Advances in Intelligent Systems and Computing

Volume 1111

The series "Advances in Intelligent Systems and Computing" contains publications on theory, applications, and design methods of Intelligent Systems and Intelligent Computing. Virtually all disciplines such as engineering, natural sciences, computer and information science, ICT, economics, business, e-commerce, environment, healthcare, life science are covered. The list of topics spans all the areas of modern intelligent systems and computing such as: computational intelligence, soft computing including neural networks, fuzzy systems, evolutionary computing and the fusion of these paradigms, social intelligence, ambient intelligence, computational neuroscience, artificial life, virtual worlds and society, cognitive science and systems, Perception and Vision, DNA and immune based systems, self-organizing and adaptive systems, e-Learning and teaching, human-centered and human-centric computing, recommender systems, intelligent control, robotics and mechatronics including human-machine teaming, knowledge-based paradigms, learning paradigms, machine ethics, intelligent data analysis, knowledge management, intelligent agents, intelligent decision making and support, intelligent network security, trust management, interactive entertainment, Web intelligence and multimedia.

The publications within "Advances in Intelligent Systems and Computing" are primarily proceedings of important conferences, symposia and congresses. They cover significant recent developments in the field, both of a foundational and applicable character. An important characteristic feature of the series is the short publication time and world-wide distribution. This permits a rapid and broad dissemination of research results.

**** Indexing: The books of this series are submitted to ISI Proceedings, EI-Compendex, DBLP, SCOPUS, Google Scholar and Springerlink ****

More information about this series at http://www.springer.com/series/11156

Hemen Dutta · Zakia Hammouch ·
Hasan Bulut · Haci Mehmet Baskonus
Editors

4th International Conference on Computational Mathematics and Engineering Sciences (CMES-2019)

Springer

Editors
Hemen Dutta
Department of Mathematics
Gauhati University
Guwahati, India

Zakia Hammouch
FSTE
Moulay Ismail University
Meknes, Morocco

Hasan Bulut
Department of Mathematics
Fırat University
Elâzığ Merkez, Turkey

Haci Mehmet Baskonus
Faculty of Education
Department of Mathematics
and Science Education
Harran University
Sanliurfa, Turkey

ISSN 2194-5357 ISSN 2194-5365 (electronic)
Advances in Intelligent Systems and Computing
ISBN 978-3-030-39111-9 ISBN 978-3-030-39112-6 (eBook)
https://doi.org/10.1007/978-3-030-39112-6

This Springer imprint is published by the registered company Springer Nature Switzerland AG
The registered company address is: Gewerbestrasse 11, 6330 Cham, Switzerland

Contents

About the Editors

Dr. Hemen Dutta has been serving in the Department of Mathematics at Gauhati University as a faculty member. He did his Master of Science in Mathematics, Postgraduate Diploma in Computer Application, Master of Philosophy in Mathematics, and Doctor of Philosophy in Mathematics. His research areas include functional analysis, mathematical modeling, etc. He has to credit more than 100 items as research papers and book chapters. He has published 10 books so far as textbooks, reference books, monographs, and edited books. He has delivered several invited talks at national and international levels. He has organized several academic events as well as associated with several conferences in different capacities. He has reviewed papers for journals and databases and also associated with editing special issues in journals. He has contributed several articles in newspaper, popular books and magazines, and science portals.

Dr. Zakia Hammouch received her M.S. in Applied Mathematics and PhD degree in Applied Mathematics and Mechanical Engineering from the University of Picardie Jules Verne, France. She received a habilitation degree from Moulay Ismail University, Morocco, in 2015. She is currently Associate Professor at Faculty of Sciences and Techniques, Moulay Ismail University, Morocco. She has published over 60 items as journal papers and chapters in books. She is an editorial member of more than 10 international journals, a reviewer of 35 international journals of mathematics and mechanical engineering, and a member of scientific committee of more than 20 international conferences. Her research interests include computational mathematics, biomathematics, nonlinear dynamics, fractional calculus, chaos theory, control and synchronization, fluid dynamics, heat and mass transfer, and soliton theory.

Dr. Hasan Bulut is currently Full Professor of mathematics in Firat University. He has published more than 200 articles journals. His research interests include stochastic differential equations, fluid and heat mechanics, finite element method, analytical methods for nonlinear differential equations, mathematical physics, numerical solutions of the partial differential equations and computer programming.

Dr. Haci Mehmet Baskonus is currently Associate Professor in Harran University. He has published more than 140 articles in various reputed and leading journals, including in SCI and SCIE indexed journals. His research interests include ordinary and partial differential equations, analytical methods for linear and nonlinear differential equations, mathematical physics, numerical solutions of the partial differential equations, fractional differential equations, and computer programming using Mathematica, Pascal, and Maple.

A Hybrid Computational Technique for Time-Fractional Newell-Whitehead-Segel Equation via Sumudu Transform

Amit Prakash[1], Vijay Verma[2], and Haci Mehmet Baskonus[3](✉)

[1] Department of Mathematics, National Institute of Technology, Kurukshetra 136119, India
amitmath@nitkkr.ac.in, amitmath0185@gmail.com

[2] Department of Mathematics, Pt. Chiranji Lal Sharma Govt. (PG) College, Karnal, Haryana, India
vijay_mtech21@rediffmail.com

[3] Faculty of Education, Department of Mathematics, Harran University, Sanliurfa, Turkey
hmbaskonus@gmail.com

Abstract. In present article, we constitute a user friendly algorithm basically expansion of homotopy perturbation method with Sumudu transform (ST), namely homotopy perturbation Sumudu transform method (HPSTM) to resolve fractional model of Newell-Whitehead-Segel equation (NWS).Thereafter, the numerical solution of the time-fractional NWS model compared with exact solution. The results attained by HPSTM may be hypothesize as a different and effective method for solving fractional model. Two tests example demonstrate the correctness as well as effectiveness of the present techniques.

Keywords: Caputo fractional derivative · Fractional Newell-Whitehead-Segel equation · Homotopy perturbation Sumudu transform method

1 Introduction

Fractional calculus (FC) is the branch of applied mathematics was introduced by Guillaume de I'Hopital as the conventional calculus before 300 years ago. In recent year, fractional differential equation has been promoted in diverse area of science and technology such as fluid mechanics, diffusion equation, biology, electro-magnetic waves, control theory visco elasticity, electrode-electrolyte heat conduction, polarization and many others physical processes, finance and biomedical engineering. Many authors describe varied techniques to reach the objective of most correct solution [7–19] and very recently, Singh et al. have suggested HPSTM [20].

In the present work, the HPSTM is tested to the fractional model of NWS equation. It is the action of the impact of the diffusion term with the nonlinear impact of the reaction term. Fractional model of NWS equation is represent as

$$u_t^\alpha = ku_{xx} + au - bu^q, t > 0, 0 < \alpha \leq 1, \tag{1}$$

H. Dutta et al. (Eds.): CMES 2019, AISC 1111, pp. 1–14, 2020.
https://doi.org/10.1007/978-3-030-39112-6_1

where a, b and $k > 0$ represent the real numbers and $represent$ the positive integers. In Eq. (1) u_t^α, u_{xx} and $au - bu^q$ illustrate the variation of $u(x, t)$ with time at a fixed location, variation of $u(x, t)$ with dimensional variable at a particular time and takes into description the impact of the source term respectively and the nonlinear distribution of temperature may be represent by function $u(x, t)$ in an infinitely narrow and lengthy rod or fluid flow as a velocity in an infinitely lengthy pipe with limited diameter.

Mainly two kinds of template are noticed, in the first roll template cylinders form by fluid stream lines and these cylinders may be turn and layout spiral like template. In the second template liquid flow is spilt into honey comb cells and layout hexagon alike template. The uniform template, stripes and hexagons occur in distinct physical structure, stripes template are show in visual cortex, on zebra skin and in a human fingerprints and hexagonal template are attained from the laser beams propagation through a nonlinear approach and in structure with chemical reaction and diffusion species [24].

Recently Nourazar et al. and Prakash et al. used homotopy perturbation method [25] and variation iteration method [26] to solved classical NWS model, respectively. Also, Prakash et al. applied Adomain decomposition method [27] and fractional variational iteration method [28] to solve fractional model of NWS equation. Kumar and Sharma [29] used homotopy analysis Sumudu transform method to solved fractional model of NWS equation. But fractional model of NWS equation has not been solved by HPSTM.

The framework of present paper is outline as follow: First section is introductory, in the Sect. 2 the key terminology of fractional calculus is examined, in Sect. 3 proposed homotopy perturbation Sumudu transforms method is discussed, in Sect. 4 two test examples of fractional model of NWS equation are given to elucidate the proposed method HPSTM and in Sect. 5 consequence of the effort is drawn.

2 Preliminaries

In present segment, we will introduce the key terminology of FC and ST used to discuss the suggested procedure.

Definition 2.1. The Sumudu transform is defined over the set of function $B = \{f(t) | \exists N, t_1, t_2 > 0, |f(t)| < Ne^{\frac{|t|}{t_j}}$ if $t \in (-1)^j \times [0, \infty)$ be the following formula

$$S[f(t)] = \int_0^\infty f(ut)e^{-t}dt, u \in (-t_1, t_2).$$

Definition 2.2. The ST of fractional derivative in Caputo sense is defined as:

$$S\left[D_x^{n\alpha} u(x,t)\right] = v^{-n\alpha} S[u(x,t)] - \sum\nolimits_{k=0}^{n-1} v^{(-n\alpha+k)} u^k(0,t), n-1 < n\alpha \le n.$$

Definition 2.3. The fractional derivative of $f(t)$, $f \in C_{-1}^n, n \in \mathrm{N}, n > 0$, in Caputo sense is defined as

$$\mathrm{D}^\alpha f(t) = I^{n-\alpha} D^n f(t) = \frac{1}{\Gamma(n-\alpha)} \int_0^t (t-x)^{n-\alpha-1} f^n(x)dx,$$

where $n - 1 < \alpha \le n$.

3 Proposed Homotopy Perturbation Sumudu Transforms Method

In present section, solution procedure of the proposed technique HPSTM is illustrated for the time-fractional nonlinear partial differential equation (NPDE). Now, consider the following time-fractional NPDE:

$$D_t^{\alpha} u(x,t) + Ru(x,t) + Nu(x,t) = f(x,t), \tag{2}$$

with initial approximation $u(x,0) = g(x)$,

where D_t^{α}, R, $N$and $f(x,t)$ illustrate Caputo fractional derivative in Caputo sense, linear differential operator, nonlinear differential operator and the source term, respectively.

After using the ST on Eq. (2), we attained

$$S(D_t^{\alpha} u(x,t)) + S(Ru(x,t)) + S(Nu(x,t)) = S((fx,t)). \tag{3}$$

Now, operating the property of ST on the Eq. (3), we obtain

$$S(u(x,t)) = g(x) + u^{\alpha} S(f(x,t)) - u^{\alpha} S(Ru(x,t) + Nu(x,t)) \tag{4}$$

After using the property of inverse ST on the Eq. (4), we obtain

$$u(x,t) = G(x,t) - S^{-1}(u^{\alpha} S(Ru(x,t) + Nu(x,t)), \tag{5}$$

where $G(x,t)$ represent the term obtained from source term and suggested initial approximation.

Now, after using the Homotopy perturbation method, the result can be demonstrate as a power series in small homotopy parameter $p \in [0,1]$ is specified in the form

$$u(x,t) = \sum_{n=0}^{\infty} p^n u_n(x,t). \tag{6}$$

Again, nonlinear term can be expressed as a He's polynomials $H_n(u)$

$$Nu(x,t) = \sum_{n=0}^{\infty} p^n H_n(u), \tag{7}$$

and He's polynomials $H_n(u)$ can be evaluated by the subsequent formula

$$H_n(u_0, u_1, u_2, \ldots\ldots\ldots u_n) = \frac{1}{\Gamma(n+1)} \frac{\partial^n}{\partial P^n} [N \sum_{n=0}^{\infty} p_i u_i(x,t)]_{p=0}. \tag{8}$$

where $n = 0, 1, 2, 3 \ldots\ldots$

Substituting (6) and (7) in (5) and using HPM by He [21, 22], we attain

$$\sum_{n=0}^{\infty} p^n u_n(x,t) = G(x,t) - p * S^{-1}\left\{u^{\alpha} S(R \sum_{n=0}^{\infty} p^n u_n(x,t) + \sum_{n=0}^{\infty} p^n H_n(u)\right\}.$$

After comparing coefficient of similar power of p on both sides, we obtain successive approximations:

$$p^0 : u_0 = G(x,t),$$
$$p^1 : u = S^{-1}\{u^\alpha S(Ru_0(x,t) + H_0(u))\},$$
$$p^2 : u_2 = S^{-1}\{u^\alpha S(Ru_1(x,t) + H_1(u))\},$$
$$p^3 : u_3 = S^{-1}\{u^\alpha S(Ru_2(x,t) + H_2(u))\},$$

Similarly, we can calculate

$$p^n : u_n = S^{-1}\left\{u^\alpha S(Ru_{n-1}(x,t) + H_{n-1}(u))\right\}.$$

Lastly, we approximate the numerical results by the series

$$u(x,t) = \lim_{n\to\infty} \sum\nolimits_{n=0}^{\infty} u_n(x,t). \tag{9}$$

4 Numerical Experiments

In present segment, we shall apply aforesaid HPSTM to two test example of time-fractional NWS equation.

Example 1. We analysis the linear model of time-fractional NWS equation

$$\mathrm{u}_\mathrm{t}^\alpha = \mathrm{u}_\mathrm{xx} - 2\mathrm{u}, \mathrm{t} > 0, 0 < \alpha \le 1, \tag{10}$$

with initial approximation

$$\mathrm{u}(\mathrm{x}, 0) = \mathrm{e}^\mathrm{x}. \tag{11}$$

Now, for particular case when $\alpha = 1$, the exact solution of the classical model of NWS equation is given as: $u(x,t) = e^{x-t}$.

After, using the suggested technique HPSTM to the previously show illustration, we obtain the successive approximations:

$$u_0 = e^x,$$
$$u_1 = -e^x \frac{t^\alpha}{\Gamma(\alpha+1)},$$
$$u_2 = e^x \frac{t^{2\alpha}}{\Gamma(2\alpha+1)},$$
$$u_3 = -e^x \frac{t^{3\alpha}}{\Gamma(3\alpha+1)},$$
$$\vdots$$
$$u_n = (-1)^n e^x \frac{t^{n\alpha}}{\Gamma(n\alpha+1)}.$$

From Figs. 1 and 2, we noticed the marvellous consistency between suggested technique HPSTM and exact solution. Here, we use eight terms of approximation to depicts the Fig. 1.

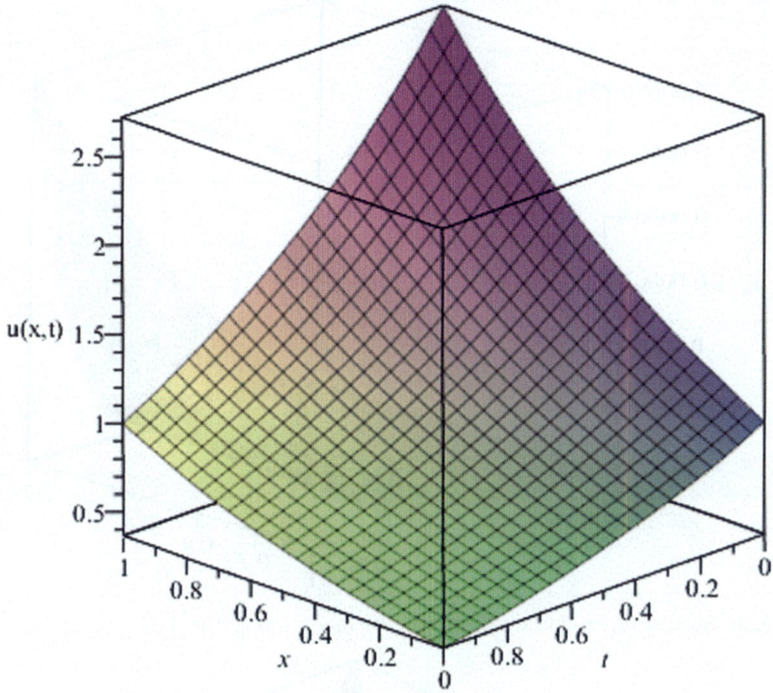

Fig. 1. Surface represents approximate solution $u(x, t)$ at $\alpha = 1$, for Ex. 1.

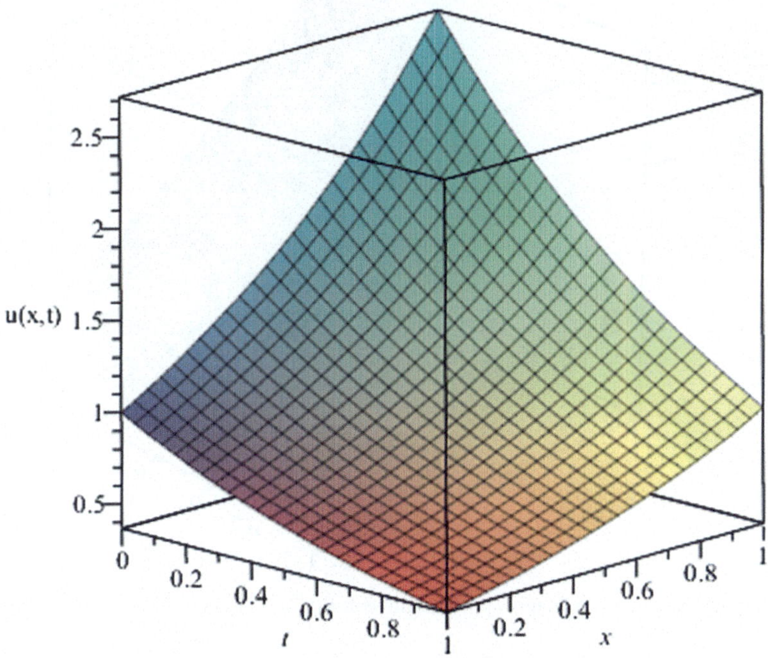

Fig. 2. Surface represents exact solution $u(x, t)$ at $\alpha = 1$, for Ex. 1.

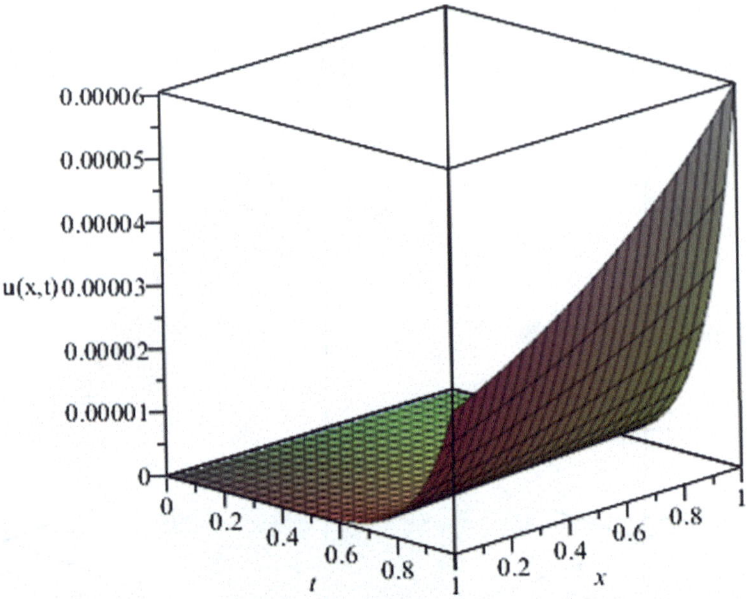

Fig. 3. Surface represents absolute error $|u_{Exact} - u_{HPSTM}|$ at $\alpha = 1$, for Ex. 1.

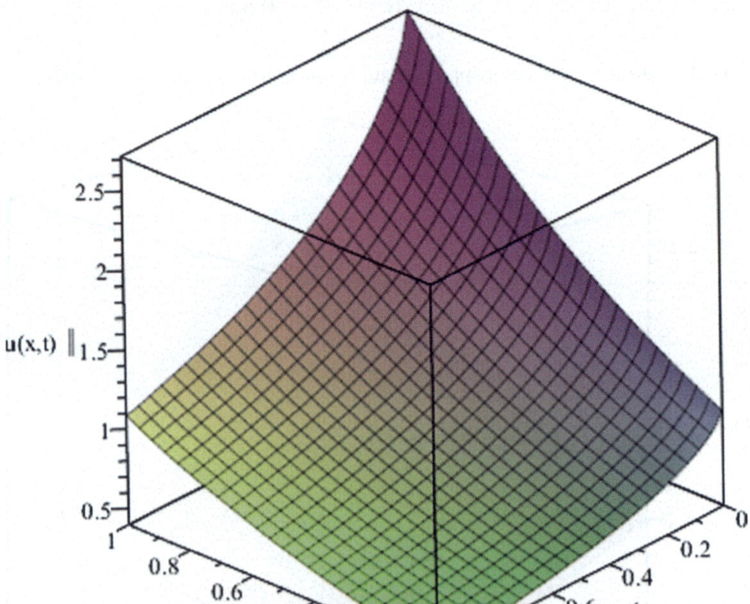

Fig. 4. Surface represents approximate solution $u(x, t)$ at $\alpha = 0.7$, for Ex. 1.

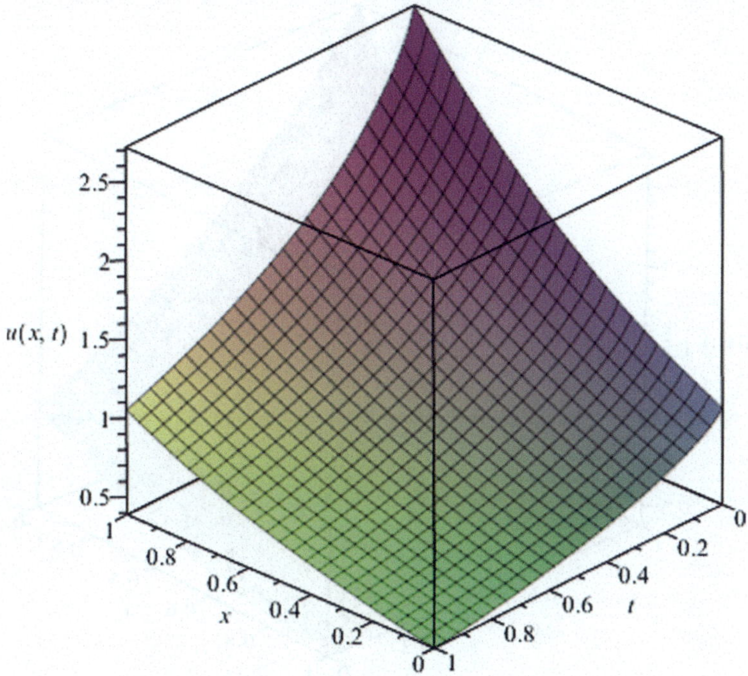

Fig. 5. Surface represents approximate solution $u(x, t)$ at $\alpha = 0.8$, for Ex. 1.

Figure 1 demonstrate the 8th order approximate result, Fig. 2 depicts exact solution. It is noticed from Figs. 1and 2 that the result attained by HPSTM is almost similar to exact solution, Fig. 3 represents the absolute error, Figs. 4, 5 and 6 shows the approximate solution for $\alpha = 0.7, 0.8\, and\, 0.9$ respectively Fig. 7 demonstrate the comparison of HPSTM solution for different values of fractional order $\alpha = 0.25, 0.50, 0.75\, and$ 1 and exact solution for $\alpha = 1\, and\, t = 0.50$. From Fig. 7 it is observed that with increase the value of x value of u increase and approach to ∞ and towards negative value of x its show asymptotic behaviour toward x axis.

Example 2 We analysis the nonlinear model of time-fractional NWS equation

$$u_t^\alpha = u_{xx} + u - u^2 = 0, t > 0, 0 < \alpha \leq 1, \tag{12}$$

with initial approximation

$$u(x, 0) = \frac{1}{\left(1 + e^{\frac{x}{\sqrt{6}}}\right)^2}. \tag{13}$$

Now, for particular case when $\alpha = 1$, the exact solution of the classical nonlinear model of NWS equation is given as: $u(x, t) = \frac{1}{\left(1+e^{\frac{x}{\sqrt{6}} - \frac{5}{6}t}\right)^2}$.

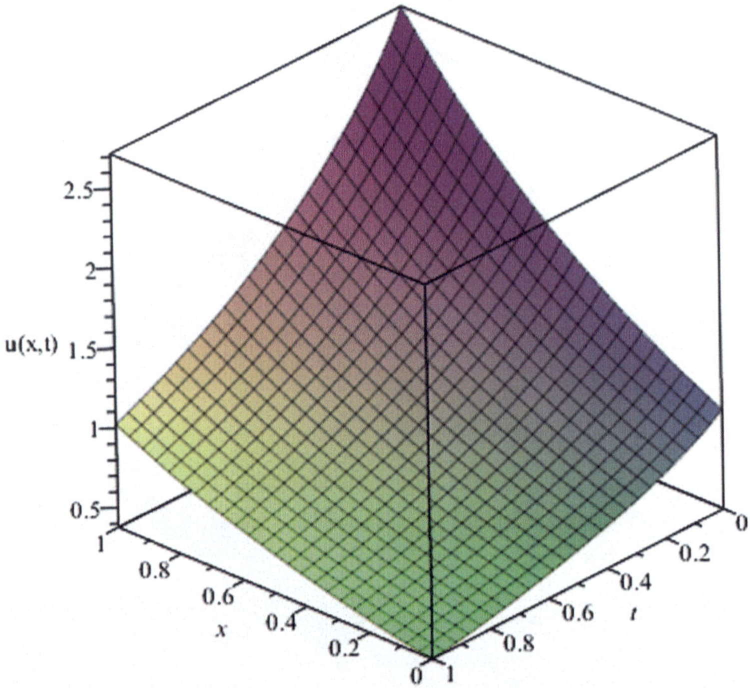

Fig. 6. Surface represents approximate solution $u(x, t)$ at $\alpha = 0.9$, for Ex. 1.

After, using the suggested technique HPSTM to the previously show illustration, we obtain the successive approximations:

$$u_0 = \frac{1}{\left(1 + e^{\frac{x}{\sqrt{6}}}\right)^2},$$

$$u_1 = \frac{5}{3} \frac{e^{\frac{x}{\sqrt{6}}}}{\left(1 + e^{\frac{x}{\sqrt{6}}}\right)^3} \frac{t^\alpha}{\Gamma(\alpha + 1)},$$

$$u_2 = \frac{25}{18} \left(\frac{e^{\frac{x}{\sqrt{6}}}\left(-1 + 2e^{\frac{x}{\sqrt{6}}}\right)}{\left(1 + e^{\frac{x}{\sqrt{6}}}\right)^4} \right) \frac{t^{2\alpha}}{\Gamma(2\alpha + 1)},$$

$$u_3 = \{ \frac{25}{18} \frac{1}{\left(1 + e^{\frac{x}{\sqrt{6}}}\right)^5} [\frac{8}{6} (e^{\frac{x}{\sqrt{6}}})^2 - 4 (e^{\frac{x}{\sqrt{6}}})^3 + \left(\frac{8}{6} (e^{\frac{x}{\sqrt{6}}})^2 - \frac{(e^{\frac{x}{\sqrt{6}}})}{6} \right) \left(1 + e^{\frac{x}{\sqrt{6}}}\right)$$

$$+ \frac{4}{6} (e^{\frac{x}{\sqrt{6}}})^2 - \frac{16}{6} (e^{\frac{x}{\sqrt{6}}})^3 + \left(2 (e^{\frac{x}{\sqrt{6}}})^2 - e^{\frac{x}{\sqrt{6}}}\right) \left(1 + e^{\frac{x}{\sqrt{6}}}\right) + \frac{\frac{-20}{6} (e^{\frac{x}{\sqrt{6}}})^3 + \frac{40}{6} (e^{\frac{x}{\sqrt{6}}})^4}{\left(1 + e^{\frac{x}{\sqrt{6}}}\right)}$$

$$-2\left(\frac{(-e^{\frac{x}{\sqrt{6}}})^1+2(e^{\frac{x}{\sqrt{6}}})^2}{\left(1+e^{\frac{x}{\sqrt{6}}}\right)}\right)]\}\frac{t^{3\alpha}}{\Gamma(3\alpha+1)}-\frac{25}{9}\frac{(e^{\frac{x}{\sqrt{6}}})^2}{\left(1+e^{\frac{x}{\sqrt{6}}}\right)^6}\frac{\Gamma(2\alpha+1)t^{3\alpha}}{\Gamma(3\alpha+1)\Gamma(\alpha+1)^2}.$$

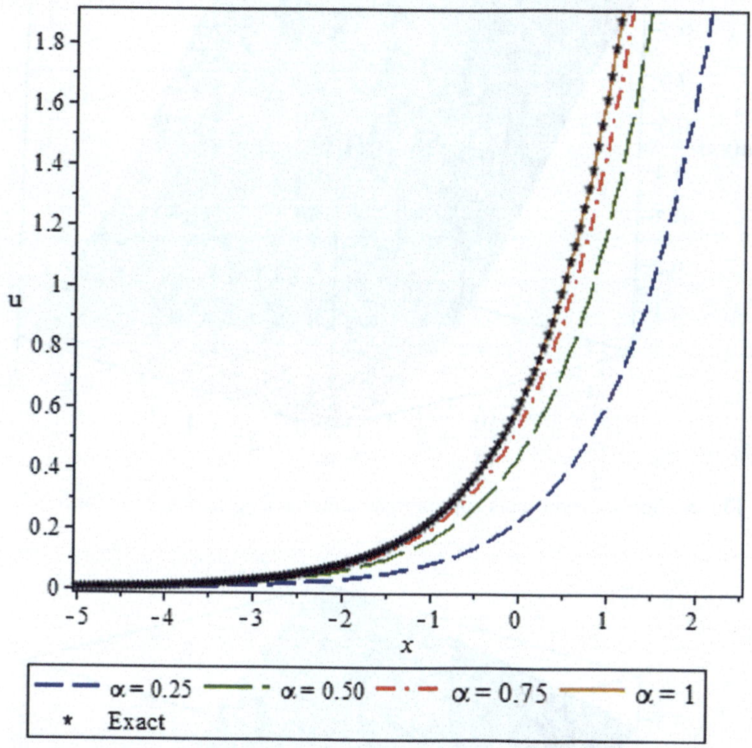

Fig. 7. Surface represent HPSTM solution for different of value α and exact sol. at $\alpha = 1$ *and* $t = 0.50$, for Ex. 1.

From Figs. 8 and 9, we noticed the marvellous consistency between suggested technique HPSTM and exact solution. Here, we use third terms of approximation to depicts the Fig. 8.

Figure 8 demonstrate the 2nd order approximate result, Fig. 9 depicts the exact solution. It is noticed from Figs. 8 and 9 that the result attained by HPSTM is nearly similar to exact solution, Fig. 10 represents the absolute error, Figs. 11, 12 and 13 shows the approximate solution for $\alpha = 0.7, 0.8$ *and* 0.9 respectively and Fig. 14 demonstrate the comparison of proposed technique HPSTM solution for $\alpha = 0.25, 0.50, 0.75, 1$ and exact solution for $\alpha = 1$ *and* $t = 0.50$. From Fig. 14 it is observed that with increase the value of x value of u decrease and approach to 0 and curve show asymptotic behaviour toward x axis.

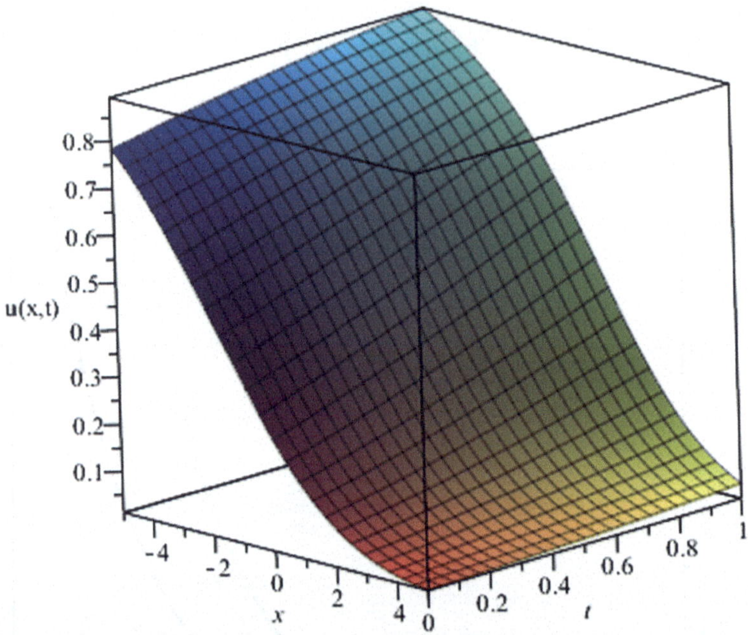

Fig. 8. Surface represents approximate solution $u(x, t)$ at $\alpha = 1$, for Ex. 2.

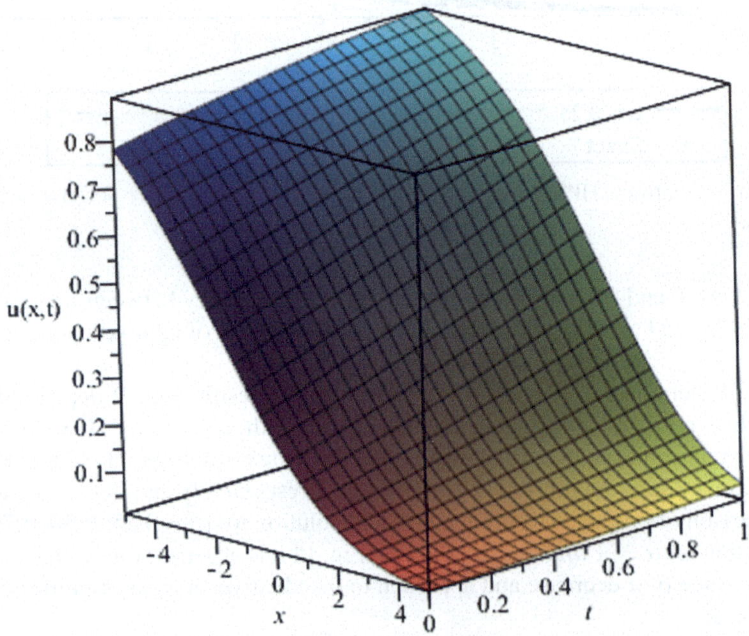

Fig. 9. Surface represents exact solution $u(x, t)$ at $\alpha = 1$, for Ex. 2.

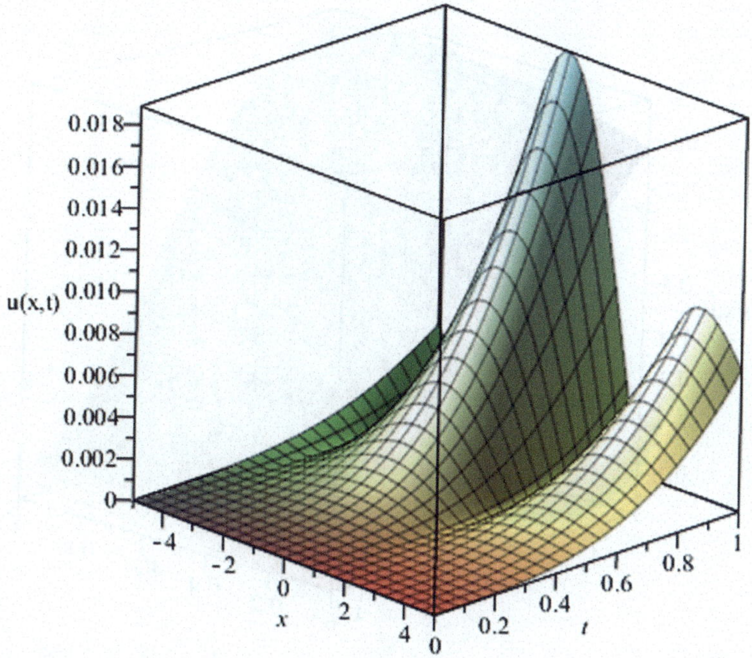

Fig. 10. Surface represents absolute error $|u_{Exact} - u_{HPSTM}|$ at $\alpha = 1$, for Ex. 2.

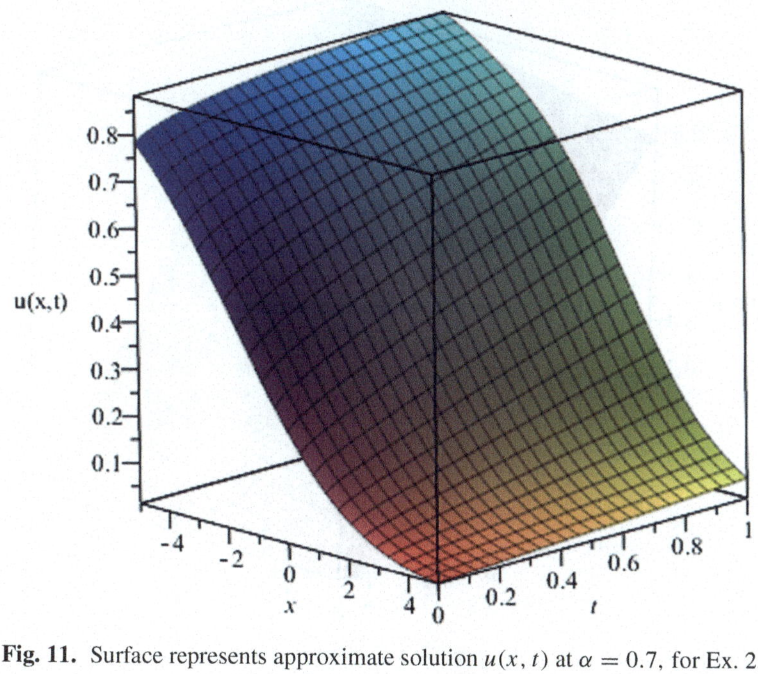

Fig. 11. Surface represents approximate solution $u(x, t)$ at $\alpha = 0.7$, for Ex. 2.

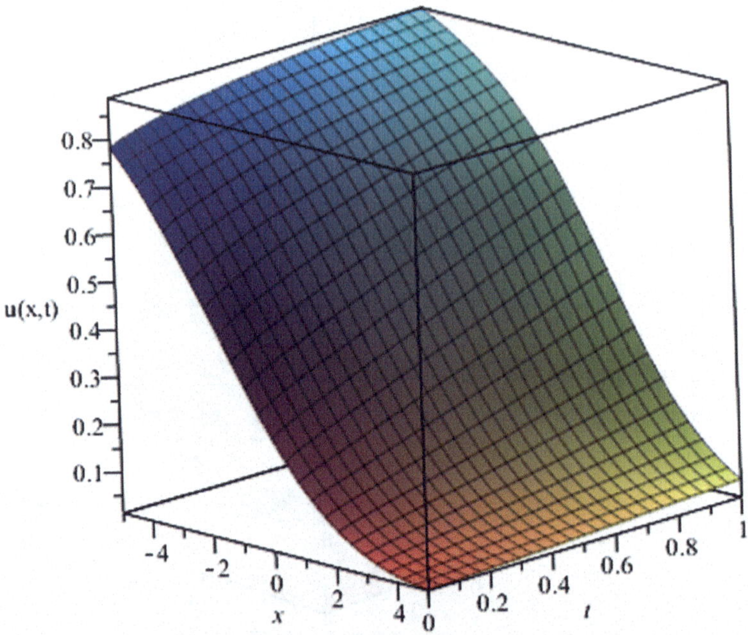

Fig. 12. Surface represents approximate solution $u(x, t)$ at $\alpha = 0.8$, for Ex. 2.

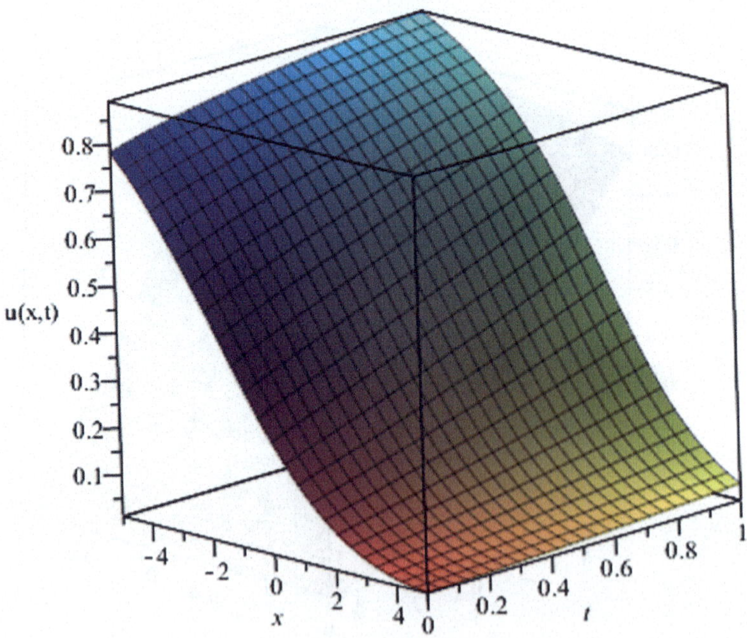

Fig. 13. Surface represents approximate solution $u(x, t)$ at $\alpha = 0.9$, for Ex. 2.

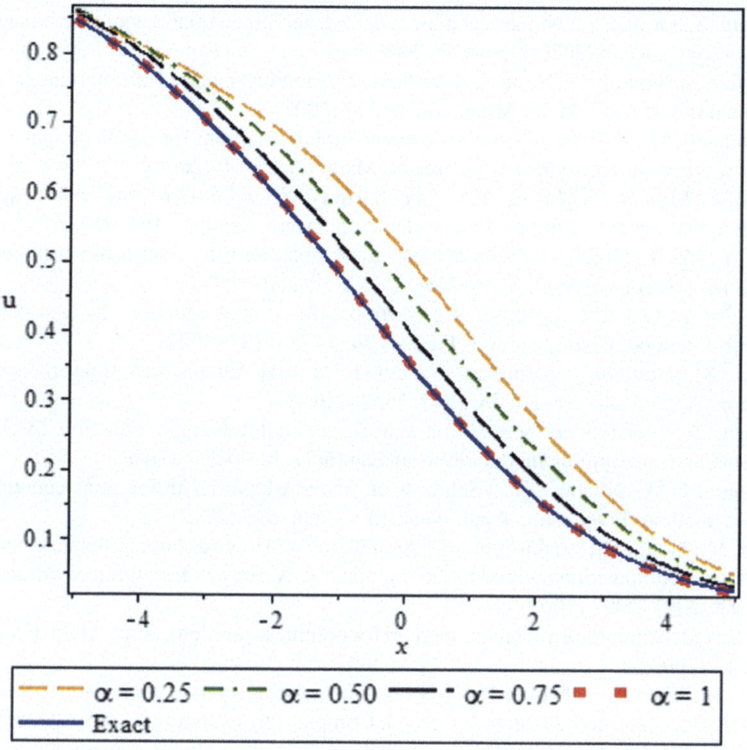

Fig. 14. Surface represent HPSTM solution for different value of α and exact sol. at $\alpha = 1$ *and* $t = 0.50$, for Ex. 2.

5 Conclusions

In the present paper, almost accurate solution of fractional model of NWS is attained with the aids of effective techniques HPSTM. The plotted graph and approximate solution demonstrate the validity and effectiveness of suggested techniques HPSTM and solution is converge towards the exact solution vary rapidly which is demonstrated numerically. We noticed the marvellous consistency between suggested techniques HPSTM and the exact solution. Moreover, efficient techniques HPSTM can be again tested to construct the accurate solution of nonlinear fractional model.

References

1. Oldham, K.B., Spanier, J.: The Fractional Calculus. Academic Press, New York (1974)
2. Podlubny, I.: Fractional differential Equation. Academic press, New York (1999)
3. Raberto, M., Scalas, E.: Waiting times and returns in high frequency financial data: an empirical study. Phys. A **314**, 749–755 (2002)
4. Magin, R.L.: Fractional Calculus in Bioengineering. Begell House Publishers, Redding (2006)
5. Ray, S.S., Bera, R.K.: Analytical solution of Bagley-Torvik equation by Adomian decomposition method. Appl. Math. Comput. **168**(1), 398–410 (2005)

6. Momani, S., Odibat, Z.: Numerical approach to differential equations of fractional orders. J. Comput. Appl. Math. **207**(1), 96–110 (2007)
7. Odibat, Z., Momani, S.: Numerical methods for nonlinear partial differential equations of fractional order. Appl. Math. Model. **32**, 28–29 (2008)
8. Meerschaert, M., Tadjeran, C.: Finite difference approximations for two sided space fractional partial differential equations. Appl. Numer. Math. **56**, 80–90 (2006)
9. Odibat, Z., Momani, S., Erturk, V.S.: Generalized differential transform method: application to differential equations of fractional order. Appl. Math. Comput. **197**, 467–477 (2008)
10. Jiang, Y., Ma, J.: Higher order finite element methods for time-fractional partial differential equations. J. Comput. Appl. Math. **235**(11), 3285–3290 (2011)
11. Arikoglu, A., Ozkol, I.: Solution of a fractional differential equations by using differential transform method. Chaos, Solitons Fractals **34**, 1473–1481 (2007)
12. Zhang, X.: Homotopy perturbation method for two dimensional time-fractional wave equation. Appl. Math. Model. **38**, 5545–5552 (2014)
13. Prakash, A.: Analytical method for space-fractional telegraph equation by homotopy perturbation transform method. Nonlinear Eng. **5**(2), 123–128 (2016)
14. Dhaigude, C.D., Nikam, V.R.: Solution of fractional partial differential equations using iterative method. Fract. Calc. Appl. Anal. **15**(4), 684–699 (2012)
15. Safari, M., Ganji, D.D., Moslemi, M.: Application of He's variational iteration method and Adomain decomposition method to the fractional KdV-Burger-Kuramoto equation. Comput. Appl. **58**, 2091–2097 (2009)
16. Liao, S.: On the homotopy analysis method for nonlinear problem. Appl. Math. Comput. **147**, 499–513 (2004)
17. Jiwari, R., Pandit, S., Mittal, R.C.: Numerical solution of two dimensional Sine-Gordon Solitons by differential quadrature method. Comput. Phys. Commun. **183**, 600–616 (2012)
18. Singh, J., Kumar, D.: Homotopy perturbation Sumudu transform method for nonlinear equation. Adv. Appl. Mech. **14**, 165–175 (2011)
19. He, J.H.: Homotopy perturbation technique. Comput. Method Appl. Mech. Eng. **178**, 257–262 (1999)
20. He, J.H.: Homotopy perturbation method: a new nonlinear analytical technique. Appl. Math. Comput. **135**, 73–79 (2003)
21. Whitham, G.B.: Variational method and application to water waves. Proc. Roy. Soc. A **299**, 6–25 (1967)
22. Broer, L.J.: Approximation equations for long water waves, appl. Sci. Res. **31**, 377–395 (1975)
23. Kaup, D.J.: A higher order water-wave equations and the method for solving it. Prog. Theor. Phys. **54**, 396–408 (1975)
24. Golovinb, N.A.: General Aspect of pattern formation, pattern formation and growth phenomena in nano-system. Alexander, pp.1–54 (2007)
25. Nourazar, S.S., Soori, M.: On the exact solution of Newell-Whitehead-Segel equation using the homotopy perturbation method. Aust. J. Basic Appl. Sci. **5**(8), 1400–1411 (2011)
26. Prakash, A., Kumar, M.: He's variation iteration method for the solution of nonlinear Newell-Whitehead-Segel equation. J. Appl. Anal. Comput. **5**(2), 123–128 (2016)
27. Prakash, A., Verma, V.: Numerical method for fractional model of Newell-Whitehead-Segel equation. Front. Phys. **7**(15), 1–10 (2019)
28. Prakash, A., Goyal, M., Gupta, S.: Fractional variational iteration method for solving time-fractional Newell-Whitehead-Segel equation. Nonlinear Eng. **8**(1), 164–171 (2019). https://doi.org/10.1515/nleng-2018-0001
29. Kumar, D., Prakash, R.: Numerical approximation of Newell-Whitehead-Segel equation of fractional order. Nonlinear Eng. **5**(2), 81–86 (2016). https://doi.org/10.1515/nleng-2015-0032

Mathematical Modeling and Stability Analysis of HIV with Contact Tracing According to the Changes in the Infected Classes

Ali Yousef[1](✉) and Fatma Bozkurt Yousef[1,2]

[1] Department of Mathematics, Kuwait College of Science and Technology, 27235 Kuwait City, Kuwait
A.yousef@kcst.edu.kw, fbozkurt@erciyes.edu.tr

[2] Faculty of Education, Department of Mathematics and Science Education, Erciyes University, 38039 Kayseri, Turkey

Abstract. In this paper, we investigate the effect of contact tracing the spread of HIV in a population. The mathematical model is given as a system of differential equations with piecewise constant arguments, where we divide the population into three sub-classes: HIV negative, HIV positive that do not know they are infected and the class with HIV positive that know they are infected. This system is analyzed using the theory of differential and difference equations. The local stability of the positive equilibrium point is investigated by using the Schur-Cohn Criteria, while for the global stability we consider an appropriate Lyapunov function. The system under consideration has shown that it has semi-cycle behaviors, but not a structure of period two. Moreover, we analyze the case for low infection rate by using the Allee effect at time t. Several examples are presented to support our theoretical findings using data from a case study in India.

Keywords: Logistic differential equations · Stability analysis · Periodic behavior · Allee effect

1 Introduction

Since 1981, HIV has spread throughout the world, and now it is a major epidemic problem worldwide [1]. Furthermore, it is an important research area in medicine as well as in applied mathematics. Mathematical models of transmission dynamics of HIV play an important role in understanding epidemiological patterns for disease control and to have a long-term prediction of HIV. Recent studies have been conducted to describe the transmission dynamics of HIV, where some of them are [2–6].

In this paper, we want to consider the study of [7, 8], where the authors analyzed the effect of contact tracing on reducing the spread of HIV/AIDS in a homogeneous population with the constant immigration of susceptible by using the theory of differential equations. However, in some biological phenomena, we need to use both continuous and discrete time. In this demography study, the model needs continuous time, since we have overlapping population growth. Moreover, the spread of transmission and to realize to

H. Dutta et al. (Eds.): CMES 2019, AISC 1111, pp. 15–35, 2020.
https://doi.org/10.1007/978-3-030-39112-6_2

be infected happens in discrete time. Theoretical studies show that differential equations with piecewise constant arguments are equivalent to integral equations and are very close to delaying differential equations [9, 10]. Cooke and Györi studied this idea and proved that a differential equation with piecewise constant arguments could be used to obtain the approximate solution of delay differential equations that contains discrete delays. Following this work, many authors have analyzed various types of biological models consisting of differential equations with piecewise constant arguments and fractional order differential equations. The reader can consult the papers [11–16] for further details.

According to the above information, our model as a system of equations with piecewise constant arguments is as follows:

$$\begin{cases} \frac{dS}{dt} = S(t)r_1(p - \alpha_1 S(t) - \beta_1 I_1([[t]]) - \beta_2 I_2([[t]])), \\ \frac{dI_1}{dt} = I_1(t)r_2(1 - \alpha_2 I_1(t) + \beta_1(1 - \varepsilon_1)S([[t]]) - \theta I_1([[t]]) + \beta_2(1 - \varepsilon_2)S([[t]])I_2([[t]])), \\ \frac{dI_2}{dt} = I_2(t)r_3(1 - \alpha_3 I_2(t) + \beta_2 \varepsilon_2 S([[t]]) + \theta I_1([[t]]) + \beta_1 \varepsilon_1 S([[t]])I_1([[t]])), \end{cases} \tag{1.1}$$

where t is the time and $[[t]]$ is the exact value of $t \geq 0$.

The model contains three populations; susceptible S, HIV positives that do not know they are infected I_1 and HIV positives that know they are infected I_2. The susceptible are composed of individuals that have not contacted the infection but can get infected through contacts (sexual, blood transfusion, etc.) with infectives. Note, r_1 is the population growth rate of the susceptible population, α_1 is the death (of natural causes) rate while, p is a rate of susceptible population per year. The susceptibles lost their class following contacts with infectives I_1 and I_2 at a rate β_1 and β_2, respectively.

I_1 is the population that has HIV positive, but they do not know it due to the invisibility of disease symptoms. In class I_1, r_2 is the population growth rate, while α_2 is the death (of natural causes) rate. The population of this class decreases by HIV test and become aware after screening at a rate θ. After contacts of classes S and I_1 in discrete time, these classes could realize they are infected. This group $(S([[t]]) \cdot I_1([[t]]))$ is given in class I_2 with rate ε_1.

It can also be possible that after tracing contact between S and I_2, a rate of the susceptible class detect in further time to be HIV positive, which is here ε_2. The population growth rate of class I_2 is given by r_3, while the death rate of natural causes is α_3.

In this paper, we analyzed in Sect. 2 the boundedness character and the non-periodic behavior of system (1.1). We obtained that the solutions show a semi-cycle behavior, which is not periodic. In Sect. 3, we investigated the local and global behavior of the system around the positive equilibrium point, which is based on specific conditions. For a prediction analysis about a small infected class of I_1 and I_2, we incorporate Allee functions at time t in Sect. 4. The conclusion part in Sect. 5 summarize the study in this paper.

2 The Boundedness Character and Analysis of Periodic Behavior

In this section, we analyze the boundedness character of the system, and we want to find conditions for periodic behavior.

For $t \in [n, n+1)$, system (1.1) is an expression of Bernoulli differential equations such as

$$\begin{cases} \frac{dS}{dt} - r_1(p - \beta_1 I_1(n) - \beta_2 I_2(n))S(t) = -\alpha_1 r_1 S(t)^2, \\ \frac{dI_1}{dt} - r_2(1 + \beta_1(1-\varepsilon_1)S(n) - \theta I_1(n) + \beta_2(1-\varepsilon_2)S(n)I_2(n))I_1(t) = -\alpha_2 r_2 I_1(t)^2, \\ \frac{dI_2}{dt} - r_3(1 + \beta_2\varepsilon_2 S(n) + \theta I_1(n) + \beta_1\varepsilon_1 S(n)I_1(n))I_2(t) = -\alpha_3 r_3 I_2(t). \end{cases} \tag{2.1}$$

Solving (2.1) for $t \in [n, n+1)$, we obtain

$$\begin{cases} S(n+1) = \frac{S(n)U_1(n)}{(U_1(n) - \alpha_1 S(n)) \cdot \exp(-r_1 U_1(n)) + \alpha_1 S(n)}, \\ I_1(n+1) = \frac{I_1(n)U_2(n)}{(U_2(n) - \alpha_2 I_1(n)) \cdot \exp(-r_2 U_2(n)) + \alpha_2 I_1(n)}, \\ I_2(n+1) = \frac{I_2(n)U_3(n)}{(U_3(n) - \alpha_3 I_2(n)) \cdot \exp(-r_3 U_3(n)) + \alpha_3 I_2(n)}, \end{cases} \tag{2.2}$$

where

$$U_1(n) = p - \beta_1 I_1(n) - \beta_2 I_2(n) \neq 0,$$
$$U_2(n) = 1 + \beta_1(1-\varepsilon_1)S(n) - \theta I_1(n) + \beta_2(1-\varepsilon_2)S(n)I_2(n) \neq 0,$$

and

$$U_3(n) = 1 + \beta_2\varepsilon_2 S(n) + \theta I_1(n) + \beta_1\varepsilon_1 S(n)I_1(n) \neq 0.$$

To analyze the stability around a positive critical point of system (1.1), we want to obtain at first the positive equilibrium point of (2.2), since (2.2) is the solution of (1.1) for $t \in [n, n+1)$. Thus, we get

$$p = \beta_1 \overline{I_1} + \beta_2 \overline{I_2} + \alpha_1 \bar{S}, \tag{2.3}$$

$$1 = -\beta_1(1-\varepsilon_1)\bar{S} + (\theta + \alpha_2)\overline{I_1} - \beta_2(1-\varepsilon_2)\bar{S}\overline{I_2} \tag{2.4}$$

and

$$1 = -\beta_2\varepsilon_2\bar{S} - \theta\overline{I_1} - \beta_1\varepsilon_1\bar{S}\overline{I_1} + \alpha_3\overline{I_2}. \tag{2.5}$$

To find the positive equilibrium point of (2.2), we have the following assumptions given in the demographic data in [7, 8];

(i) $\beta_1 = 3\beta_2$,
(ii) $\varepsilon_1 = \varepsilon_2$,
(iii) $\varepsilon_1 + \varepsilon_2 < 1$,
(iv) $\varepsilon_1 = \varepsilon_2 < 0.75$.

Considering (2.4) and (2.5), we obtain

$$\overline{I_1} = \frac{(1-\varepsilon_1)}{3\varepsilon_1}\overline{I_2}, \tag{2.6}$$

where we can write

$$-\beta_2\varepsilon_1(3-4\varepsilon_1)\bar{S}+((2\theta+\alpha_2)\beta_2(1-\varepsilon_1)-\alpha_3\varepsilon_1)\overline{I_2}=0 \tag{2.7}$$

and

$$\alpha_1\varepsilon_1\bar{S}+\beta_2\overline{I_2}=\varepsilon_1 p. \tag{2.8}$$

Thus, we have the equilibrium point $\left(\bar{S},\overline{I_1},\overline{I_2}\right)$ as follows;

$$\begin{aligned}
\bar{S} &= \frac{p}{\alpha_1}-\frac{\beta_2^2 p(3-4\varepsilon_1)}{\alpha_1\left(\beta_2^2(3-4\varepsilon_1)+\alpha_1((2\theta+\alpha_2)\beta_2-\varepsilon_1((2\theta+\alpha_2)\beta_2+\alpha_3))\right)},\\
\overline{I_1} &= \frac{\beta_2\varepsilon_1 p(3-4\varepsilon_1)(1-\varepsilon_1)}{3\varepsilon_1\left(\beta_2^2(3-4\varepsilon_1)+\alpha_1((2\theta+\alpha_2)\beta_2-\varepsilon_1((2\theta+\alpha_2)\beta_2+\alpha_3))\right)},\\
\overline{I_2} &= \frac{\beta_2\varepsilon_1 p(3-4\varepsilon_1)}{\beta_2^2(3-4\varepsilon_1)+\alpha_1((2\theta+\alpha_2)\beta_2-\varepsilon_1((2\theta+\alpha_2)\beta_2+\alpha_3))},
\end{aligned}$$

where $\varepsilon_1<\frac{(2\theta+\alpha_2)\beta_2}{(2\theta+\alpha_2)\beta_2+\alpha_3}$ and $p>\frac{\beta_2^2 p(3-4\varepsilon_1)}{\beta_2^2(3-4\varepsilon_1)+\alpha_1((2\theta+\alpha_2)\beta_2-\varepsilon_1((2\theta+\alpha_2)\beta_2+\alpha_3))}$.

The Jacobian matrix for the positive equilibrium point $\bar{X}=\left(\bar{S},\overline{I_1},\overline{I_2}\right)$ is

$$J\left(\bar{X}\right)=\begin{pmatrix} a_{11} & a_{12} & a_{13}\\ a_{21} & a_{22} & a_{23}\\ a_{31} & a_{32} & a_{33}\end{pmatrix} \tag{2.9}$$

where

$$\begin{aligned}
a_{11} &= e^{-r_1\overline{U_1}},\ a_{12}=\frac{3\beta_2\left(e^{-r_1\overline{U_1}}-1\right)}{\alpha_1},\ a_{13}=\frac{\beta_2\left(e^{-r_1\overline{U_1}}-1\right)}{\alpha_1},\\
a_{21} &= \frac{\beta_2(1-\varepsilon_1)(3+\overline{I_2})\left(1-e^{-r_2\overline{U_2}}\right)}{\alpha_2},\ a_{22}=\frac{-\theta+(\theta+\alpha_2)e^{-r_2\overline{U_2}}}{\alpha_2},\\
a_{23} &= \frac{\beta_2(1-\varepsilon_1)\bar{S}\left(1-e^{-r_2\overline{U_2}}\right)}{\alpha_2},\ a_{31}=\frac{\beta_2\varepsilon_1\left(1+3\overline{I_2}\right)\left(1-e^{-r_3\overline{U_3}}\right)}{\alpha_3},\\
a_{32} &= \frac{\left(\theta+3\beta_2\varepsilon_1\bar{S}\right)\left(1-e^{-r_3\overline{U_3}}\right)}{\alpha_3},\ a_{33}=e^{-r_3\overline{U_3}}.
\end{aligned}$$

The characteristic equation of (2.9) around the equilibrium point is given by

$$\begin{aligned}
&\lambda^3+(-a_{11}-a_{22}-a_{33})\lambda^2+(a_{11}a_{22}+a_{11}a_{33}+a_{22}a_{33}-a_{13}a_{31}-a_{23}a_{32}-a_{12}a_{21})\lambda\\
&+(a_{13}a_{31}a_{22}+a_{23}a_{32}a_{11}+a_{12}a_{21}a_{33}-a_{11}a_{22}a_{33}-a_{21}a_{32}a_{13}-a_{31}a_{12}a_{23})=0.
\end{aligned} \tag{2.10}$$

Theorem 2.1. Assume that $(S(n),I_1(n),I_2(n))_{n=0}^{\infty}$ is a positive solution of system (2.2). Then the following statements are true.

(i) If

$$\begin{cases} p > \beta_1 I_1(n) + \beta_2 I_2(n) > \alpha_1 S(n), \\ 1 > 1 + \beta_1(1-\varepsilon_1)S(n) - \theta I_1(n) + \beta_2(1-\varepsilon_2)S(n)I_2(n) > \alpha_2 I_1(n) > 0\,, \\ 1 > 1 + \beta_2\varepsilon_2 S(n) + \theta I_1(n) + \beta_1\varepsilon_1 S(n)I_1(n) > \alpha_3 I_2(n), \end{cases} \tag{2.11}$$

then the solution of system (2.2) increases monotonically.

(ii) If

$$\begin{cases} \alpha_1 S(n) > p - \beta_1 I_1(n) - \beta_2 I_2(n) > 0, \\ \alpha_2 I_1(n) > 1 + \beta_1(1-\varepsilon_1)S(n) - \theta I_1(n) + \beta_2(1-\varepsilon_2)S(n)I_2(n) > 0\,, \\ \alpha_3 I_2(n) > 1 + \beta_2\varepsilon_2 S(n) + \theta I_1(n) + \beta_1\varepsilon_1 S(n)I_1(n) > 0, \end{cases} \tag{2.12}$$

then the solution of system (2.2) decreases monotonically.

Proof

(i) Assume that $(S(n), I_1(n), I_2(n))_{n=0}^{\infty}$ is a positive solution of system (2.2). If (2.11) holds, then

$$\begin{cases} \dfrac{S(n+1)}{S(n)} = \dfrac{U_1(n)}{(U_1(n) - \alpha_1 S(n)) \cdot \exp(-r_1 U_1(n)) + \alpha_1 S(n)} > 1, \\ \dfrac{I_1(n+1)}{I_1(n)} = \dfrac{U_2}{(U_2(n) - \alpha_2 I_1(n)) \cdot \exp(-r_2 U_2(n)) + \alpha_2 I_1(n)} > 1, \\ \dfrac{I_2(n+1)}{I_2(n)} = \dfrac{U_3}{(U_3(n) - \alpha_3 I_2(n)) \cdot \exp(-r_3 U_3(n)) + \alpha_3 I_2(n)} > 1, \end{cases}$$

which implies that the solution increases.

(ii) The proof is similar to Theorem 2.1/(i) and thus omitted.

Theorem 2.2. Assume that $(S(n), I_1(n), I_2(n))_{n=0}^{\infty}$ is a positive solution of system (2.2). Then the following statements are true.

(i) Let Theorem 2.1/(i) hold. Then

$$S(n) \in \left(0, \frac{p}{\alpha_1}\right),\ I_1(n+1) \in \left(0, \frac{1}{\alpha_2}\right) \text{ and } I_2(n+1) \in \left(0, \frac{1}{\alpha_3}\right). \tag{2.13}$$

(ii) Let Theorem 2.1/(ii) hold. If $U_1 > \frac{1}{r_1}\ln 2$, $U_2 > \frac{1}{r_2}\ln 2$ and $U_3 > \frac{1}{r_3}\ln 2$, then system (2.2) is bounded from above with

$$\begin{cases} S(n) < S(0)\exp\left(r_1\alpha_1 \sum_{i=0}^{n-1} S(i)\right), \\ I_1(n) < I_1(0)\exp\left(r_2\alpha_2 \sum_{i=0}^{n-1} I_1(i)\right), \\ I_2(n) < I_2(0)\exp\left(r_2\alpha_2 \sum_{i=0}^{n-1} I_2(i)\right). \end{cases} \tag{2.14}$$

Proof

(i) Assume that the conditions in Theorem 2.1./(i) hold. Then we obtain

$$
\begin{aligned}
S(n+1) &= \frac{S(n)U_1(n)}{(U_1(n)-\alpha_1 S(n))\cdot \exp(-r_1U_1(n))+\alpha_1 S(n)}\\
&= \frac{S(n)\cdot U_1(n)}{U_1(n)\cdot \exp(-r_1U_1(n))+\alpha_1 S(n)\cdot(1-\exp(-r_1U_1(n)))}\\
&< \frac{p}{\alpha_1\cdot(1-\exp(-r_1U_1(n)))} < \frac{p}{\alpha_1\cdot(1-\exp(-r_1\alpha_1 S(n)))} < \frac{p}{\alpha_1}\\
I_1(n+1) &= \frac{I_1(n)U_2(n)}{U_2(n)\cdot \exp(-r_2U_2(n))+\alpha_2 I_1(n)(1-\exp(-r_2U_2(n)))}\\
&< \frac{U_2(n)}{\alpha_2(1-\exp(-r_2U_2(n)))} < \frac{1}{\alpha_2(1-\exp(-r_2\alpha_2 I_1(n)))} < \frac{1}{\alpha_2}\\
I_2(n+1) &= \frac{I_2(n)U_3(n)}{U_3(n)\cdot \exp(-r_3U_3)+\alpha_3 I_2(n)(1-\exp(-r_3U_3))}\\
&< \frac{U_3(n)}{\alpha_3 I_2(n)(1-\exp(-r_3U_3(n)))} < \frac{1}{\alpha_3(1-\exp(-r_3\alpha_3 I_2(n)))} < \frac{1}{\alpha_3}.
\end{aligned}
$$

This completes the proof of part (i).

(ii) Assume that the conditions in Theorem 2.1./(ii) hold. Then we have

$$
\begin{aligned}
S(n+1) &= \frac{S(n)\cdot U_1(n)}{U_1(n)\cdot \exp(-r_1U_1(n))+\alpha_1 S(n)\cdot(1-\exp(-r_1U_1(n)))}\\
&< \frac{S(n)}{1-\exp(-r_1U_1(n))} = \frac{S(n)\cdot \exp(r_1U_1(n))}{\exp(r_1U_1(n))-1} < \frac{S(n)\exp(r_1\alpha_1 S(n))}{\exp(r_1U_1(n))-1}\\
&< S(n)\exp(r_1\alpha_1 S(n)).
\end{aligned}
$$

For $n = 1, \ldots$ we get

$$S(n) < S(0)\exp\left(r_1\alpha_1 \sum_{i=0}^{n-1} S(i)\right).$$

Similarly, we obtain

$$I_1(n) < I_1(0)\exp\left(r_2\alpha_2 \sum_{i=0}^{n-1} I_1(i)\right) \text{ and } I_2(n) < I_2(0)\exp\left(r_2\alpha_2 \sum_{i=0}^{n-1} I_2(i)\right).$$

This completes the proof of part (ii).

Theorem 2.3. Let $(S(n), I_1(n), I_2(n))_{n=0}^{\infty}$ be a positive solution of system (2.2), which consists a single semi cycle. Furthermore, assume (2.11) holds. If

$$U_1(n-1) < \frac{1}{r_1}\ln\left(\frac{\bar{S}}{S(n-1)}\right),\quad U_2(n-1) < \frac{1}{r_2}\ln\left(\frac{\overline{I_1}}{I_1(n-1)}\right) \text{ and}$$

$$U_3(n-1) < \frac{1}{r_3}\ln\left(\frac{\overline{I_2}}{I_2(n-1)}\right), \tag{2.15}$$

for all $n \geq 1$, then $(S(n), I_1(n), I_2(n))_{n=0}^{\infty}$ converges monotonically to the positive equilibrium point $\bar{x} = (\bar{S}, \overline{I_1}, \overline{I_2})$.

Proof. Suppose that $0 < S(n-1) < \bar{S}$ for all $n \geq 1$. Note that in this case, we have

$$S(n-1) < S(n) = \frac{S(n-1)(p - \beta_1 I_1(n-1) - \beta_2 I_2(n-1))}{(p - \beta_1 I_1(n-1) - \beta_2 I_2(n-1) - \alpha_1 S(n-1)) \cdot e^{-r_1(p-\beta_1 I_1(n-1)-\beta_2 I_2(n-1))} + \alpha_1 S(n-1)},$$

where

$$\left(1 - e^{-r_1(p-\beta_1 I_1(n-1)-\beta_2 I_2(n-1))}\right) \cdot (p - \beta_1 I_1(n-1) - \beta_2 I_2(n-1) - \alpha_1 S(n-1)) > 0$$

holds for the conditions in (2.11). Moreover, from

$$S(n) = \frac{S(n-1)(p - \beta_1 I_1(n-1) - \beta_2 I_2(n-1))}{(p - \beta_1 I_1(n-1) - \beta_2 I_2(n-1) - \alpha_1 S(n-1)) \cdot e^{-r_1(p-\beta_1 I_1(n-1)-\beta_2 I_2(n-1))} + \alpha_1 S(n-1)} < \bar{S},$$

we can write

$$\begin{aligned} &S(n-1)\left(p - \beta_1 I_1(n-1) - \beta_2 I_2(n-1) - \alpha_1 \bar{S}\right) \\ &< \bar{S}(p - \beta_1 I_1(n-1) - \beta_2 I_2(n-1) - \alpha_1 S(n-1)) \cdot e^{-r_1(p-\beta_1 I_1(n-1)-\beta_2 I_2(n-1))}, \end{aligned}$$

which holds for

$$\begin{aligned} &p - \beta_1 I_1(n-1) - \beta_2 I_2(n-1) - \alpha_1 \bar{S} \\ &< p - \beta_1 I_1(n-1) - \beta_2 I_2(n-1) - \alpha_1 S(n-1) \end{aligned}$$

and

$$S(n-1) < \bar{S} \cdot e^{-r_1(p-\beta_1 I_1(n-1)-\beta_2 I_2(n-1))} \Rightarrow U_1(n-1) < \frac{1}{r_1} \ln\left(\frac{\bar{S}}{S(n-1)}\right). \quad (2.16)$$

Thus we obtain

$$0 < S(n-1) < S(n) < \bar{S} \text{ for all } n \geq 1.$$

In a similar way as before, we can prove that

$$0 < I_1(n-1) < I_1(n) < \overline{I_1} \text{ for all } n \geq 1$$

and

$$0 < I_2(n-1) < I_2(n) < \overline{I_2} \text{ for all } n \geq 1.$$

The proof is completed.

Theorem 2.4. Let $(S(n), I_1(n), I_2(n))_{n=0}^{\infty}$ be a positive solution of (2.2). System (2.2) has no positive solutions of prime period two.

Proof. Let

$$\ldots, \theta, \mu, \theta, \mu, \ldots \tag{2.17}$$

be period two solutions of the $(S(n))_{n=0}^{\infty}$ such that $\theta \neq \mu$. Then, we have

$$\theta = \frac{\mu(p - \beta_1 I_1(n) - \beta_2 I_2(n))}{(p - \beta_1 I_1(n) - \beta_2 I_2(n) - \alpha_1 \mu) \cdot e^{-r_1(p - \beta_1 I_1(n) - \beta_2 I_2(n))} + \alpha_1 \mu} \tag{2.18}$$

and

$$\mu = \frac{\theta(p - \beta_1 I_1(n) - \beta_2 I_2(n))}{(p - \beta_1 I_1(n) - \beta_2 I_2(n) - \alpha_1 \theta) \cdot e^{-r_1(p - \beta_1 I_1(n) - \beta_2 I_2(n))} + \alpha_1 \theta}. \tag{2.19}$$

Note that

$$\mu(p - \beta_1 I_1(n) - \beta_2 I_2(n)) - \theta\{(p - \beta_1 I_1(n) - \beta_2 I_2(n) - \alpha_1 \mu) \cdot e^{-r_1(p - \beta_1 I_1(n) - \beta_2 I_2(n))}\} \\ = \alpha_1 \theta \mu \tag{2.20}$$

and

$$\theta(p - \beta_1 I_1(n) - \beta_2 I_2(n)) - \mu\{(p - \beta_1 I_1(n) - \beta_2 I_2(n) - \alpha_1 \theta) \cdot e^{-r_1(p - \beta_1 I_1(n) - \beta_2 I_2(n))}\} \\ = \alpha_1 \theta \mu. \tag{2.21}$$

From (2.20) and (2.21), we can write

$$(\mu - \theta)(p - \beta_1 I_1(n) - \beta_2 I_2(n))\left(1 + e^{-r_1(p - \beta_1 I_1(n) - \beta_2 I_2(n))}\right) = 0. \tag{2.22}$$

Since $U_1(n) \neq 0$ and $1 + e^{-r_1(p - \beta_1 I_1(n) - \beta_2 I_2(n))} \neq 0$, (2.22) can holds if $\mu = \theta$, which contradicts with our assumption. This completes the proof.

3 Local and Global Stability Analysis

In this section, we analyze the local stability by using the Schur-Cohn Criteria. We consider the case, where the infection is active, and the awareness of protection is low.

To show the local asymptotic stability of the positive equilibrium point, we use the Schur-Cohn criterion.

Theorem 3.1. [17] The characteristic polynomial

$$P(\lambda) = \lambda^3 + a_2 \lambda^2 + a_1 \lambda + a_0 , \tag{3.1}$$

has all its roots inside the unit disk ($|\lambda| < 1$) if and only if

(i) $P(1) = 1 + a_2 + a_1 + a_0 > 0$ and $(-1)^3 P(-1) = 1 - a_2 + a_1 - a_0 > 0$.
(ii) $D_2^{+} = 1 + a_1 - a_0^2 - a_0 a_2 > 0$

(iii) $D_2^- = 1 - a_1 - a_0^2 + a_0 a_2 > 0.$

Theorem 3.2. Let $\bar{X} = (\bar{S}, \overline{I_1}, \overline{I_2})$ be the positive equilibrium point of system (2.2). Assume
$\frac{\alpha_1}{\alpha_3} > \frac{3\beta_2(3+\overline{I_2})}{(\theta+3\beta_2\varepsilon_1\bar{S})\bar{S}}$, $\beta_2 > \frac{(1-\varepsilon_1)(\theta+3\beta_2\varepsilon_1\bar{S})\bar{S}\alpha_1}{\varepsilon_1(1+3\overline{I_2})\alpha_2+3(1-\varepsilon_1)(3+\overline{I_2})\alpha_3}$ and $\frac{\alpha_2}{\alpha_1} > \frac{(1-\varepsilon_1)(\theta+3\beta_2\varepsilon_1\bar{S})}{\beta_2\varepsilon_1(1+3\overline{I_2})}$
hold.

If

$$r_1 \in \left(\ln\left(\frac{\beta_2^2\varepsilon_1(1+3\overline{I_2})\alpha_2 + 3\beta_2^2(1-\varepsilon_1)(3+\overline{I_2})\alpha_3}{\beta_2^2\varepsilon_1(1+3\overline{I_2})\alpha_2 + 3\beta_2^2(1-\varepsilon_1)(3+\overline{I_2})\alpha_3 - \beta_2(1-\varepsilon_1)(\theta+3\beta_2\varepsilon_1\bar{S})\bar{S}\alpha_1}\right)^{\frac{1}{\overline{U_1}}}, \ln\left(\frac{\beta_2^2\varepsilon_1(1+3\overline{I_2})\alpha_2}{\beta_2^2\varepsilon_1(1+3\overline{I_2})\alpha_2 - \beta_2(1-\varepsilon_1)(\theta+3\beta_2\varepsilon_1\bar{S})\bar{S}\alpha_1}\right)^{\frac{1}{\overline{U_1}}} \right) \tag{3.2}$$

$$r_2 \in \left(\ln\left(\frac{3\beta_2(1-\varepsilon_1)\bar{S}+\theta+\alpha_2}{3\beta_2(1-\varepsilon_1)\bar{S}+\theta}\right)^{\frac{1}{\overline{U_2}}}, \ln\left(\frac{\theta+\alpha_2}{\theta}\right)^{\frac{1}{\overline{U_2}}} \right), \tag{3.3}$$

and

$$\ln\left(\frac{\beta_2(1-\varepsilon_1)(\theta+3\beta_2\varepsilon_1\bar{S})\bar{S}\alpha_1}{\beta_2(1-\varepsilon_1)(\theta+3\beta_2\varepsilon_1\bar{S})\bar{S}\alpha_1 - 3\beta_2^2(1-\varepsilon_1)(3+\overline{I_2})\alpha_3}\right)^{\frac{1}{\overline{U_3}}} < r_3 \tag{3.4}$$

then the positive equilibrium point of system (2.2) is local asymptotically stable.

Proof. We consider at first Theorem 2.1./(i), where we have

$$\begin{aligned} 1 &+ (-a_{11} - a_{22} - a_{33}) + (a_{11}a_{22} + a_{11}a_{33} + a_{22}a_{33} - a_{13}a_{31} - a_{23}a_{32} - a_{12}a_{21}) \\ &+ (a_{13}a_{31}a_{22} + a_{23}a_{32}a_{11} + a_{12}a_{21}a_{33} - a_{11}a_{22}a_{33} - a_{21}a_{32}a_{13} - a_{31}a_{12}a_{23}) > 0 \end{aligned} \tag{3.5}$$

and

$$\begin{aligned} 1 &- (-a_{11} - a_{22} - a_{33}) + (a_{11}a_{22} + a_{11}a_{33} + a_{22}a_{33} - a_{13}a_{31} - a_{23}a_{32} - a_{12}a_{21}) \\ &- (a_{13}a_{31}a_{22} + a_{23}a_{32}a_{11} + a_{12}a_{21}a_{33} - a_{11}a_{22}a_{33} - a_{21}a_{32}a_{13} - a_{31}a_{12}a_{23}) > 0. \end{aligned} \tag{3.6}$$

Considering (3.5) and (3.6), we obtain

$$1 + a_{11}a_{22} + a_{11}a_{33} + a_{22}a_{33} > a_{13}a_{31} + a_{23}a_{32} + a_{12}a_{21}. \tag{3.7}$$

The above inequality (3.7) can be written in the following form;

$$-\frac{\beta_2^2\varepsilon_1(1+3\overline{I_2})\left(1-e^{-r_1\overline{U_1}}\right)\left(1-e^{-r_3\overline{U_3}}\right)}{\alpha_1\alpha_3}$$

$$
\begin{aligned}
&+\frac{\beta_2(1-\varepsilon_1)(\theta+3\beta_2\varepsilon_1\bar{S})\bar{S}\left(1-e^{-r_2\overline{U_2}}\right)\left(1-e^{-r_3\overline{U_3}}\right)}{\alpha_2\alpha_3}\\
&-\frac{3\beta_2^2(1-\varepsilon_1)(3+\bar{I_2})\left(1-e^{-r_1\overline{U_1}}\right)\left(1-e^{-r_2\overline{U_2}}\right)}{\alpha_1\alpha_2}\\
&<1+\left(\frac{-\theta+(\theta+\alpha_2)e^{-r_2\overline{U_2}}}{\alpha_2}\right)e^{-r_1\overline{U_1}}+e^{-(r_1\overline{U_1}+r_3\overline{U_3})}\\
&+\left(\frac{-\theta+(\theta+\alpha_2)e^{-r_2\overline{U_2}}}{\alpha_2}\right)e^{-r_3\overline{U_3}}
\end{aligned}
\tag{3.8}
$$

From

$$
\frac{-\theta+(\theta+\alpha_2)e^{-r_2\overline{U_2}}}{\alpha_2}>0, \tag{3.9}
$$

we obtain

$$
r_2<\frac{1}{\overline{U_2}}\ln\left(\frac{\theta+\alpha_2}{\theta}\right). \tag{3.10}
$$

Furthermore from (3.8), if

$$
\begin{aligned}
&-\frac{\beta_2^2\varepsilon_1(1+3\bar{I_2})\left(1-e^{-r_1\overline{U_1}}\right)\left(1-e^{-r_3\overline{U_3}}\right)}{\alpha_1\alpha_3}\\
&+\frac{\beta_2(1-\varepsilon_1)(\theta+3\beta_2\varepsilon_1\bar{S})\bar{S}\left(1-e^{-r_2\overline{U_2}}\right)\left(1-e^{-r_3\overline{U_3}}\right)}{\alpha_2\alpha_3}\\
&-\frac{3\beta_2^2(1-\varepsilon_1)(3+\bar{I_2})\left(1-e^{-r_1\overline{U_1}}\right)\left(1-e^{-r_2\overline{U_2}}\right)}{\alpha_1\alpha_2}<0,
\end{aligned}
\tag{3.11}
$$

then

$$
\begin{aligned}
&-\beta_2^2\varepsilon_1(1+3\bar{I_2})\alpha_2\left(1-e^{-r_1\overline{U_1}}-e^{-r_3\overline{U_3}}+e^{-r_3\overline{U_3}-r_1\overline{U_1}}\right)\\
&+\beta_2(1-\varepsilon_1)(\theta+3\beta_2\varepsilon_1\bar{S})\bar{S}\alpha_1\left(1-e^{-r_2\overline{U_2}}-e^{-r_3\overline{U_3}}+e^{-r_3\overline{U_3}-r_2\overline{U_2}}\right)\\
&-3\beta_2^2(1-\varepsilon_1)(3+\bar{I_2})\alpha_3\left(1-e^{-r_2\overline{U_2}}-e^{-r_1\overline{U_1}}+e^{-r_1\overline{U_1}-r_2\overline{U_2}}\right)<0.
\end{aligned}
\tag{3.12}
$$

Rearrangement of (3.12), we get

$$
\begin{aligned}
&\left(\beta_2^2\varepsilon_1(1+3\bar{I_2})\alpha_2+3\beta_2^2(1-\varepsilon_1)(3+\bar{I_2})\alpha_3\right)e^{-r_1\overline{U_1}}\\
&+\beta_2(1-\varepsilon_1)(\theta+3\beta_2\varepsilon_1\bar{S})\bar{S}\alpha_1e^{-r_3\overline{U_3}-r_2\overline{U_2}}+\\
&\left(\beta_2^2\varepsilon_1(1+3\bar{I_2})\alpha_2-\beta_2(1-\varepsilon_1)(\theta+3\beta_2\varepsilon_1\bar{S})\bar{S}\alpha_1\right)e^{-r_3\overline{U_3}}
\end{aligned}
$$

$$
\begin{aligned}
&< \Big(\beta_2^2\varepsilon_1\big(1+3\overline{I_2}\big)\alpha_2 + 3\beta_2^2(1-\varepsilon_1)\big(3+\overline{I_2}\big)\alpha_3 - \beta_2(1-\varepsilon_1)\big(\theta+3\beta_2\varepsilon_1\bar{S}\big)\bar{S}\alpha_1\Big) \\
&+ \beta_2^2\varepsilon_1\big(1+3\overline{I_2}\big)\alpha_2 e^{-r_3\overline{U_3}-r_1\overline{U_1}} + \\
&\Big(\beta_2(1-\varepsilon_1)\big(\theta+3\beta_2\varepsilon_1\bar{S}\big)\bar{S}\alpha_1 - 3\beta_2^2(1-\varepsilon_1)\big(3+\overline{I_2}\big)\alpha_3\Big)e^{-r_2\overline{U_2}} \\
&+ 3\beta_2^2(1-\varepsilon_1)\big(3+\overline{I_2}\big)\alpha_3 e^{-r_1\overline{U_1}-r_2\overline{U_2}}
\end{aligned}
\tag{3.13}
$$

Considering (3.13), we have

$$
\begin{aligned}
&\beta_2(1-\varepsilon_1)\big(\theta+3\beta_2\varepsilon_1\bar{S}\big)\bar{S}\alpha_1 e^{-r_3\overline{U_3}-r_2\overline{U_2}} \\
&< \big(\beta_2(1-\varepsilon_1)\big(\theta+3\beta_2\varepsilon_1\bar{S}\big)\bar{S}\alpha_1 - 3\beta_2^2(1-\varepsilon_1)\big(3+\overline{I_2}\big)\alpha_3\big)e^{-r_2\overline{U_2}}
\end{aligned}
$$

where we obtain

$$
r_3 > \frac{1}{\overline{U_3}}\ln\left(\frac{\big(\theta+3\beta_2\varepsilon_1\bar{S}\big)\bar{S}\alpha_1}{\big(\theta+3\beta_2\varepsilon_1\bar{S}\big)\bar{S}\alpha_1 - 3\beta_2\big(3+\overline{I_2}\big)\alpha_3}\right). \tag{3.14}
$$

for

$$
\frac{\alpha_1}{\alpha_3} > \frac{3\beta_2\big(3+\overline{I_2}\big)}{\big(\theta+3\beta_2\varepsilon_1\bar{S}\big)\bar{S}}. \tag{3.15}
$$

Furthermore, from (3.13) we get

$$
\begin{aligned}
&(\beta_2^2\varepsilon_1\big(1+3\overline{I_2}\big)\alpha_2 + 3\beta_2^2(1-\varepsilon_1)\big(3+\overline{I_2}\big)\alpha_3)e^{-r_1\overline{U_1}} < (\beta_2^2\varepsilon_1\big(1+3\overline{I_2}\big)\alpha_2 \\
&+ 3\beta_2^2(1-\varepsilon_1)\big(3+\overline{I_2}\big)\alpha_3 - \beta_2(1-\varepsilon_1)\big(\theta+3\beta_2\varepsilon_1\bar{S}\big)\bar{S}\alpha_1)
\end{aligned}
\tag{3.16}
$$

where we have

$$
r_1 > \frac{1}{\overline{U_1}}\ln\left(\frac{\beta_2^2\varepsilon_1\big(1+3\overline{I_2}\big)\alpha_2 + 3\beta_2^2(1-\varepsilon_1)\big(3+\overline{I_2}\big)\alpha_3}{\beta_2^2\varepsilon_1\big(1+3\overline{I_2}\big)\alpha_2 + 3\beta_2^2(1-\varepsilon_1)\big(3+\overline{I_2}\big)\alpha_3 - \beta_2(1-\varepsilon_1)\big(\theta+3\beta_2\varepsilon_1\bar{S}\big)\bar{S}\alpha_1}\right) \tag{3.17}
$$

and

$$
\beta_2 > \frac{(1-\varepsilon_1)\big(\theta+3\beta_2\varepsilon_1\bar{S}\big)\bar{S}\alpha_1}{\varepsilon_1\big(1+3\overline{I_2}\big)\alpha_2 + 3(1-\varepsilon_1)\big(3+\overline{I_2}\big)\alpha_3}. \tag{3.18}
$$

At last, from the next inequality

$$
\Big(\beta_2^2\varepsilon_1\big(1+3\overline{I_2}\big)\alpha_2 - \beta_2(1-\varepsilon_1)\big(\theta+3\beta_2\varepsilon_1\bar{S}\big)\bar{S}\alpha_1\Big)e^{-r_3\overline{U_3}} < \beta_2^2\varepsilon_1\big(1+3\overline{I_2}\big)\alpha_2 e^{-r_3\overline{U_3}-r_1\overline{U_1}}, \tag{3.19}
$$

we have

$$
r_1 < \frac{1}{\overline{U_1}}\ln\left(\frac{\beta_2^2\varepsilon_1\big(1+3\overline{I_2}\big)\alpha_2}{\beta_2^2\varepsilon_1\big(1+3\overline{I_2}\big)\alpha_2 - \beta_2(1-\varepsilon_1)\big(\theta+3\beta_2\varepsilon_1\bar{S}\big)\bar{S}\alpha_1}\right), \tag{3.20}
$$

where

$$\frac{\alpha_2}{\alpha_1} > \frac{(1-\varepsilon_1)(\theta + 3\beta_2\varepsilon_1\bar{S})}{\beta_2\varepsilon_1(1+3\bar{I}_2)}. \tag{3.21}$$

Considering (3.17) and (3.20), we get

$$\begin{aligned}
&\frac{1}{U_1}\ln\left(\frac{\beta_2^2\varepsilon_1(1+3\bar{I}_2)\alpha_2+3\beta_2^2(1-\varepsilon_1)(3+\bar{I}_2)\alpha_3}{\beta_2^2\varepsilon_1(1+3\bar{I}_2)\alpha_2+3\beta_2^2(1-\varepsilon_1)(3+\bar{I}_2)\alpha_3-\beta_2(1-\varepsilon_1)(\theta+3\beta_2\varepsilon_1\bar{S})\bar{S}\alpha_1}\right) < r_1 \\
&< \frac{1}{U_1}\ln\left(\frac{\beta_2^2\varepsilon_1(1+3\bar{I}_2)\alpha_2}{\beta_2^2\varepsilon_1(1+3\bar{I}_2)\alpha_2-\beta_2(1-\varepsilon_1)(\theta+3\beta_2\varepsilon_1\bar{S})\bar{S}\alpha_1}\right).
\end{aligned}$$

The conditions for (i) are obtained.

Considering (ii) and (iii), we have

$$1 - c_1^2 > |-c_2 - c_1c_3|, \tag{3.22}$$

where

$$\begin{aligned}
c_1 &= a_{11}a_{22}a_{33} + a_{21}a_{32}a_{13} + a_{31}a_{12}a_{23} - a_{13}a_{31}a_{22} - a_{23}a_{32}a_{11} - a_{12}a_{21}a_{33}, \\
c_2 &= a_{13}a_{31} + a_{23}a_{32} + a_{12}a_{21} - a_{11}a_{22} - a_{11}a_{33} - a_{22}a_{33}, \\
c_3 &= a_{11} + a_{22} + a_{33}.
\end{aligned}$$

The inequality (3.22) can be written as

$$c_1(c_1 - c_3) - c_2 < c_1(c_1 + c_3) + c_2 < 1. \tag{3.23}$$

In this case, we have to show the condition

$$c_1(c_1 + c_3) + c_2 < 0 + 1 \tag{3.24}$$

for $c_1 < 0$, $c_2 < 1$ and $c_1 + c_3 > 0$. Let us consider now

$$c_2 < 1 \Rightarrow a_{13}a_{31} + a_{23}a_{32} + a_{12}a_{21} < 1 + a_{11}a_{22} + a_{11}a_{33} + a_{22}a_{33}, \tag{3.25}$$

and

$$\begin{aligned}
&c_1 + c_3 > 0 \\
&\Rightarrow -a_{11}a_{22}a_{33} - a_{21}a_{32}a_{13} - a_{31}a_{12}a_{23} < a_{11} + a_{22} + a_{33} - a_{13}a_{31}a_{22} \\
&\qquad\qquad -a_{23}a_{32}a_{11} - a_{12}a_{21}a_{33}
\end{aligned} \tag{3.26}$$

and

$$c_1 < 0 \Rightarrow a_{11}a_{22}a_{33} + a_{21}a_{32}a_{13} + a_{31}a_{12}a_{23} < a_{13}a_{31}a_{22} + a_{23}a_{32}a_{11} + a_{12}a_{21}a_{33}. \tag{3.27}$$

From (3.26) and (3.27), we obtain

$$a_{11} + a_{22} + a_{33} > 0, \tag{3.28}$$

which holds for

$$e^{-r_1\overline{U_1}} + \frac{-\theta + (\theta + \alpha_2)e^{-r_2\overline{U_2}}}{\alpha_2} + e^{-r_3\overline{U_3}} > 0 \Rightarrow r_2 < \frac{1}{\overline{\overline{U_2}}}\ln\left(\frac{\theta + \alpha_2}{\theta}\right). \tag{3.29}$$

Condition (3.25) is already proven in (i).
Furthermore, if the following conditions hold, then (3.27) is available;
I. $a_{31}a_{12}a_{23} < a_{13}a_{31}a_{22} \Rightarrow a_{12}a_{23} < a_{13}a_{22}$, where we obtain

$$r_2 > \frac{1}{\overline{\overline{U_2}}}\ln\left(\frac{3\beta_2(1-\varepsilon_1)\bar{S} + \theta + \alpha_2}{3\beta_2(1-\varepsilon_1)\bar{S} + \theta}\right). \tag{3.30}$$

Considering (3.29) and (3.30), we obtain

$$r_2 \in \left(\frac{1}{\overline{\overline{U_2}}}\ln\left(\frac{3\beta_2(1-\varepsilon_1)\bar{S} + \theta + \alpha_2}{3\beta_2(1-\varepsilon_1)\bar{S} + \theta}\right), \frac{1}{\overline{\overline{U_2}}}\ln\left(\frac{\theta + \alpha_2}{\theta}\right)\right) \tag{3.31}$$

II. $a_{21}a_{32}a_{13} < a_{12}a_{21}a_{33} \Rightarrow a_{32}a_{13} < a_{12}a_{33}$, where we get

$$r_3 > \frac{1}{\overline{\overline{U_3}}}\ln\left(\frac{\theta + 3\left(\beta_2\varepsilon_1\bar{S} + \alpha_3\right)}{\theta + 3\beta_2\varepsilon_1\bar{S}}\right). \tag{3.32}$$

From (3.14) and (3.32), we have

$$\frac{1}{\overline{\overline{U_3}}}\ln\left(\frac{\theta + 3\left(\beta_2\varepsilon_1\bar{S} + \alpha_3\right)}{\theta + 3\beta_2\varepsilon_1\bar{S}}\right) <$$
$$\frac{1}{\overline{\overline{U_3}}}\ln\left(\frac{\beta_2(1-\varepsilon_1)\left(\theta + 3\beta_2\varepsilon_1\bar{S}\right)\bar{S}\alpha_1}{\beta_2(1-\varepsilon_1)\left(\theta + 3\beta_2\varepsilon_1\bar{S}\right)\bar{S}\alpha_1 - 3\beta_2^2(1-\varepsilon_1)\left(3+\overline{I_2}\right)\alpha_3}\right) < r_3 \tag{3.33}$$

III. $a_{11}a_{22}a_{33} < a_{23}a_{32}a_{11} \Rightarrow a_{22}a_{33} < a_{23}a_{32}$, where we obtain that it holds for (3.32).

This completes the proof.

Theorem 3.3. Let $\bar{X} = \left(\bar{S}, \overline{I_1}, \overline{I_2}\right)$ be the positive equilibrium point of system (2.2) and assume the conditions in Theorem 3.2. hold.

(i) Let Theorem 2.1./(i) holds. Then the positive equilibrium point of system (2.2) is global asymptotically stable, if

$$0 < U_1(n) < \ln\left(\frac{2\bar{S} - S(n)}{S(n)}\right) \text{ and } S(n) < \bar{S},$$
$$0 < U_2(n) < \ln\left(\frac{2\overline{I_1} - I_1(n)}{I_1(n)}\right) \text{ and } I_1(n) < \overline{I_1}$$

and

$$0 < U_3(n) < \ln\left(\frac{2\overline{\mathrm{I}_2} - \mathrm{I}_2(n)}{\mathrm{I}_2(n)}\right) \quad \text{and } \mathrm{I}_2(n) < \overline{\mathrm{I}_2}.$$

(ii) Let Theorem 2.1./(ii) holds. Then the positive equilibrium point of system (2.2) is global asymptotically stable, if

$$S(n) > 2\bar{S}, \quad \mathrm{I}_1(n) > 2\overline{\mathrm{I}_1} \quad \text{and } \mathrm{I}_2(n) > 2\overline{\mathrm{I}_2}.$$

Proof. Let $\bar{X} = \left(\bar{\mathrm{S}}, \overline{\mathrm{I}_1}, \overline{\mathrm{I}_2}\right)$ be the positive equilibrium point of system (2.2) and let us consider a Lyapunov function $V(n)$ defined by

$$V(n) = \left(X(n) - \bar{X}\right)^2, \; n = 0, 1, 2, \ldots \tag{3.34}$$

where $X(n) = (S(n), \mathrm{I}_1(n), \mathrm{I}_2(n))$ and $\bar{X} = \left(\bar{\mathrm{S}}, \overline{\mathrm{I}_1}, \overline{\mathrm{I}_2}\right)$.

The change along the solutions of the system is

$$\begin{aligned} \Delta V(n) &= V(n+1) - V(n) \\ &= \left(X(n+1) - \bar{X}\right)^2 - \left(X(n) - \bar{X}\right)^2 \\ &= (X(n+1) - X(n))\left(X(n+1) + X(n) - 2\bar{X}\right). \end{aligned} \tag{3.35}$$

From the first equation of system (2.2), we have

$$\begin{aligned} \Delta V_1(n) =& (S(n+1) - S(n))\left(S(n+1) + S(n) - 2\bar{S}\right) \\ =& \left(\frac{\mathrm{S(n)}U_1(n)}{(U_1(n) - \alpha_1\mathrm{S(n)}) \cdot \exp(-\mathrm{r}_1 U_1(\mathrm{n})) + \alpha_1 \mathrm{S(n)}} - S(n)\right) \\ & \left(\frac{\mathrm{S(n)}U_1(n)}{(U_1(n) - \alpha_1\mathrm{S(n)}) \cdot \exp(-\mathrm{r}_1 U_1(\mathrm{n})) + \alpha_1 \mathrm{S(n)}} + S(n) - 2\bar{S}\right) \\ =& \frac{1}{(K(n))^2} \cdot (\mathrm{S(n)}U_1(n) - K(n) \cdot S(n)) \cdot \left(\mathrm{S(n)}U_1(n) + \left(S(n) - 2\bar{S}\right)K(n)\right), \end{aligned}$$

where $K(n) = (U_1(n) - \alpha_1 \mathrm{S(n)}) \cdot \exp(-\mathrm{r}_1 U_1(\mathrm{n})) + \alpha_1 \mathrm{S(n)}$.

(i) In this case, if (2.11) in Theorem 2.1/(i) holds, then

$$\mathrm{S(n)}U_1(n) - K(n) \cdot S(n) = S(n)(U_1(n) - \alpha_1 \mathrm{S(n)})\left(1 - e^{-\mathrm{r}_1 U_1(\mathrm{n})}\right) > 0. \tag{3.36}$$

So, we have to obtain

$$\mathrm{S(n)}U_1(n) + \left(S(n) - 2\bar{S}\right)K(n) < 0, \tag{3.37}$$

to get $\Delta V_1(n) < 0$, which holds for

$$0 < U_1(n) < \ln\left(\frac{2\bar{S} - S(n)}{S(n)}\right) \quad \text{and } S(n) < \bar{S}. \tag{3.38}$$

This implies that $\lim_{n\to\infty} S(n) = \bar{S}$. Similarly, we can obtain the conditions

$$0 < U_2(n) < \ln\left(\frac{2\overline{I_1} - I_1(n)}{I_1(n)}\right) \quad \text{and } I_1(n) < \overline{I_1} \tag{3.39}$$

for $\Delta V_2(n) < 0$ and

$$0 < U_3(n) < \ln\left(\frac{2\overline{I_2} - I_2(n)}{I_2(n)}\right) \quad \text{and } I_2(n) < \overline{I_2}. \tag{3.40}$$

This completes part (i).

(ii) If (2.12) in Theorem 2.1/(ii) holds, then

$$S(n)U_1(n) - K(n)\cdot S(n) = S(n)(U_1(n) - \alpha_1 S(n))\left(1 - e^{-r_1 U_1(n)}\right) > 0, \tag{3.41}$$

since $U_1(n) < 0$. Additionally,

$$\begin{aligned} &S(n)U_1(n) + \left(S(n) - 2\bar{S}\right)K(n) \\ &= U_1(n)\left[S(n) + \left(S(n) - 2\bar{S}\right)e^{-r_1 U_1(n)}\right] - \alpha_1 S(n)\left(S(n) - 2\bar{S}\right)\left(e^{-r_1 U_1(n)} - 1\right) < 0, \end{aligned}$$

if $S(n) > 2\bar{S}$. In a similar way, we can obtain $I_1(n) > 2\overline{I_1}$ and $I_2(n) > 2\overline{I_2}$, which completes the proof.

Example 3.1. In this work, we consider the HIV transmission case in India using the data of [7, 8]. The initial conditions are given as follows;

$S(0) = 100,000,000$ adult population that were recorded in 1990
$I_1(0) = 500,000$ assumed that were found after 1990 and were already infected that time
$I_2(0) = 200,000$ number of HIV positives at the end of 1990.

The blue graph denotes the susceptible class $S(n)$, the red graph is the HIV positives that do not know that they are infected, namely $I_1(n)$, and the green graph shows the class that know they are infected, which is $I_2(n)$. In Fig. 1, we see the behavior of these classes according to the given data, where $r_1 \in [0.012, 0.05]$ and $r_1 = 4\cdot r_2 = 12\cdot r_3$. In Fig. 2, we increased the rate of infections of class I_1 and I_2 to consider the case where both populations continue to infect people. We choose $\beta_1 = 1.5$ and $\beta_2 = 0.5$, where all other parameters and initial conditions are fixed. In Fig. 3, show the existence of the non-negative equilibrium point for I_1 and I_2. Figure 4 shows the per capita growth rate of each population class, while $r_1 \in [0.012, 0.05]$ and $r_1 = 4\cdot r_2 = 12\cdot r_3$ (Table 1).

Table 1. The parameters for $S(n)$, $I_1(n)$ and $I_2(n)$ populations (see [7, 8])

Parameter			Parameter		
p	Yearly immigrated adult persons	3, 000, 000	r_1	The population growth rate of susceptible class S	0.012
α_1	Death (of natural causes) for class S	0.0748	r_2	The population growth rate of class I_1	0.003
α_2	Death (of natural causes) for class I_1	0.0919	r_3	The population growth rate of class I_2	0.001
α_3	Death (of natural causes) for class I_2	0.0920	ε_1	Awareness of class S after tracing contact with I_1	0.01
β_1	The rate of infection from class I_1	1.3440	ε_2	Awareness of class S after tracing contact with I_2	0.01
β_2	The rate of infection from class I_2	0.4480	θ	Aware of infection in class I_1 by HIV test	0.015

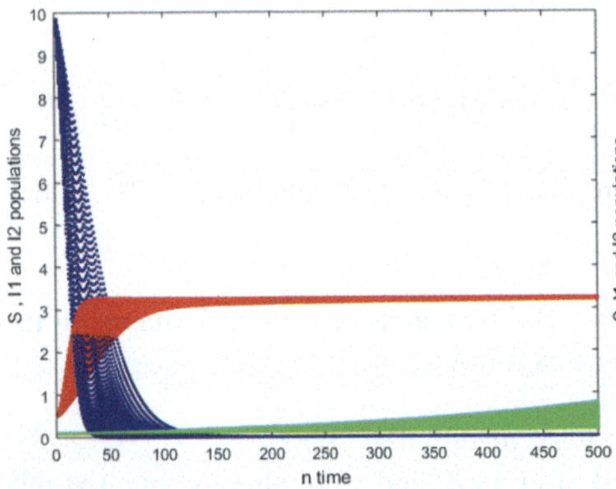

Fig. 1. Dynamical behavior of S(n), $I_1(n)$ $I_2(n)$.

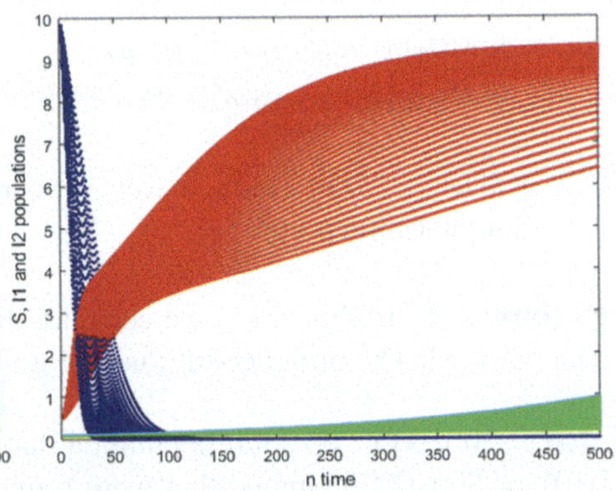

Fig. 2. Dynamical behavior of S(n), $I_1(n)$ and and $I_2(n)$, where $\beta_1 = 1.5$ and $\beta_2 = 0.5$

4 Stability Analysis for a Model with Allee Effect

Significant research for population models is obtained by Allee [18], who demonstrated that the Allee effect occurs when the population growth rate is reduced at low population size. It is well known that the logistic model assumes that per-capita growth rate declines monotonically when the density increase; however, it is shown that for population subjected to an Allee effect, per-capita growth rate gives a humped curve increasing at low density, up to a maximum intermediate density and then declines again. Many theoretical and laboratory studies have demonstrated the importance of the Allee effect in the dynamics of small populations, see for example [19–23].

In this section, we use Allee functions to I_1 and I_2 to analyse the spread of infection when the population of the infected classes are low.

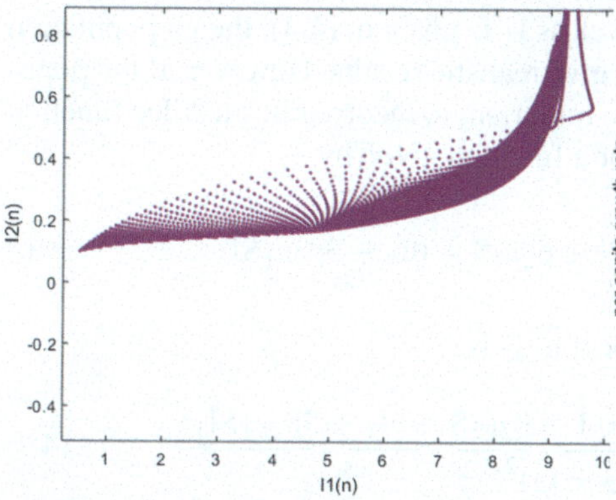

Fig. 3. Existence of non-negative equilibrium growth point of the infected classes

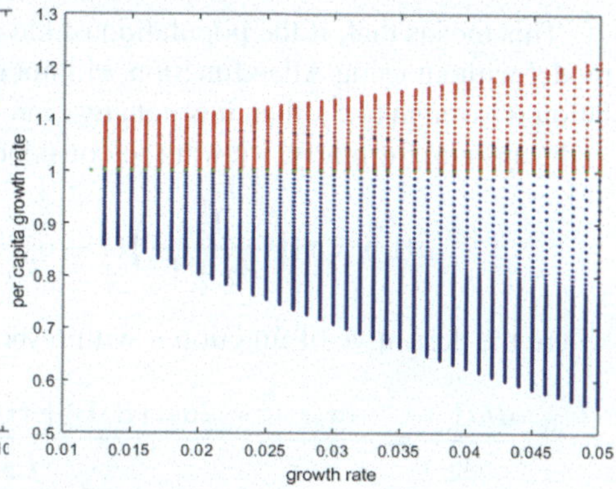

Fig. 4. Dynamical behavior of per-capita rate of each population

Let

$$\begin{cases} \frac{1}{S}\frac{dS}{dt} = r_1(p - \alpha_1 S - 3\beta_2 I_1 - \beta_2 I_2), \\ \frac{1}{I_1}\frac{dI_1}{dt} = r_2\left(\frac{I_1}{E_1+I_1}\right)(1 - (\alpha_2+\theta)I_1 + 3\beta_2(1-\varepsilon_1)S + \beta_2(1-\varepsilon_1)SI_2), \\ \frac{1}{I_2}\frac{dI_2}{dt} = r_3\left(\frac{I_2}{E_2+I_2}\right)(1 - \alpha_3 I_2 + \beta_2\varepsilon_2 S + \theta I_1 + 3\beta_2\varepsilon_1 SI_1), \end{cases} \tag{4.1}$$

where $a(I_1) = \frac{I_1}{E_1+I_1}$ and $a(I_2) = \frac{I_2}{E_2+I_2}$ are Allee functions, while E_1 and E_2 are Allee constants. By defining

$$g(I_1) = r_2\left(\frac{I_1}{E_1+I_1}\right)(1 - (\alpha_2+\theta)I_1 + \beta_1(1-\varepsilon_1)S + \beta_2(1-\varepsilon_2)SI_2), \tag{4.2}$$

moreover, taking the derivative with respect to I_1, we obtain

$$\frac{dg(I_1)}{dI_1} = \frac{-r_2(\alpha_2+\theta)I_1^2 - 2E_1 r_2(\alpha_2+\theta)I_1 + E_1 r_2(1 + 3\beta_2(1-\varepsilon_1)S + \beta_2(1-\varepsilon_1)SI_2)}{(E_1+I_1)^2} \tag{4.3}$$

By considering the sign of (4.3), we obtain that g is an increasing function, if

$$I_1 \in \left(0, \frac{-E_1(\alpha_2+\theta) + \sqrt{E_1^2(\alpha_2+\theta)^2 + (\alpha_2+\theta)E_1(1 + 3\beta_2(1-\varepsilon_1)S + \beta_2(1-\varepsilon_1)SI_2)}}{\alpha_2+\theta}\right), \tag{4.4}$$

and a decreasing function, if

$$I_1 \in \left(\frac{-E_1(\alpha_2+\theta) + \sqrt{E_1^2(\alpha_2+\theta)^2 + (\alpha_2+\theta)E_1(1 + 3\beta_2(1-\varepsilon_1)S + \beta_2(1-\varepsilon_1)SI_2)}}{\alpha_2+\theta}, \infty\right). \tag{4.5}$$

This means that, if the population of the class I_1 is given in (4.4), then a population model without using Allee function will not give realistic results. However, if the population is given in (4.5), then it is not any more important to incorporate an Allee function to the system. Similarly, we want to consider a function given by

$$h(I_2) = r_3\left(\frac{I_2}{E_2 + I_2}\right)(1 - \alpha_3 I_2 + \beta_2\varepsilon_2 S + \theta I_1 + 3\beta_2\varepsilon_1 S I_1), \tag{4.6}$$

where the derivative of function h with respect to I_2 is

$$\frac{dh(I_2)}{dI_2} = \frac{-\alpha_3 r_3 I_2^2 - 2\alpha_3 r_3 E_2 I_2 + r_3(1 + \beta_2\varepsilon_2 S + \theta I_1 + 3\beta_2\varepsilon_1 S I_1)}{(E_2 + I_2)^2}. \tag{4.7}$$

If

$$I_2 \in \left(0, \frac{-\alpha_3 E_2 + \sqrt{\alpha_3^2 E_2^2 + \alpha_3(1 + \beta_2\varepsilon_2 S + \theta I_1 + 3\beta_2\varepsilon_1 S I_1)}}{\alpha_3}\right), \tag{4.8}$$

then h is monotonic increasing and if

$$I_2 \in \left(\frac{-\alpha_3 E_2 + \sqrt{\alpha_3^2 E_2^2 + \alpha_3(1 + \beta_2\varepsilon_2 S + \theta I_1 + 3\beta_2\varepsilon_1 S I_1)}}{\alpha_3}, \infty\right), \tag{4.9}$$

then it is decreasing, which means that I_2 has to be in the interval (4.8).

Furthermore, for $t \in [n, n + 1)$, system (2.2) with Allee effect can be obtained as

$$\begin{cases} S(n+1) = \frac{S(n)U_1(n)}{(U_1(n) - \alpha_1 S(n))\cdot \exp(-r_1 U_1(n)) + \alpha_1 S(n)}, \\ I_1(n+1) = \frac{I_1(n)U_2(n)}{(U_2(n) - \alpha_2 I_1(n))\cdot \exp(-r_2 a(I_1(n))U_2(n)) + \alpha_2 I_1(n)}, \\ I_2(n+1) = \frac{I_2(n)U_3(n)}{(U_3(n) - \alpha_3 I_2(n))\cdot \exp(-r_3 a(I_2(n))U_3(n)) + \alpha_3 I_2(n)}, \end{cases} \tag{4.10}$$

where $a(I_1(n)) = \frac{I_1(n)}{E_1 + I_1(n)}$ and $a(I_2([[t]])) = \frac{I_2(n)}{E_2 + I_2(n)}$.

The following theorems are given without proof since it is similar to Sect. 3.

Theorem 4.1. Let $\bar{X} = (\bar{S}, \overline{I_1}, \overline{I_2})$ be the positive equilibrium point of system (4.10). Let $\bar{X} = (\bar{S}, \overline{I_1}, \overline{I_2})$ be the positive equilibrium point of system (2.2). Assume that $\frac{\alpha_1}{\alpha_3} > \frac{3\beta_2(3 + \overline{I_2})}{(\theta + 3\beta_2\varepsilon_1\bar{S})\bar{S}}$, $\beta_2 > \frac{(1-\varepsilon_1)(\theta + 3\beta_2\varepsilon_1\bar{S})\bar{S}\alpha_1}{\varepsilon_1(1 + 3\overline{I_2})\alpha_2 + 3(1-\varepsilon_1)(3 + \overline{I_2})\alpha_3}$ and $\frac{\alpha_2}{\alpha_1} > \frac{(1-\varepsilon_1)(\theta + 3\beta_2\varepsilon_1\bar{S})}{\beta_2\varepsilon_1(1 + 3\overline{I_2})}$ hold. If

$$\begin{aligned} &\frac{1}{\overline{U_1}} \ln\left(\frac{\beta_2^2\varepsilon_1(1+3\overline{I_2})\alpha_2 + 3\beta_2^2(1-\varepsilon_1)(3+\overline{I_2})\alpha_3}{\beta_2^2\varepsilon_1(1+3\overline{I_2})\alpha_2 + 3\beta_2^2(1-\varepsilon_1)(3+\overline{I_2})\alpha_3 - \beta_2(1-\varepsilon_1)(\theta+3\beta_2\varepsilon_1\bar{S})\bar{S}\alpha_1}\right) < r_1 \\ &< \frac{1}{\overline{U_1}} \ln\left(\frac{\beta_2^2\varepsilon_1(1+3\overline{I_2})\alpha_2}{\beta_2^2\varepsilon_1(1+3\overline{I_2})\alpha_2 - \beta_2(1-\varepsilon_1)(\theta+3\beta_2\varepsilon_1\bar{S})\bar{S}\alpha_1}\right), \\ &r_2 \in \left(\frac{1}{a(\overline{I_1}n))\overline{U_2}} \ln\left(\frac{3\beta_2(1-\varepsilon_1)\bar{S}+\theta+\alpha_2}{3\beta_2(1-\varepsilon_1)\bar{S}+\theta}\right), \frac{1}{a(\overline{I_1}n))\overline{U_2}} \ln\left(\frac{\theta+\alpha_2}{\theta}\right)\right), \end{aligned}$$

and

$$\frac{1}{a(\overline{I_2}n))\overline{U_2}}\ln\left(\frac{\beta_2(1-\varepsilon_1)(\theta+3\beta_2\varepsilon_1\bar{S})\bar{S}\alpha_1}{\beta_2(1-\varepsilon_1)(\theta+3\beta_2\varepsilon_1\bar{S})\bar{S}\alpha_1-3\beta_2^2(1-\varepsilon_1)(3+\overline{I_2})\alpha_3}\right)<r_3,$$

then the positive equilibrium point of system (4.10) is local asymptotically stable.

Theorem 4.2. Let $\bar{X}=(\bar{S},\overline{I_1},\overline{I_2})$ be the positive equilibrium point of system (4.10) and assume that the conditions in Theorem 2.1/(i) and Theorem 4.1 hold. Then the positive equilibrium point of system (4.10) is global asymptotically stable, if

$$0<U_1(n)<\ln\left(\frac{2\bar{S}-S(n)}{S(n)}\right)\quad\text{and } S(n)<\bar{S},$$

$$0<U_2(n)<\frac{1}{a(I_1(n))}\ln\left(\frac{2\overline{I_1}-I_1(n)}{I_1(n)}\right)\quad\text{and } I_1(n)<\overline{I_1}$$

and

$$0<U_3(n)<\frac{1}{a(I_2(n))}\ln\left(\frac{2\overline{I_2}-I_2(n)}{I_2(n)}\right)\quad\text{and } I_2(n)<\overline{I_2}.$$

Example 4.1. The initial condition for the susceptible class is as given in Example 3.1, namely $S(0)=100,000,000$. From (4.4), we obtain

$$I_1\in\left(0,\frac{-E_1(\alpha_2+\theta)+\sqrt{E_1^2(\alpha_2+\theta)^2+(\alpha_2+\theta)E_1(1+3\beta_2(1-\varepsilon_1)S+\beta_2(1-\varepsilon_1)SI_2)}}{\alpha_2+\theta}\right)$$
$$\approx(0,24939)$$

for $E_1=0.5$, where we choose $I_1(0)=24,000$. Furthermore,

$$I_2\in\left(0,\frac{-\alpha_3E_2+\sqrt{\alpha_3^2E_2^2+\alpha_3(1+\beta_2\varepsilon_2S+\theta I_1+3\beta_2\varepsilon_1SI_1)}}{\alpha_3}\right)\approx(0,2207)$$

for $E_2=0.5$, where $I_2(0)=2200$.

Figure 5 shows the population of each class and the per capita growth rate of each population according to the data of Fig. 1. Using the initial values given above, we obtain Fig. 6, where we can see the increase to a humped curve and after that decrease.

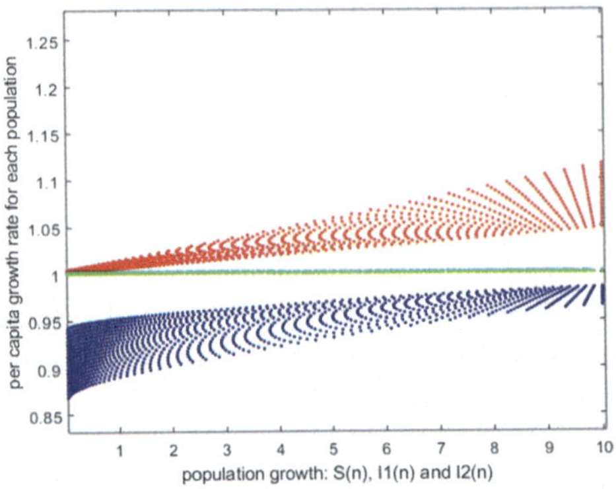

Fig. 5. Graph without Allee effect

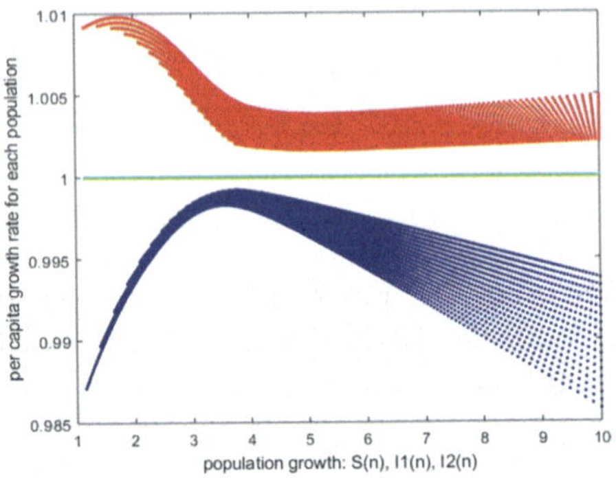

Fig. 6. Graph with Allee effect

5 Conclusion

In this paper, we construct a system of differential equations with piecewise constant arguments to analyze the spread of infection in an overlapping population, where the infection happens in discrete time. In this system, the population is divided into three sub-classes as mentioned before. In Sect. 2, we obtained conditions for a semi-cycle behavior and showed that the behavior is dependent on the attitude of each class to the HIV infection. Furthermore, we show that the behavior around the equilibrium point cannot be in a period two structure. In Sect. 3, we consider the local and global stability around the positive equilibrium point under specific conditions by using the Schur-Cohn Criteria and Lyapunov function, respectively. Example 3.1 shows the spread of each class according to the data of [7, 8]. The demographic information of HIV shows in Fig. 1 that after 80 years a significant increase of ($I_1(t)$) and a decrease in ($S(t)$) will happen. We want to see the case, if a human in India become more unaware in infecting the classes, which show us a dramatic graph in Fig. 2. In Fig. 3, we want to show the existence of non-negative equilibrium points in the infected classes that are ($I_1(t)$) and ($I_2(t)$), while Fig. 4 shows the growth rate of each population according to the per capita growth rate. Due to the immigration to the susceptible class, we expect that ($S(t)$) will be always positive, since India does not have a closed population, and has an endemic spread.

Furthermore, if the infection can be detected earlier by screening and the awareness of the HIV infected class do not transmit in any contact tracing, then the infection rate will be reduced significantly. This prediction was analyzed in Sect. 4, where we consider ($I_1(t)$) and ($I_2(t)$) in a small population. Figures 5 and 6 show the differences of the populations in large and small populations.

Finally, the most effective way to reduce the infection in India is to educate humans about the infection and to show the consequences of doing sex without protection or any other kinds of transmission risk.

References

1. Dalal, N., Greenhalgh, D., Mao, X.: A stochastic model for internal HIV dynamics. J. Math. Anal. Appl. **341**, 1084–1101 (2008)
2. Arazoza, H.D., Lounes, R.: A non-linear model for a sexually transmitted disease with contact tracing. IMA J. Math. Appl. Med. **19**, 221–234 (2002)
3. Busenberg, K., Cooke, K., Ying-Hen, H.: A model for HIV in Asia. Math. Biosci. **128**, 185–210 (1995)
4. Doyle, M., Greenhalgh, D.: Asymmetry and multiple endemic equilibria in a model for HIV transmission in a heterosexual population. Math. Comput. Model. **29**, 43–61 (1999)
5. Hsieh, Y.H., Sheu, S.P.: The effect of density-dependent treatment and behavior change on the dynamics of HIV transmission. J. Math. Biol. **43**, 69–80 (2001)
6. Ma, Z., Liu, J., Li, J.: Stability analysis for differential infectivity epidemic models. Nonlinear Anal.: Real World Appl. **4**, 841–856 (2003)
7. Naresh, R., Tripathi, A., Sharma, D.: A nonlinear HIV/AIDS model with contact tracing. Appl. Math. Comput. **217**, 9575–9591 (2011)
8. Tripathi, A., Naresh, R., Sharma, D.: Modelling the effect of screening of unaware infectives on the spread of HIV infection. Appl. Math. Comput. **184**, 1053–1068 (2007)
9. Cooke, K.L., Györi, I.: Numerical approximation of the solutions of delay differential equations on an infinite interval using piecewise constant arguments. Comput. Math Appl. **28**, 81–92 (1994)
10. Akhmet, M.: Nonlinear Hybrid Continuous-Discrete Time Models. Atlantis Press, Paris (2011)
11. Bozkurt, F.: Modeling a tumor growth with piecewise constant arguments. Discrete Dyn. Nat. Soc. **2013**, 1–8 (2013). Article ID 841764
12. Bozkurt, F., Hajji, M.A.: Stability and density analysis of glioblastoma (GB) with piecewise constant arguments. Wulfenia J. **23**(2), 305–320 (2016)
13. Bozkurt, F., Peker, F.: Mathematical modelling of HIV epidemic and stability analysis. Adv. Differ. Eqn. **95**, 1–17 (2014)
14. Gopalsamy, K., Liu, P.: Persistence and global stability in a population model. J. Math. Anal. Appl. **224**, 59–80 (1998)
15. Veeresha, P., Parakasha, D.G., Baskonus, H.M.: New numerical surfaces to the mathematical model of cancer chemotherapy effect in Caputo fractional derivatives. Chaos **29**, 013119 (2019)
16. Veeresha, P., Parakasha, D.G., Baskonus, H.M.: Solving smoking epidemic model of fractional order using a modified homotopy analysis transform method. Math. Sci. **13**(2), 1–14 (2019)
17. Li, X., Mou, C., Niu, W., Wang, D.: Stability analysis for discrete biological models using algebraic methods. Math. Comput. Sci. **5**, 247–262 (2011)
18. Allee, W.C.: Animal Aggregations: A Study in General Sociology. University of Chicago Press, Chicago (1931)
19. Asmussen, M.A.: Density-Dependent Selection II. The Allee effect. Am. Nat. **114**, 796–809 (1979)
20. Courchamp, F., Berec, L.: Gascoigne: Allee Effects in Ecology and Conservation. Oxford University Press, Oxford (2008)
21. Stephens, P.A., Sutherland, W.J., Freckleton, R.P.: What is Allee effect? Oikos **87**, 185–190 (1999)
22. Lande, R.: Extinction threshold in demographic models of territorial populations. Am. Nat. **130**(4), 624–635 (1987)
23. Allen, L.J.S.: An Introduction to Mathematical Biology. Prentice Hall, Pearson (2007)

Fractional Optimal Economic Control Problem Described by the Generalized Fractional Order Derivative

Abdou Thiao[1(✉)] and Ndolane Sene[2(✉)]

[1] Département d'Economies appliquées, Faculté des Sciences Economiques et Gestion, Université Cheikh Anta Diop de Dakar, Dakar, Senegal
thiaoabdou76@gmail.com

[2] Laboratoire Lmdan, Département de Mathématiques de la Décision, Faculté des Sciences Economiques et Gestion, Université Cheikh Anta Diop de Dakar, Fann, 5683 Dakar, Senegal
ndolanesene@yahoo.fr

Abstract. In this chapter, a game theory model has been addressed in the context of the fractional calculus. We established three conditions under which the continuous time dynamical game model related to the semi-renewable resource electricity can be solved. The fractional optimal control theory has been used for getting the optimal control of the model. The generalized fractional derivative has been used in our studies. For an illustration of our results, we minimize the playoff functional in electricity for the miners in our country, constraint by a fractional bilinear equation involving in the left generalized fractional order operator. The Nash equilibrium point has been provided for our proposed model and has been discussed, interpreted economically. Furthermore, our main results have been illustrated graphically.

Keywords: Fractional optimal control problem · Mittag-Leffler fractional operator · Nonsingular kernels operators

Highlight

▷ The fractional optimal control problem for continuous-time dynamical game model related to the semi-renewable resource electricity described by the Mittag-Leffler fractional operator has been presented.

▷ The optimal control of the proposed model has been provided.

▷ The Laplace transform of the generalized fractional derivative has been used for solving the fractional bilinear equation.

▷ The Nash equilibrium of the proposed model has been provided. Economic interpretation of our obtain Nash equilibrium has been analyzed. The graphical representations illustrate the main results of our chapter.

H. Dutta et al. (Eds.): CMES 2019, AISC 1111, pp. 36–48, 2020.
https://doi.org/10.1007/978-3-030-39112-6_3

1 Introduction

Fractional calculus has many applications in biology, in sciences and engineering, in physics and mechanics and many other fields. Recently, many papers appear and address many applications of fractional calculus in real word problems. [34] has proposed the analytical and the numerical solution of the fractional diffusion equation described by the generalized fractional derivative. In [31,33], Sene has proposed the analytical solution of the fractional diffusion equation described by the Atangana-Baluenu fractional derivative using the Fourier Laplace transform and the Fourier sine transform. In [17,18], Santos has proposed the statistical interpretations of the fractional diffusion of the particle. In [17,18], these contributions, Santos has used the mean square displacement to classify the fractional diffusion equations in different classes. The discrete versions of the fractional derivatives have been proposed recently by Thabet et al. in [1–4]. Statistical application in fractional calculus has been started by Atangana et al. in [9]. Applications of fractional calculus in biology can be found in [7,23]. Many other applications of fractional calculus exist and can be found in [20,21,25,28,29,32,35].

The fractional optimal control has been recently addressed in some recent works. In [6], Agarwal et al. have proposed the fractional optimal control problems with several state and control variables. In [11], Bahaa has introduced the fractional optimal control problem for a differential system with delay. In [11], Bahaa et al. have presented the necessary and sufficient condition for the fractional optimal problems described by the fractional derivative with nonsingular kernel. In [12], Bahaa et al. have studied the fractional optimal control of initial value problems on time scales. In [27], Pooseh et al. have studied the fractional optimal control problems in which the dynamic control system are described by the fractional order derivatives and the terminal time is free. In [24], Lotfi et al. have proposed a numerical technique for solving fractional optimal control problems. There exist many other applications of fractional calculus in control theory, see in [13,37]. In this paper, we try to determine the Nash Equilibrium using the fractional optimal control theory. We consider the continuous-time dynamical game model related to the semi-renewable resource electricity described by the generalized fractional derivative. The optimal control has been got for this problem. The optimal control obtained for the considered model is called the Nash Equilibrium point due to the fact a game theory model has been studied in our chapter. The game theory is one of the mathematical fields which interests many mathematicians. Game theory is a systematic study of strategic interactions among rational individuals. In all games, the decision for all players depends on the strategy adopted by the other players. The game theory finds many applications in the economic model, in Finance and many other fields. Cournot proposed the first work in game theory at 1838 [15], later other work in this field has been proposed by Bertrand at 1883 [14]. Von Neumann and Morgenstern proposed the main important work considered as the beginning of the game theory model at 1944 [26].

The paper is structured as follows. In Sect. 2, we propose a review on fractional derivative operators. In Sect. 3, we propose the game theory model considered in this Section. In Sect. 4, we determine the optimal control of our game theory model known as the Nash Equilibrium. In Sect. 5, we discuss our main results proposed in the previous sections. In Section, we give the concluding remarks.

2 Review on Fractional Derivative Operators

In this section, we recall the definitions of the existing fractional derivatives. We recall the Laplace transforms, introduce the Mittag-Leffler function with two parameters, and propose the property related to the integration by parts. The integration by parts is one of the essential property in fractional optimal control. It plays an essential role in establishing the different conditions for optimality.

We begin with the Mittag-Leffler function. The Mittag–Leffler function with two parameters is represented as the following series

$$E_{\alpha,\beta}(z) = \sum_{k=0}^{\infty} \frac{z^k}{\Gamma(\alpha k+\beta)}, \tag{1}$$

where $\alpha > 0$, $\beta \in \mathbb{R}$, $z \in \mathbb{C}$ and the function $\Gamma(...)$ represents the Euler Gamma function.

The fractional derivative with Mittag-Leffler kernel [8] for a given function u in Riemann-Liouville sense, of order $\alpha \in (0,1)$ is defined by

$$D_{\alpha}^{ABR} u(y,t) = \frac{B(\alpha)}{1-\alpha}\frac{d}{dt}\int_0^t u(y(s),s) E_{\alpha}\left(-\frac{\alpha}{1-\alpha}(t-s)^{\alpha}\right) ds, \tag{2}$$

for all $t > 0$.

The fractional derivative with Mittag-Leffler kernel [8] for a function u in Caputo sense, of order $\alpha \in (0,1)$ is defined as the following form

$$D_{\alpha}^{ABC} u(y,t) = \frac{B(\alpha)}{1-\alpha}\int_0^t u'(y(s),s) E_{\alpha}\left(-\frac{\alpha}{1-\alpha}(t-s)^{\alpha}\right) ds, \tag{3}$$

for all $t > 0$.

Let's the function u, the Caputo-Fabrizio fractional derivative of the function u of order $\alpha \in (0,1)$ is expressed in the form [8]

$$D_{\alpha}^{CF} u(y,t) = \frac{M(\alpha)}{1-\alpha}\int_0^t u'(y(s),s) \exp\left(-\frac{\alpha}{1-\alpha}(t-s)\right) ds, \tag{4}$$

for all $t > 0$.

The Riemann-Liouville integral [8] for a given function u, of order $\alpha \in (0,1)$ is represented as the form

$$I^{\alpha} u(y,t) = \frac{1}{\Gamma(\alpha)}\int_a^t (t-s)^{\alpha-1} u(y(s),s) ds, \tag{5}$$

for all $t > 0$.

The Atangana-Baleanu integral [8] for a given function u, of order $\alpha \in (0,1)$ is defined as the form

$$I_{\alpha}^{AB} u(y,t) = \frac{1-\alpha}{B(\alpha)} u(y,t) + \frac{\alpha}{B(\alpha)} I_{\alpha}^{RL} u(y,t), \tag{6}$$

for all $t > 0$.

Let's the function u, the left generalized derivative of the function u of order $\alpha \in (0,1)$ is expressed in the form [22]

$$\left(D^{\alpha,\rho} u\right)(y,t) = \frac{1}{\Gamma(1-\alpha)} \left(t^{1-\rho} \frac{d}{dt}\right) \int_0^t \left(\frac{t^\rho - s^\rho}{\rho}\right)^{-\alpha} u(y(s),s) \frac{ds}{s^{1-\rho}}, \tag{7}$$

for all $t > 0$.

Let's the function u, the left generalized derivative in Caputo sense of the function u of order $\alpha \in (0,1)$ is expressed in the form [22]

$$\left(D_c^{\alpha,\rho} u\right)(y,t) = \frac{1}{\Gamma(1-\alpha)} \int_0^t \left(\frac{t^\rho - s^\rho}{\rho}\right)^{-\alpha} u'(y(s),s) ds, \tag{8}$$

for all $t > 0$.

We recall the Laplace transform of the Atangana-Baleanu-Caputo derivative [8] and the Atangana-Baleanu-Riemann derivative. The following Laplace transform is used for solving the fractional differential equations described by the fractional derivatives with Mittag-Leffler functions. We have the following forms

$$\mathcal{L}\left\{D_{\alpha}^{ABC} u\right\}(s) = \frac{B(\alpha)}{1-\alpha} \frac{s^\alpha \mathcal{L}\{u\} - s^{\alpha-1} u(0)}{s^\alpha + \frac{\alpha}{1-\alpha}}. \tag{9}$$

Here $\mathcal{L}$ represents the classical Laplace transform.

$$\mathcal{L}\left\{D_{\alpha}^{ABR} u\right\}(s) = \frac{B(\alpha)}{1-\alpha} \frac{s^\alpha \mathcal{L}\{u\}}{s^\alpha + \frac{\alpha}{1-\alpha}}. \tag{10}$$

Let's recall the Laplace transform used for solving a class of the fractional differential equations described by the left generalized fractional derivative. The ρ-Laplace transform was introduced in [22]. The ρ-Laplace transform of the generalized fractional derivative in the Caputo sense is expressed in the following form

$$\mathcal{L}_\rho\left\{\left(D_c^{\alpha,\rho} f\right)(t)\right\} = s^\alpha \mathcal{L}_\rho\{f(t)\} - s^{\alpha-1} f(0), \tag{11}$$

The ρ-Laplace transform of function f is given in the form

$$\mathcal{L}_\rho\{g(t)\}(s) = \int_0^\infty e^{-s\frac{t^\rho}{\rho}} g(t) \frac{dt}{t^{1-\rho}}. \tag{12}$$

We finish this section by recalling the integration by parts related to generalized fractional derivative recently introduced by Thabet et al. in [5]; we have the following expression

$$\int_a^b f(t) D_a^{\alpha,\rho} g(t) dt = \int_a^b g(t) D_b^{\alpha,\rho} f(t) dt. \tag{13}$$

3 Presentation of the Game Theory Model

In this section, we present the continuous-time dynamical game model related to the semi-renewable resource electricity in our country. Let's the set of finite players $i \in I = \{1, 2, 3, ..., n\}$. We denote by u_i the electricity consummation of the miners or player i. In our game model [36] the profile of the player i is represented by the following continuous function

$$f(x, u_i, t) = \left[P - C - \sum_{i=1}^{n} u_i\right] u_i, \tag{14}$$

where P is considered as the price of the bitcoin, and C represents the price of the electricity in our country. Our objective in this paper is to get the optimal strategy u_i of the player i, which maximize the total profile of the players

$$\begin{aligned} J(x, \bar{u}) &= \int_0^1 e^{-rt} \sum_{i=1}^{n} \left[P - C - \sum_{i=1}^{n} u_i\right] u_i dt \\ &= \int_0^1 e^{-rt} \sum_{i=1}^{n} \left[P - C - n\bar{u}\right] u_i dt \\ &= \int_0^1 e^{-rt} \left[P - C - n\bar{u}\right] \sum_{i=1}^{n} u_i dt \\ &= \int_0^1 n e^{-rt} \left[P - C - n\bar{u}\right] \bar{u} dt, \end{aligned} \tag{15}$$

under which the strategy satisfies the following constraint

$$D_0^{\alpha,\rho} x = \xi x - \sum_{i=1}^{n} u_i = \xi x - n\bar{u}. \tag{16}$$

Note that here $x(u_i) = u_i$. In this paper, we consider the game is cooperative. That is, all players cooperate and decide to consume the electricity respecting the price of the electricity fixed by the government and in return. Furthermore, the players get the bitcoin market price as profit so, they maximize their profit, and the profit is identically shared among all the players. As explained in the next lines.

The above-presented model is a game theory model. In games theory, there exist two different types of games: cooperative game and non-cooperative game. A game is called cooperative when all the players can cooperate between them and make sign contrats or make agreements between them. A game is called non-cooperative when the profits of each economic player do not depend on his own decision but the decision of the other players in the game. A fundamental property in a non-cooperative game is the players cannot sign contrast between them. In our paper, we consider the case on which the game is cooperative. There exist two types of a cooperative game, namely transferable game and non-transferable game. A cooperative game is said to be non-transferable when all the utilities of the players in the game cannot be summed and distributed between them. A cooperative game is said to be transferable when the utilities of all the players in the game can be summed and distributed between all the players individually. When the strategy of the game is cooperative and transferable all players try to maximize their own profit and the profits obtained by the other players in the game. We said in this case; the players have a common market. In our chapter, we consider the game is cooperative and transferable. The Nash Equilibrium is one of the essential concepts in game theory nowadays. In many situations, the profile obtained by the player depends on his strategy and the strategy proposed by the other players in the game. That is when a player chooses a strategy, and he will have in mind the approach of the other players in the game. Summarizing, the Nash Equilibrium is an action profile a^* with the property that no player i can do better by choosing an action different from a_i^*, giving that every player j adheres to the action a_j^*. The concept of Nash Equilibrium is now fundamental in mathematics, optimal control, economics, and game theory. The Nash Equilibrium is a situation under which each player doesn't want to change his strategy, having in mind the action of the other players.

4 Optimization of the Continuous Time Dynamical Game Model

In this section, we address the optimal control which maximizes the payoff function of the continuous time dynamical game model in cooperative strategy related to the semi-renewable resource electricity in our country described by

$$J(x,\bar{u}) = \int_0^1 ne^{-rt}\left[P - C - n\bar{u}\right]\bar{u}dt, \tag{17}$$

where P denotes the price of the bitcoin, C represents the electricity price fixed by the government, $n\bar{u} = \sum_{i=1}^{n} u_i$ and r denotes the discount rate. Furthermore, the optimal control satisfies the fractional differential equation with exogenous input

$$D_0^{\alpha,\rho}x = \xi x - n\bar{u}, \tag{18}$$

where ξ represents the regeneration rate of the electricity. For simplification of our problem, let's the following function

$$f(x,u_i,t) = ne^{-rt}\left[P - C - n\bar{u}\right]\bar{u} \text{ and } g(x,u_i,t) = \xi x - n\bar{u}. \tag{19}$$

The function f represents the payoff function of miner i. Thus, the optimization problem is defined by Eqs. (17) and (18). The problem consists of finding the optimal control which represents the Nash equilibrium in our game problem of the playoff function defined by

$$J(x,\bar{u}) = \int_0^1 f(x,\bar{u},t)dt, \tag{20}$$

and satisfies the fractional differential equation defined by

$$D_0^{\alpha,\rho} x = g(x, \bar{u}, t), \tag{21}$$

under initial boundary condition defined by $x(0) = \eta$. In other words, the problem consists of finding optimal control for the game, maximizing the total profit of miner i represented by Eq. (17) satisfying the constraint (18). Many methods exist to solve the problem defined by Eqs. (17) and (18). In our studies, we use fractional optimal control theory. Let's the Hamiltonian function defined by the following equation

$$H(x, \bar{u}, \lambda, t) = f(x, \bar{u}, t) - \lambda^T g(x, \bar{u}, t), \tag{22}$$

where λ denotes the Lagrange multiplier. Note the Eq. (17) can be expressed in the term of the Hamiltonian function defined by

$$J(x, \bar{u}) = \int_0^1 \left[H(x, \bar{u}, \lambda, t) - \lambda^T D_0^{\alpha,\rho} x\right] dt. \tag{23}$$

The necessaries and the sufficient conditions for optimality of the problem defined by Eqs. (17) and (18) are described in the following Theorem [6, 11].

Theorem 1. *Consider the Nash equilibrium $\bar{u}$ and the state $x(t)$ be the optimal points for the cooperative game related to the semi-renewable resource electricity for our country in Eqs. (17) and (18), then there exist a Lagrange multiplier λ such that the following relationships are held:*

$$D_0^{\alpha,\rho} x = \frac{\partial H}{\partial \lambda} \quad \textit{and} \quad D_1^{\alpha,\rho} \lambda = \frac{\partial H}{\partial x} \quad \textit{and} \quad \frac{\partial H}{\partial \bar{u}} = 0. \tag{24}$$

The proof of this Thereom is classic in fractional calculus, Banaa et al. [11] provided the proof in the context of the Riemann-Liouville fractional derivative, early stated by Agrawal et al. in [6]. Here we recall the same Theorem and prove it in the context of the generalized fractional derivative recently introduced in Fractional Calculus. In other words, the novelty of this works is the use of the generalized fractional derivative and the application of the optimal control. This work provided too the relation existing between the game theory model and fractional calculus.

Proof: We adopt the same proof as in [6, 11]. The game theory model Eqs. (17)–(18) in term of the Hamiltonian function is described by

$$J(x, \bar{u}) = \int_0^1 \left[H(x, \bar{u}, \lambda, t) - \lambda^T D_0^{\alpha,\rho} x\right] dt. \tag{25}$$

Taking the variation of the total profit function, we obtain the following relationship

$$\delta J(x, \bar{u}) = \int_0^1 \left\{ \left[\frac{\partial H}{\partial x}\right]^T \delta x - \lambda^T D_0^{\alpha,\rho} \delta x + \left[\frac{\partial H}{\partial \lambda} - D_0^{\alpha,\rho} x\right] \delta\lambda + \left[\frac{\partial H}{\partial \bar{u}}\right]^T \delta\bar{u} \right\} dt. \tag{26}$$

Recalling the integration by parts in term of the generalized fractional derivative stated by Thabet et al. in [5, 22], we have the following expression

$$\int_0^1 \lambda^T D_0^{\alpha,\rho} \delta x dt = \int_0^1 \left(D_1^{\alpha,\rho}\lambda\right)^T \delta x dt. \tag{27}$$

Replacing Eqs. (27) into (26) and imposing the variation of the playoff function null to obtain the maximum, we have respectively

$$D_0^{\alpha,\rho} x = \frac{\partial H}{\partial \lambda} \quad \text{and} \quad D_1^{\alpha,\rho}\lambda = \frac{\partial H}{\partial x} \quad \text{and} \quad \frac{\partial H}{\partial \bar{u}} = 0. \tag{28}$$

In this new paragraph, we replace the function f and g by their values and use the optimality conditions established in Theorem 1 to get the Nash equilibrium of the continuous time dynamical game model related to the semi-renewable resource electricity in our country. The explicit form of the Hamiltonian function of the game model is given by

$$H(x, \bar{u}, \lambda, t) = ne^{-rt}\left[P - C - n\bar{u}\right]\bar{u} - \lambda^T\left[\xi x - n\bar{u}\right]. \tag{29}$$

We begin our investigation by the last optimality condition given in Eq. (28), we have the following equation

$$\frac{\partial H}{\partial \bar{u}} = ne^{-rt}\left[-2n\bar{u} + P - C\right] + n\lambda = 0. \tag{30}$$

From Eq. (30), it follows the value of the Nash equilibrium in the cooperative game of the continuous time dynamical model related to the semi-renewable resource electricity in our country is given by the following expression

$$\bar{u} = \frac{P - C + \lambda e^{rt}}{2n}. \tag{31}$$

We can observe the Nash equilibrium obtained in this section depends on the Lagrange multiplicator λ. The second step of the resolution of our problem consists of determining the state obtained by solving the first equation in Eq. (24), that is

$$D_0^{\alpha,\rho} x = \frac{\partial H}{\partial \lambda} = \xi x - n\bar{u}. \tag{32}$$

Applying the ρ-Laplace transform to both sides of Eq. (32), we have the following relation

$$\begin{aligned} q^\alpha \tilde{x} - x_0 &= \xi\tilde{x} - n\bar{u}(q) \\ q^\alpha \tilde{x} - \xi\tilde{x} &= x_0 - n\bar{u}(q) \\ \tilde{x} &= \frac{x_0}{q^\alpha - \xi} - \frac{n\bar{u}(q)}{q^\alpha - \xi}. \end{aligned} \tag{33}$$

Applying the inverse of the ρ-Laplace transform to both sides of Eq. (33), we obtain the state defined by

$$x(t) = \eta\left(\frac{t^\rho}{\rho}\right)^{\alpha-1} E_{\alpha,\alpha}\left(\xi\left(\frac{t^\rho}{\rho}\right)^\alpha\right) + \int_0^t \left(\frac{t^\rho - s^\rho}{\rho}\right)^{\alpha-1} E_{\alpha,\alpha}\left(\xi\left(\frac{t^\rho - s^\rho}{\rho}\right)^\alpha\right) n\bar{u}(s)ds. \tag{34}$$

The last step of the resolution concerns the determination of the Lagrange multiplicator λ, using second equation into Eq. (24), we have the following equation to solve

$$D_1^{\alpha,\rho}\lambda = \frac{\partial H}{\partial x} = \xi\lambda + ne^{-rt}\left[-2n\bar{u}\frac{\partial \bar{u}}{\partial x} + (P-C)\frac{\partial \bar{u}}{\partial x}\right]. \tag{35}$$

We can observe the second term of Eq. (35) depends on time, for simplification, we let the following function

$$v(t) = ne^{-rt}\left[-2n\bar{u}\frac{\partial \bar{u}}{\partial x} + (P-C)\frac{\partial \bar{u}}{\partial x}\right]. \tag{36}$$

But $\frac{\partial \bar{u}}{\partial x} = 0$ in our model, finaly Eq. (35) becomes the following

$$D_1^{\alpha,\rho}\lambda = \xi\lambda. \tag{37}$$

Applying the ρ-Laplace transform to both sides of Eq. (37), we have the following relationships

$$\begin{aligned} q^\alpha\bar{\lambda} - \lambda(1) &= \xi\bar{\lambda} \\ q^\alpha\bar{\lambda} - \xi\bar{\lambda} &= \lambda(1) \\ \bar{\lambda} &= \frac{\lambda(1)}{q^\alpha - \xi}. \end{aligned} \tag{38}$$

Applying the inverse of the ρ-Laplace transform to both sides of Eq. (38), we obtain the following solution for Eq. (37)

$$\lambda(t) = \lambda(1)\left(\frac{t^\rho}{\rho}\right)^{\alpha-1} E_{\alpha,\alpha}\left(\xi\left(\frac{t^\rho}{\rho}\right)^\alpha\right). \tag{39}$$

Finally, the triplet $(x, \bar{u}, \lambda)$ is the optimal solution which maximizes the total profit of the miner i of the game theory model related to the semi-renewable resource electricity in our country.

5 Discussion and Graphical Representations

In this section, we represent and discuss the optimal control, which maximizes the total profit of the player i of the game theory model related to the semi-renewable resource electricity in our country. We depict the optimal control $\bar{u}$ of the problems (17) and (18). We consider $n = 2000$ players in Dakar, $\alpha = 1$, different values for the order ρ. Furthermore $r = 0.03$ and $\xi = 0.02$. The price of the bitcoin in Dakar is fixed to $P = 8302,57\ USD$, and then the price of the electricity of a player in Dakar is $C = 6000\ USD$ for all players equivalent to 3 USD per player.

In Fig. 1a, we depict the values of the Nash equilibrium in time when we consider $n = 2000$ players. We can observe the Nash equilibrium increase with

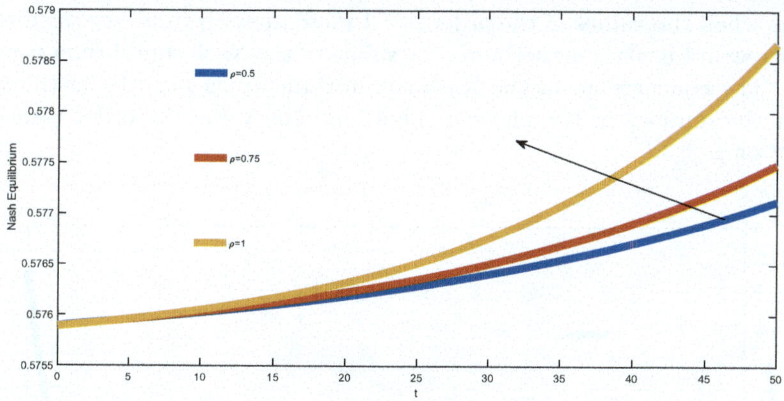

(a)

Fig. 1. Nash equilibrium evolution in times.

the time when the values of the order $\rho \leq 1$ increase, see the direction of the arrow in Fig. 1a.

In Fig. 2a, we depict the values of the Nash equilibrium in time when we consider $n = 2000$ players. We can observe the Nash equilibrium decrease with the time when the values of the order $\rho \leq 1$ decrease, see the direction of the arrow in Fig. 2a.

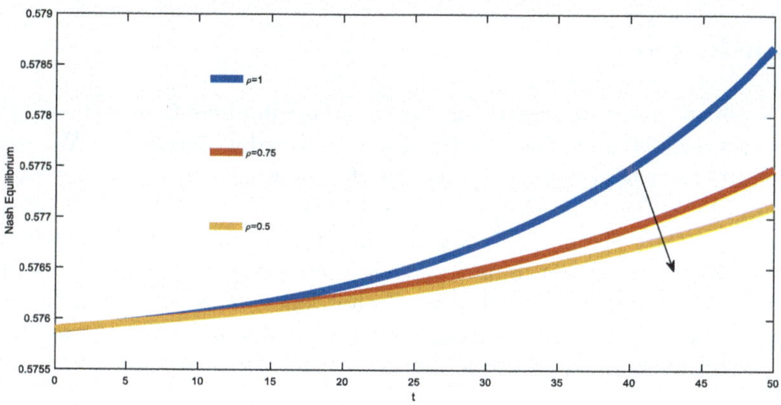

(a)

Fig. 2. Nash equilibrium evolution in times.

In Fig. 3a, we depict the values of the Nash equilibrium in time when we consider $n = 2000$ players. We can observe the Nash equilibrium decrease with the time when the values of the order $\rho \geq 1$ increase with time, see the direction of the arrow in Fig. 3a. Furthermore, the values of the Nash equilibrium converges to zero. The convergence of the Nash equilibrium to zero can be justified from the fact the number of the player is high, and the value the order respect the assumption $\rho \geq 1$.

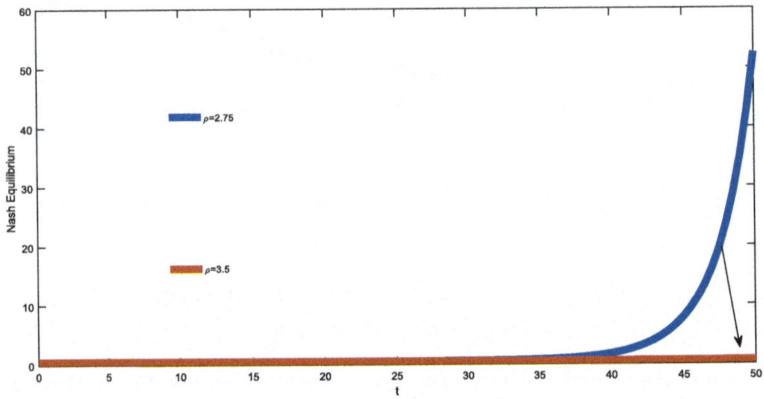

(a)

Fig. 3. Nash equilibrium evolution in times.

6 Conclusion

In this paper, we have discussed the fractional optimal control of the continuous game theory model described by the fractional order derivative. We establish three conditions which permit to obtain the optimal control of our presented model. For Dakar based on the price of the electricity and the price of bitcoin, we have proposed the optimal control which maximizes the profit for 2000 players. The order ρ has a significant impact on the values of the Nash equilibrium. Future works, we will extend the studies in all the Senegal. What will happen when the fractional derivative with Mittag-Leffler kernel is used. For the forthcoming paper, we will focus on these directions.

References

1. Abdeljawad, T.: Different type kernel h-fractional differences and their fractional h-sums. Chaos Soli. Fract. **116**, 146–156 (2018)
2. Abdeljawad, T., Al-Mdallal, Q.M.: Discrete Mittag-Leffler kernel type fractional difference initial value problems and Gronwall's inequality. J. Comput. Appl. Math. **339**, 218–230 (2018)

3. Abdeljawad, T., Baleanu, D.: Discrete fractional differences with non-singular discrete Mittag-Leffler kernels. Adv. Differ. Equ. **2016**, 232 (2016)
4. Abdeljawad, T., Baleanu, D.: On fractional derivatives with generalized Mittag-Leffler kernels. Adv. Diff. Equat. **2018**, 468 (2018)
5. Abdeljawad, T.: Fractional operators with boundary points dependent kernels and integration by parts. Discrt. conti.dyn. syt. S 13, 351–375 (2020)
6. Agrawal, O.P., Defterli, O., Baleanu, D.: Fractional optimal control problems with several state and control variables. J. Vibra. Contr. **16**, 1967 (2010)
7. Arqub, O.A., Ajou, A.E.: Solution of the fractional epidemic model by homotopy analysis method. J. King Saud Univ. Scien. **25**, 73–81 (2013)
8. Atangana, A., Baleanu, D.: New fractional derivatives with nonlocal and non-singular kernel: theory and application to heat transfer model, arXiv preprint arXiv:1602.03408 (2016)
9. Atangana, A., Aguilar, J.F.G.: Fractional derivatives with no-index law property: application to chaos and statistics. Chaos Soli. Fract. **114**, 516–535 (2018)
10. Bahaa, G.M.: Fractional optimal control problem for differential system with delay argument. Advan. Diff. Equat. **2017**, 69 (2017)
11. Bahaa, G.M., Atangana, A.: Necessary and sufficient optimality conditions for fractional problems. Involving Atangana Baleanus Derivatives: Trends and Applications in Science and Engineering (2019)
12. Bahaa, G.M., Torres, D.F.M.: Time-fractional optimal control of initial value problems on time scales. arXiv:1904.07684v1 [math.OC] (2019)
13. Baskonus, H.M., Mekkaoui, T., Hammouch, Z., Bulut, H.: Active control of a chaotic fractional order economic system. Entropy **17**, 5771–5783 (2015)
14. Bertrand, J.: Revue des ouvrages d'Augustin Cournot et de Léon Walras. J. des Savants **499** (1883)
15. Cournot, A.: Recherches sur les Principes Mathematiques de la Theorie des Richesses (1883)
16. Caputo, M., Fabrizio, M.: A new definition of fractional derivative without singular kernel. Progr. Fract. Differ. Appl **1**(2), 1–15 (2015)
17. Santos, M.D., Gomez, I.S.: A fractional Fokker-Planck equation for non-singular kernel operators. J. Stat. Mech: Theory Exp. **2018**, 123205 (2018)
18. Santos, M.D.: Fractional prabhakar derivative in diffusion equation with non-static stochastic resetting. Physics **1**, 40–58 (2019)
19. Santos, M.D.: Non-Gaussian distributions to random walk in the context of memory kernels. Fractal Fract. **2**, 20 (2018)
20. Hashemi, M.S., Baleanu, D., Haghighi, M.P.: Solving the time fractional diffusion equation using a lie group integrator. Therm. Sci. **19**, 77–83 (2015)
21. Henry, B.I., Langlands, T.A.M., Straka, P.: An Introduction to Fractional Diffusion. World Scientific Review (2009)
22. Fahd, J., Abdeljawad, T.: A modified Laplace transform for certain generalized fractional operators. Results Nonlinear Anal. **2**, 88–98 (2018)
23. Khan, M.A., Hammouch, Z., Baleanu, D.: Modeling the dynamics of hepatitis E via the Caputo-Fabrizio derivative. Math. Model. Nat. Phenom. **14**(3), 19 (2019)
24. Lotfi, A., Dehghan, M., Yousefi, S.A.: A numerical technique for solving fractional optimal control problems. Comput. Math. Appl. **62**, 1055–1067 (2011)
25. Meerschaert, M.M., Tadjeran, C.: Finite difference approximations for fractional advection-dispersion flow equations. J. Comput. Appl. Math. **172**, 65–67 (2004)
26. Newmann, J.V., Morgenstern, O.: Theory of Games and Economic Behaviour. Princeton University Press, Princeton (1944)

27. Pooseh, S., Almeida, R., Torres, D.F.M.: Fractional order optimal control problems with free terminal time, arXiv preprint arXiv:1302.1717 (2013)
28. Al-Refai, M., Abdeljawad, T.: Analysis of the fractional diffusion equations with fractional derivative of non-singular kernel. Adv. Differ. Equ. **2017**, 315 (2017)
29. Khader, M.M.: On the numerical solutions for the fractional diffusion equation. Commun. Nonlinear Sci. Numer. Simulat. **16**, 2535–2542 (2011)
30. Tasbozan, O., Esen, A., Yagmurlu, N.M., Ucar, Y.: A numerical solution to fractional diffusion equation for force-free case. Abstr. Appl. Anal. **2013**(6) (2013)
31. Sene, N.: Analytical solutions of Hristov diffusion equations with non-singular fractional derivatives. Choas **29**, 023112 (2019)
32. Sene, N.: Solutions of fractional diffusion equations and Cattaneo-Hristov diffusion models. Int. J. Appli. Anal. **17**(2), 191–207 (2019)
33. Sene, N.: Stokes' first problem for heated flat plate with Atangana-Baleanu fractional derivative. Chaos, Solitons Fractals **117**, 68–75 (2018)
34. Sene, N.: Analytical solutions and numerical schemes of certain generalized fractional diffusion models. Eur. Phys. J. Plus **134**, 199 (2019)
35. Sene, N.: Homotopy perturbation ρ- laplace transform method and its application to the fractional diffusion equation and the fractional diffusion-reaction equation. Fractal Fract. **3**, 14 (2019)
36. Singh, R., Dwivedi, A.D., Srivastava, G.: Energy consumption in Bitcoin Mining: A Game Theoretic Analysis (2019)
37. Hammouch, Z., Mekkaoui, T.: Control of a new chaotic fractional-order system using Mittag-Leffler stability. Nonli. Stud. **22**(4), 1–13 (2015)

An Efficient Technique for Coupled Fractional Whitham-Broer-Kaup Equations Describing the Propagation of Shallow Water Waves

P. Veeresha[1], D. G. Prakasha[2(✉)], and Haci Mehmet Baskonus[3]

[1] Department of Mathematics, Karnatak University, Dharwad 580 003, India
viru0913@gmail.com
[2] Department of Mathematics, Faculty of Science, Davangere University, Shivagangothri, Davangere 577 007, India
prakashadg@gmail.com
[3] Department of Mathematics and Science Education, Faculty of Education, Harran University, Sanliurfa, Turkey
hmbaskonus@gmail.com

Abstract. In the present work, the approximated analytical solution for the time-fractional coupled Whitham-Broer-Kaup (WBK) equations describing the propagation of shallow water waves are obtained with the aid of an efficient computational technique called, q-Homotopy analysis transform method (briefly, q-HATM). To demonstrate the reliability and efficiency of the proposed technique, two examples are illustrated. The homotopy polynomials are hired in order to handle the nonlinear terms and the suggested algorithm provides the auxiliary parameters $\hbar$ and n, which help us to control and adjust the convergence region of the obtained series solution. Numerical simulation has been carried out in terms of absolute error. The obtained results revels that, the proposed algorithm is highly methodical and very efficient to solve coupled nonlinear differential system.

Keywords: q-homotopy analysis transform method · Fractional Whitham-Broer-Kaup equations · Laplace transform

1 Introduction

The derivatives of fractional order were debut in 1695, as in the question of the extension of meaning. In recent years, many researchers are attentive to study the fractional calculus due to its ability to provide an exact description for various types of non-linear phenomena. Fractional order differential equations are the generalization of traditional differential equations having nonlocal and genetic consequence in the material properties, which are studied and described by many pioneers including Caputo [1], Miller and Ross [2], Podlubny [3], Liao [4], and others. Nowadays, fractional partial differential equations have gained popularity in developing procedure for non-linear models and investigation of dynamical systems [5]. The problems relating to applications of fractional differential equations are situated in various connected branches of science and engineering; like fluid and continuum mechanics [6],

H. Dutta et al. (Eds.): CMES 2019, AISC 1111, pp. 49–75, 2020.
https://doi.org/10.1007/978-3-030-39112-6_4

electrodynamics [7], nanotechnology [8], ocean engineering [9], cosmology [10] and many other branches [11–16].

The solutions of the nonlinear differential equations of arbitrary order play a vital role in describing the nature and characteristics of complex problems arised in science and technology. Moreover, it is very difficult to obtain the analytical solutions for these differential equations. On other hand, last three decades be the witness for initiating and studying the number of numerical techniques. There have been many such techniques are available in literature and one of them is q-homotopy analysis transform method (briefly, q-HATM). This method was introduced by Singh et al. [17], to study the linear and nonlinear differential equation of integer and fractional order exist is different areas of science. The q-HATM is an elegant amalgamation of q-homotopy analysis method and Laplace transform.

In this paper, we present the numerical solution of nonlinear coupled fractional Whitham-Broer-Kaup (WBK) equations with the help of q-HATM. The WBK equations were studied by Whitham [18], Broer [19] and Kaup [20], and these equations describe the propagation of shallow water waves with different dispersion relations. The WBK equations arise in hydrodynamics to describe the propagation of waves in dissipative and nonlinear media. They are advisable for problems arise in the leakage of water in porous subsurface stratum and widely used in ocean and coastal engineering. Moreover, these equations are the foundation of numerous models utilize to portray the unconfined subsurface like, drainage and groundwater flow problems.

Consider coupled nonlinear WBK equations of fractional order [21]:

$$\begin{aligned} &D_t^\alpha u + u u_x + v_x + b u_{xx} = 0, \\ &D_t^\alpha v + u v_x + v u_x + a u_{xxx} - b v_{xx} = 0, \end{aligned} \qquad 0<\alpha\leq 1, \tag{1}$$

where $u = u(x,t)$ is the horizontal velocity and $v = v(x,t)$ be the height that deviating from equilibrium position of the liquid. Further, a and b are constants which are representing different diffusion powers i.e., if $a = 1$ and $b = 0$, then Eq. (1) reduces to the modified Boussinesq equation, and if $a = 0$ and $b = 1$, then the system represents classical long wave equation. The solutions for coupled fractional WBK equations were obtained and studied by many authors through different numerical techniques like, Adomian decomposition method [22], variation iteration method [23], optimal homotopy asymptotic method [24], coupled fractional reduced differential transform method [21], residual power series method [25] and Laplace Adomian decomposition method [26].

2 Preliminaries

We recall some definitions and properties of fractional calculus and Laplace transform, which are used in the sequel:

Definition 1. Let a function $f(t) \in C_\mu(\mu \geq -1)$, then the Riemann-Liouville integral of fractional order $(\alpha > 0)$ is given as [3]:

$$J^{\alpha}f(t) = \frac{1}{\Gamma(\alpha)}\int_{0}^{t}(t-\vartheta)^{\alpha-1}f(\vartheta)d\vartheta,$$
$$J^{0}f(t) = f(t) \tag{2}$$

Definition 2. The fractional derivative of $f \in C_{-1}^{n}$ in the Caputo [1] sense is defined as

$$D_{t}^{\alpha}f(t) = \begin{cases} \frac{d^{n}f(t)}{dt^{n}}, & \alpha = n \in N, \\ \frac{1}{\Gamma(n-\alpha)}\int_{0}^{t}(t-\vartheta)^{n-\alpha-1}f^{(n)}(\vartheta)d\vartheta, & n-1<\alpha<n, n \in \mathbb{N}. \end{cases} \tag{3}$$

Definition 3. The Laplace transform (LT) of a Caputo fractional derivative $D_{t}^{\alpha}f(t)$ is represented as [1, 2]

$$L\left[D_{t}^{\alpha}f(t)\right] = s^{\alpha}F(s) - \sum_{r=0}^{n-1} s^{\alpha-r-1}f^{(r)}(0^{+}), (n-1<\alpha\leq n), \tag{4}$$

where $F(s)$ is symbolize the LT of the function $f(t)$.

3 Fundamental Idea of q-HATM

To present the fundamental idea of proposed method [27–35], we consider a general fractional order nonlinear non-homogeneous partial differential equation of the form:

$$D_{t}^{\alpha}\mathcal{U}(x,t) + R\mathcal{U}(x,t) + N\mathcal{U}(x,t) = f(x,t), \quad n-1<\alpha\leq n, \tag{5}$$

where $D_{t}^{\alpha}\mathcal{U}(x,t)$ denotes the Caputo's fractional derivative of the function $\mathcal{U}(x,t)$, R and N specifies the linear and nonlinear differential operator, respectively, and $f(x,t)$ represents the source term. Now, by employing the LT on Eq. (5), we get

$$s^{\alpha}L[\mathcal{U}(x,t)] - \sum_{k=0}^{n-1} s^{\alpha-k-1}\mathcal{U}^{(k)}(x,0) + L[R\mathcal{U}(x,t)] + L[N\mathcal{U}(x,t)] = L[f(x,t)]. \tag{6}$$

On simplifying Eq. (6), we have

$$L[\mathcal{U}(x,t)] - \frac{1}{s^{\alpha}}\sum_{k=0}^{n-1} s^{\alpha-k-1}\mathcal{U}^{k}(x,0) + \frac{1}{s^{\alpha}}\{L[R\mathcal{U}(x,t)] + L[N\mathcal{U}(x,t)] - L[f(x,t)]\} = 0. \tag{7}$$

According to homotopy analysis method [4], the nonlinear operator defined as

$$N[\varphi(x,t;q)] = L[\varphi(x,t;q)] - \frac{1}{s^{\alpha}}\sum_{k=0}^{n-1} s^{\alpha-k-1}\varphi^{(k)}(x,t;q)(0^{+})$$
$$+ \frac{1}{s^{\alpha}}\{L[R\varphi(x,t;q)] + L[N\varphi(x,t;q)] - L[f(x,t)]\}, \tag{8}$$

where $q \in \left[0, \frac{1}{n}\right]$, and $\varphi(x,t;q)$ is real function of x, t and q.

We construct a homotopy for non-zero auxiliary function $H(x,t)$ as follows:

$$(1 - nq)L[\varphi(x,t;q) - \mathcal{U}_0(x,t)] = \hbar q H(x,t) N[\varphi(x,t;q)], \tag{9}$$

where L be a symbol of the Laplace transform, $\hbar \neq 0$ is an auxiliary parameter, $q \in \left[0, \frac{1}{n}\right] (n \geq 1)$ is the embedding parameter, $\mathcal{U}_0(x,t)$ is an initial guess of $\mathcal{U}(x,t)$ and $\varphi(x,t;q)$ is an unknown function. The following results hold for $q = 0$ and $q = \frac{1}{n}$:

$$\varphi(x,t;0) = \mathcal{U}_0(x,t), \ \varphi\left(x,t;\frac{1}{n}\right) = \mathcal{U}(x,t), \tag{10}$$

respectively. Thus, by amplifying q from 0 to $\frac{1}{n}$, the solution $\varphi(x,t;q)$ converge from $\mathcal{U}_0(x,t)$ to the solution $\mathcal{U}(x,t)$. Expanding the function $\varphi(x,t;q)$ in series form by employing Taylor theorem near to q, one can get

$$\varphi(x,t;q) = \mathcal{U}_0(x,t) + \sum\nolimits_{m=1}^{\infty} \mathcal{U}_m(x,t) q^m, \tag{11}$$

where

$$\mathcal{U}_m(x,t) = \frac{1}{m!} \frac{\partial^m \varphi(x,t;q)}{\partial q^m}\Big|_{q=0}. \tag{12}$$

On choosing the auxiliary linear operator, the initial guess $\mathcal{U}_0(x,t)$, the auxiliary parameter $n, \hbar$ and $H(x,t)$, the series (11) converges at $q = \frac{1}{n}$, then it gives one of the solutions of the original nonlinear equation of the form

$$\mathcal{U}(x,t) = \mathcal{U}_0(x,t) + \sum\nolimits_{m=1}^{\infty} \mathcal{U}_m(x,t) \left(\frac{1}{n}\right)^m. \tag{13}$$

Now, differentiating the *zero-th* order deformation Eq. (9) m-times with respect to q and then dividing by $m!$ and finally taking $q = 0$, which yields

$$L[\mathcal{U}_m(x,t) - k_m \mathcal{U}_{m-1}(x,t)] = \hbar H(x,t) \Re_m\left(\vec{\mathcal{U}}_{m-1}\right), \tag{14}$$

where the vectors are defined as

$$\vec{\mathcal{U}}_m = \{\mathcal{U}_0(x,t), \mathcal{U}_1(x,t), \ldots, \mathcal{U}_m(x,t)\}. \tag{15}$$

Applying the inverse Laplace transform on Eq. (14), it provide the following recursive equation

$$\mathcal{U}_m(x,t) = k_m \mathcal{U}_{m-1}(x,t) + \hbar L^{-1}\left[H(x,t) \Re_m\left(\vec{\mathcal{U}}_{m-1}\right)\right], \tag{16}$$

where

$$\Re_m\left(\vec{\mathcal{U}}_{m-1}\right) = \frac{1}{(m-1)!}\frac{\partial^{m-1}N[\varphi(x,t;q)]}{\partial q^{m-1}}|_{q=0} \tag{17}$$

and

$$\mathsf{k}_m = \begin{cases} 0, m \leq 1, \\ n, m > 1. \end{cases} \tag{18}$$

Finally, on solving Eq. (16) we obtain the components of the q-HATM series solution.

4 Numerical Examples

To demonstrate the efficiency and applicability of the proposed algorithm, we consider two examples as an illustration.

Example 4.1. Consider the coupled WBK equations of time-fractional order [21]:

$$\begin{cases} D_t^\alpha u = -u\frac{\partial u}{\partial x} - \frac{\partial v}{\partial x} - b\frac{\partial^2 u}{\partial x^2}, \\ D_t^\alpha v = -u\frac{\partial v}{\partial x} - v\frac{\partial u}{\partial x} - a\frac{\partial^3 u}{\partial x^3} + b\frac{\partial^2 v}{\partial x^2}, \end{cases} \quad 0<\alpha\leq 1, \tag{19}$$

with initial conditions

$$u(x,0) = \omega - 2B\ell\ coth[\ell(x+c)],\ v(x,0) = -2B(B+b)\ell^2 csch^2[\ell(x+c)] \tag{20}$$

where ω, ℓ and c are arbitrary constants and $B = \sqrt{a+b^2}$. Now by performing LT on both sides of Eq. (19) and make use of conditions provided in Eq. (20), we have

$$\begin{aligned} L[u(x,t)] - \tfrac{\omega - 2B\ell coth[\ell(x+c)]}{s} + \tfrac{1}{s^\alpha}L\left\{u\tfrac{\partial u}{\partial x} + \tfrac{\partial v}{\partial x} + b\tfrac{\partial^2 u}{\partial x^2}\right\} = 0, \\ L[v(x,t)] + \tfrac{2B(B+b)\ell^2 csch^2[\ell(x+c)]}{s} + \tfrac{1}{s^\alpha}L\left\{u\tfrac{\partial v}{\partial x} + v\tfrac{\partial u}{\partial x} + a\tfrac{\partial^3 u}{\partial x^3} - b\tfrac{\partial^2 v}{\partial x^2}\right\} = 0. \end{aligned} \tag{21}$$

Define the non-linear operator as

$$\begin{aligned} N^1[\varphi_1(x,t;q),\varphi_2(x,t;q)] &= L[\varphi_1(x,t;q)] - \frac{\omega - 2B\ell\ coth[\ell(x+c)]}{s} \\ &+ \frac{1}{s^\alpha}L\left\{\varphi_1(x,t;q)\frac{\partial\varphi_1(x,t;q)}{\partial x} + \frac{\partial\varphi_2(x,t;q)}{\partial y} + b\frac{\partial^2\varphi_1(x,t;q)}{\partial x^2}\right\}, \\ N^2[\varphi_1(x,t;q),\varphi_2(x,t;q)] &= L[\varphi_2(x,t;q)] + \frac{2B(B+b)\ell^2 csch^2[\ell(x+c)]}{s} \\ &+ \frac{1}{s^\alpha}L\left\{\varphi_1(x,t;q)\frac{\partial\varphi_2(x,t;q)}{\partial x} + \varphi_2(x,t;q)\frac{\partial\varphi_1(x,t;q)}{\partial x}\right. \\ &\left. + a\frac{\partial^3\varphi_1(x,t;q)}{\partial x^3} - b\frac{\partial^2\varphi_2(x,t;q)}{\partial x^2}\right\}. \end{aligned} \tag{22}$$

By proposed algorithm, the deformation equation of *m-th* order for $H(x,t) = 1$, is given as

$$\begin{aligned} L[u_m(x,t) - \mathsf{k}_m u_{m-1}(x,t)] &= h\Re_{1,m}[\vec{u}_{m-1}, \vec{v}_{m-1}], \\ L[v_m(x,t) - \mathsf{k}_m v_{m-1}(x,t)] &= h\Re_{2,m}[\vec{u}_{m-1}, \vec{v}_{m-1}]. \end{aligned} \tag{23}$$

where

$$\begin{aligned} \Re_{1,m}[\vec{u}_{m-1}, \vec{v}_{m-1}] &= L[u_{m-1}(x,t)] - \left(1 - \frac{\mathsf{k}_m}{n}\right) \frac{\omega - 2B\ell \, coth[\ell(x+c)]}{s} \\ &+ \frac{1}{s^\alpha} L\left\{ \sum_{i=0}^{m-1} u_i \frac{\partial u_{m-1-i}}{\partial x} + \frac{\partial v_{m-1}}{\partial x} + b \frac{\partial^2 u_{m-1}}{\partial x^2} \right\}, \\ \Re_{2,m}[\vec{u}_{m-1}, \vec{v}_{m-1}] &= L[v_{m-1}(x,t)] + \left(1 - \frac{\mathsf{k}_m}{n}\right) \frac{2B(B+b)\ell^2 csch^2[\ell(x+c)]}{s} \\ &+ \frac{1}{s^\alpha} L\left\{ \sum_{i=0}^{m-1} u_i \frac{\partial v_{m-1-i}}{\partial x} + \sum_{i=0}^{m-1} v_i \frac{\partial u_{m-1-i}}{\partial x} + a \frac{\partial^3 u_{m-1}}{\partial x^3} - b \frac{\partial^2 v_{m-1}}{\partial x^2} \right\}. \end{aligned} \tag{24}$$

By applying inverse LT on Eq. (24), we get

$$\begin{aligned} u_m(x,t) &= \mathsf{k}_m u_{m-1}(x,t) + \hbar L^{-1}\left\{\Re_{1,m}[\vec{u}_{m-1}, \vec{v}_{m-1}]\right\}, \\ v_m(x,t) &= \mathsf{k}_m v_{m-1}(x,t) + \hbar L^{-1}\left\{\Re_{2,m}[\vec{u}_{m-1}, \vec{v}_{m-1}]\right\}. \end{aligned} \tag{25}$$

On solving above system of equations, we have

$$\begin{aligned} u_0(x,t) &= \omega - 2B\ell \, coth[\ell(x+c)], \\ v_0(x,t) &= -2B(B+b)\ell^2 csch^2[\ell(x+c)], \\ u_1(x,t) &= \frac{2hB\ell^2 \omega \, csch^2[\ell(x+c)] t^\alpha}{\Gamma[\alpha+1]}, \\ v_1(x,t) &= \frac{-2hB\ell^3 csch^4[\ell(x+c)](-4(a+b^2-B^2)\ell - 2(a+b^2-B^2)\ell \, cosh[2\ell(x+c)] - (b+B)\omega sinh[2\ell(x+c)])t^\alpha}{\Gamma[\alpha+1]}, \\ u_2(x,t) &= \frac{2(n+\hbar)\hbar B\ell^2 \omega \, csch^2[\ell(x+c)] t^\alpha}{\Gamma[\alpha+1]} + \frac{2B\hbar^2 \ell^3 t^{2\alpha}}{\Gamma[2\alpha+1]} (-20a\ell^2 - 20b^2\ell^2 + 20B^2\ell^2 + \omega^2 \\ &- (4a\ell^2 + 4b^2\ell^2 - 4B^2\ell^2 + \omega^2) cosh[2\ell(x+c)]) coth[\ell(x+c)] csch^4[\ell(x+c)], \\ v_2(x,t) &= \frac{-2(n+\hbar)\hbar B\ell^3 csch^4[\ell(x+c)](-4(a+b^2-B^2)\ell - 2(a+b^2-B^2)\ell cosh[2\ell(x+c)] - (b+B)\omega \, sinh[2\ell(x+c)])t^\alpha}{\Gamma[\alpha+1]} \\ &- \frac{B\hbar^2 \ell^4 csch[\ell(x+c)]^6 t^{2\alpha}}{\Gamma[2\alpha+1]} ((132b\ell^2 - 60B\ell^2)(a+b^2-B^2) - 3\omega^2(b+B) \\ &+ 2cosh[2\ell(x+c)](4a(13b-7B)\ell^2 + (b+B)(52b^2\ell^2 - 80bB\ell^2 + 28B^2\ell^2 + \omega^2)) \\ &+ cosh[4\ell(x+c)](4a(b-B)\ell^2 + (b+B)(4b^2\ell^2 - 8bB\ell^2 + 4B^2\ell^2 + \omega^2)) \\ &+ 4\ell\omega(10 \, sinh[2\ell(x+c)] + sinh[4\ell(x+c)])(a+b^2-B^2)) \end{aligned}$$

$$
\begin{aligned}
u_3(x,t) &= \frac{2(n+\hbar)^2\hbar B\ell^2\omega\, csch^2[\ell(x+c)]t^\alpha}{\Gamma[\alpha+1]} + \frac{4(n+\hbar)B\hbar^2\ell^3 t^{2\alpha}}{\Gamma[2\alpha+1]}(-20a\ell^2 - 20b^2\ell^2 + 20B^2\ell^2 + \omega^2 \\
&- (4a\ell^2 + 4b^2\ell^2 - 4B^2\ell^2 + \omega^2)cosh[2\ell(x+c)])coth[\ell(x+c)]csch^4[\ell(x+c)] \\
&+ \frac{B\hbar^3\ell^4 t^{3\alpha} csch^7[\ell(x+c)]}{2\Gamma[\alpha+1]^2\Gamma[3\alpha+1]}(-8B\ell\omega^2\Gamma[2\alpha+1]sinh[\ell(x+c)]sinh[2\ell(x+c)] \\
&+ \Gamma[\alpha+1]^2(-8B\ell(260a\ell^2 + 260b^2\ell^2 - 260B^2\ell^2 + \omega^2)cosh[\ell(x+c)] \\
&+ 8B\ell(-62a\ell^2 - 62b^2\ell^2 + 62B^2\ell^2 + \omega^2)cosh[3\ell(x+c)] - 16aB\ell^3 cosh[5\ell(x+c)] \\
&- 16b^2B\ell^3 cosh[5\ell(x+c)] + 16B^3\ell^3 cosh[5\ell(x+c)] + 480a\ell^2\omega\, sinh[\ell(x+c)] \\
&+ 480b^2\ell^2\omega\, sinh[\ell(x+c)] - 480B^2\ell^2\omega\, sinh[\ell(x+c)] - 8\omega^3 sinh[\ell(x+c)] \\
&+ 300a\ell^2\omega\, sinh[3\ell(x+c)] + 300b^2\ell^2\omega sinh[3\ell(x+c)] - 300B^2\ell^2\omega\, sinh[3\ell(x+c)] \\
&+ \omega^3 sinh[3\ell(x+c)] + 12a\ell^2\omega\, sinh[5\ell(x+c)] + 12b^2\ell^2\omega\, sinh[5\ell(x+c)] \\
&- 12B^2\ell^2\omega\, sinh[5\ell(x+c)] + \omega^3 sinh[5\ell(x+c)])),
\end{aligned}
$$

$$
\begin{aligned}
v_3(x,t) &= \frac{-2(n+\hbar)^2\hbar B\ell^3 csch^4[\ell(x+c)](-4(a+b^2-B^2)\ell - 2(a+b^2-B^2)\ell cosh[2\ell(x+c)] - (b+B)\omega sinh[2\ell(x+c)])t^\alpha}{\Gamma[\alpha+1]} \\
&- \frac{(n+\hbar)\hbar^2 B\ell^4 csch[\ell(x+c)]^6 t^{2\alpha}}{\Gamma[2\alpha+1]}((132b\ell^2 - 60B\ell^2)(a+b^2-B^2) - 3\omega^2(b+B) \\
&+ 2cosh[2\ell(x+c)](4a(13b-7B)\ell^2 + (b+B)(52b^2\ell^2 - 80bB\ell^2 + 28B^2\ell^2 + \omega^2)) \\
&+ cosh[4\ell(x+c)](4a(b-B)\ell^2 + (b+B)(4b^2\ell^2 - 8bB\ell^2 + 4B^2\ell^2 + \omega^2)) \\
&+ 4\ell\omega(10\, sinh[2\ell(x+c)] + sinh[4\ell(x+c)])(a+b^2-B^2)) \\
&- \frac{B\hbar^3\ell^5 csch[\ell(x+c)]^8 t^{3\alpha}}{2\Gamma[1+\alpha]^2\Gamma[3\alpha+1]}(16B\ell\omega\Gamma[2\alpha+1]sinh[\ell(x+c)](16(a+b^2-B^2)\ell cosh[\ell(x+c)] \\
&+ 2(a+b^2-B^2)\ell cosh[3\ell(x+c)] + (b+B)\omega(2sinh[\ell(x+c)] + sinh[3\ell(x+c)])) \\
&- \Gamma[\alpha+1]^2(2\ell(8(604a^2\ell^2 + a(8\ell^2(151b^2 - 65bB - 70B^2) - 15\omega^2) \\
&+ (b+B)(4\ell^2(151b^3 - 281b^2B + 141bB^2 - 11B^3) + \omega^2(13B - 15b)) \\
&+ (4764a^2\ell^2 + a(4\ell^2(2382b^2 - 1090bB - 1060B^2) + 45\omega^2) \\
&+ (b+B)(4\ell^2(1191b^3 - 2281b^2B + 1221bB^2 - 131B^3) + 45b\omega^2 - 37B\omega^2))cosh[2\ell(x+c)]) \\
&+ 16\ell(60a^2\ell^2 + a(4\ell^2(30b^2 - 17bB - 11B^2) + 9\omega^2) \\
&+ (b+B)(4\ell^2(15b^3 - 32b^2B + 21bB^2 - 4B^3) + \omega^2(9b - 8B)))cosh[4\ell(x+c)] \\
&+ 2\ell(a+b^2-B^2)(4a\ell^2 + 4b^2\ell^2 - 8bB\ell^2 + 4B^2\ell^2 + 3\omega^2)cosh[6\ell(x+c)] \\
&- \omega(-4a(735b - 283B)\ell^2 - (b+B)(4\ell^2(735b^2 - 1018bB + 283B^2) - 19\omega^2))sinh[2\ell(x+c)] \\
&+ 8\omega(4a(21b - 11B)\ell^2 + (b+B)(4\ell^2(21b^2 - 32bB + 11B^2) + \omega^2))sinh[4\ell(x+c)] \\
&+ \omega(12a(b-B)\ell^2 + (b+B)(4\ell^2(3b^2 - 6bB + 3B^2\ell^2) + \omega^2))sinh[6\ell(x+c)])) \\
&\vdots
\end{aligned}
$$

In this approach, the rest of the iterative terms can be obtained. Then, the family of ***q***-HATM series solution of the of Eq. (19) is

$$
\begin{aligned}
u(x,t) &= u_0(x,t) + \sum_{m=1}^{\infty} u_m(x,t)\left(\frac{1}{n}\right)^m, \\
v(x,t) &= v_0(x,t) + \sum_{m=1}^{\infty} v_m(x,t)\left(\frac{1}{n}\right)^m.
\end{aligned} \tag{26}
$$

If we set $\alpha = 1, h = -1$, and $n = 1$, then the obtained solutions $\sum_{m=1}^{N} u_m(x,t)\left(\frac{1}{n}\right)^m$ and $\sum_{m=1}^{N} v_m(x,t)\left(\frac{1}{n}\right)^m$, respectively converges to the exact solutions

$u(x,t) = \omega - 2B\ell coth[\ell(x + c - \omega t)]$ and $v(x,t) = -2B(B+b)\ell^2 csch^2[\ell(x + c - \omega t)]$ of the classical order coupled WBK equations as $N \to \infty$.

Example 4.2. Consider the time-fractional coupled WBK Eq. (1) for $a = 3$ and $b = 1$ [26, 36]:

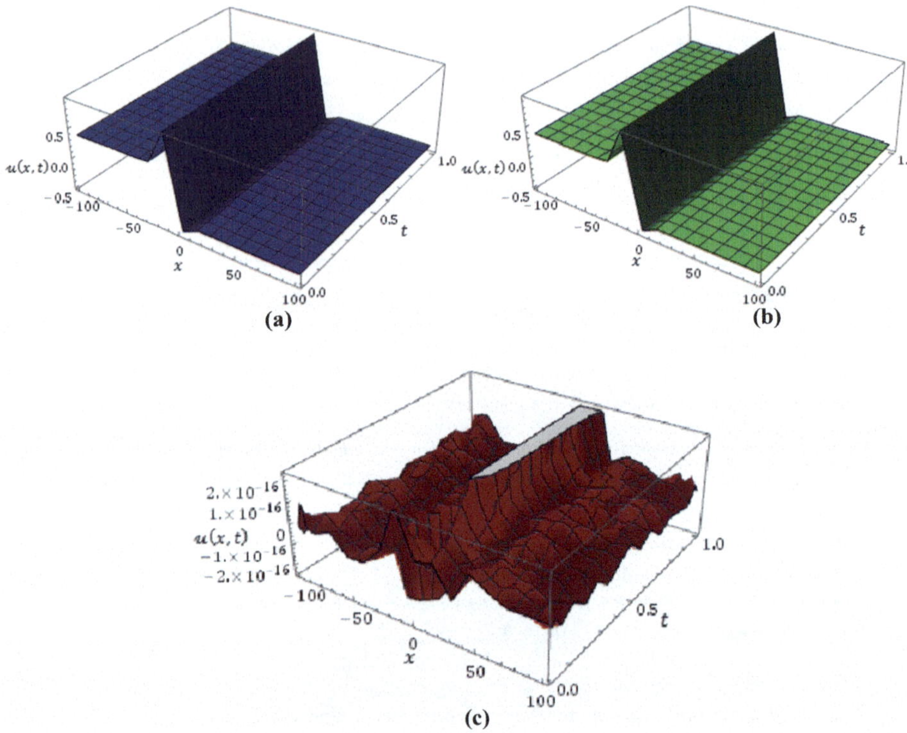

Fig. 1. **(a)** Surface of approximate solution **(b)** Surface of exact solution **(c)** Surface of absolute error $= |u_{exa.} - u_{app.}|$ for Example 4.1 at $n = 1, \alpha = 1, \hbar = -1, \omega = 0.005, a = 1.5, b = 1.5, \ell = 0.1$ and $c = 10$.

$$\begin{cases} D_t^\alpha u = -u\frac{\partial u}{\partial x} - \frac{\partial v}{\partial x} - \frac{\partial^2 u}{\partial x^2}, \\ D_t^\alpha v = -u\frac{\partial v}{\partial x} - v\frac{\partial u}{\partial x} + \frac{\partial^2 v}{\partial x^2} - 3\frac{\partial^3 u}{\partial x^3}, \end{cases} \quad 0 < \alpha \leq 1, \tag{27}$$

with initial conditions

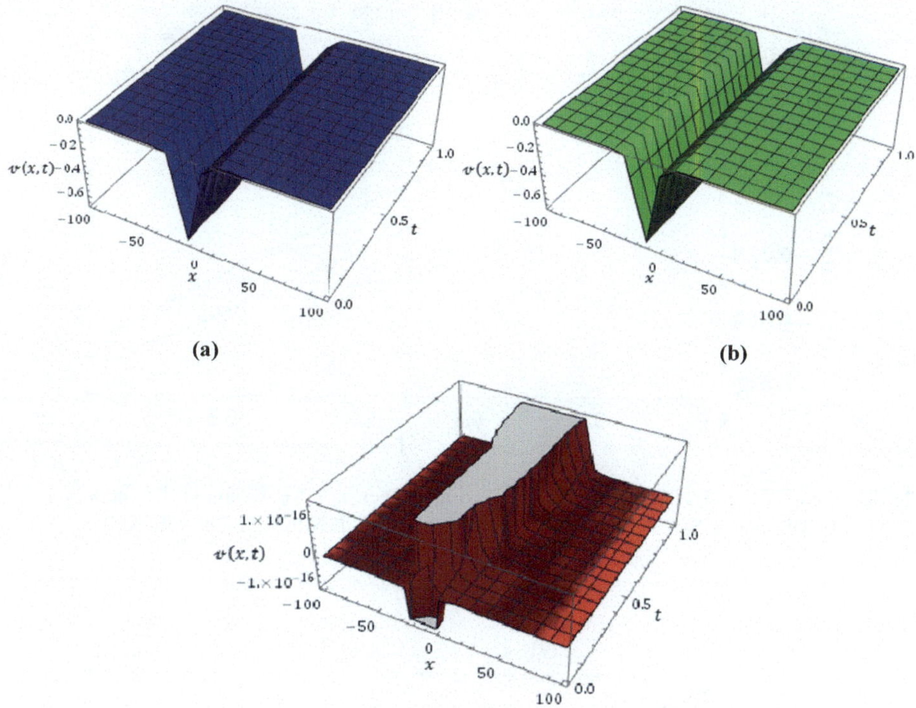

(c)

Fig. 2. **(a)** Surface of approximate solution **(b)** Surface of exact solution **(c)** Surface of absolute errore error $= \left|v_{exa.} - v_{app.}\right|$ at $n = 1, \alpha = 1, \omega = 0.005, a = 1.5, b = 1.5, \hbar = -1, \ell = 0.1$ and $c = 10$ for Example 4.1.

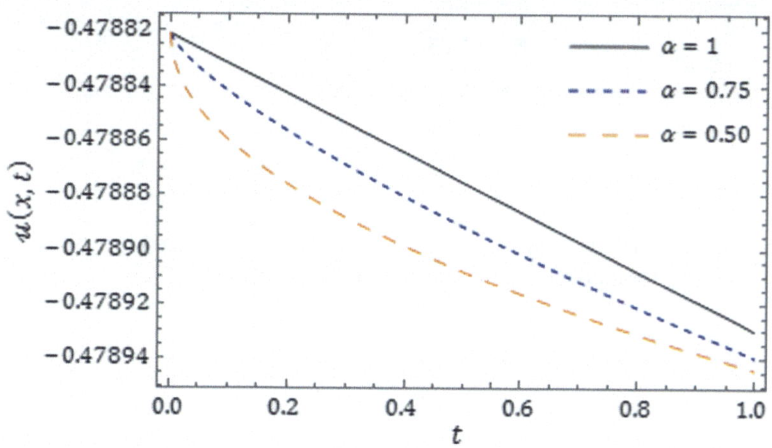

Fig. 3. Plot of q-HATM solution $u(x,t)$ for Example 4.1 with respect to t when $n = 1, \hbar = -1, \omega = 0.005$, $a = 1.5, b = 1.5, \ell = 0.1$, $c = 10$ and $x = 1$ with diverse values of α.

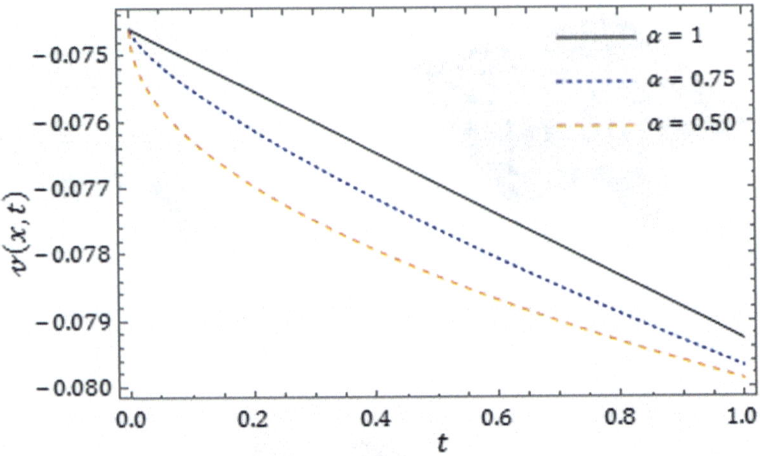

Fig. 4. Nature of q-HATM solution $v(x,t)$ with respect to t for Example 4.1, at $n=1, \hbar=-1, \omega=0.005$, $a=1.5, b=1.5, \ell=0.1$, $c=10$ and $x=1$ with different values of α.

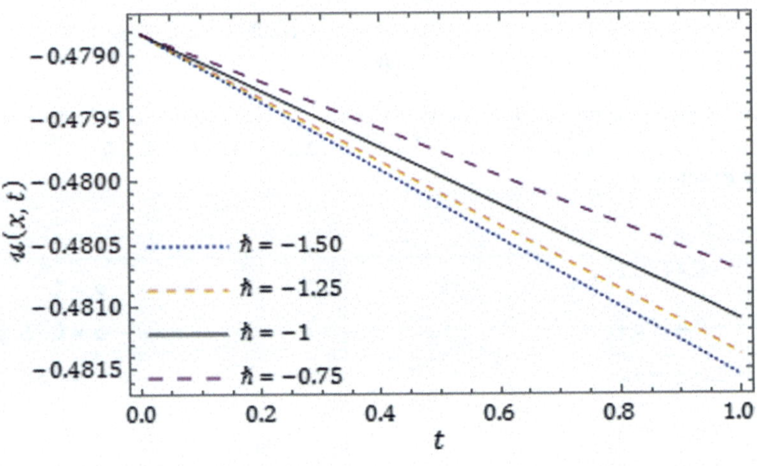

Fig. 5. Nature of q-HATM solution $u(x,t)$ at $n=5, \alpha=1, \omega=0.005, \ell=0.1, a=1.5, b=1.5, c=10$ and $x=1$ with diverse values of $\hbar$ for Example 4.1.

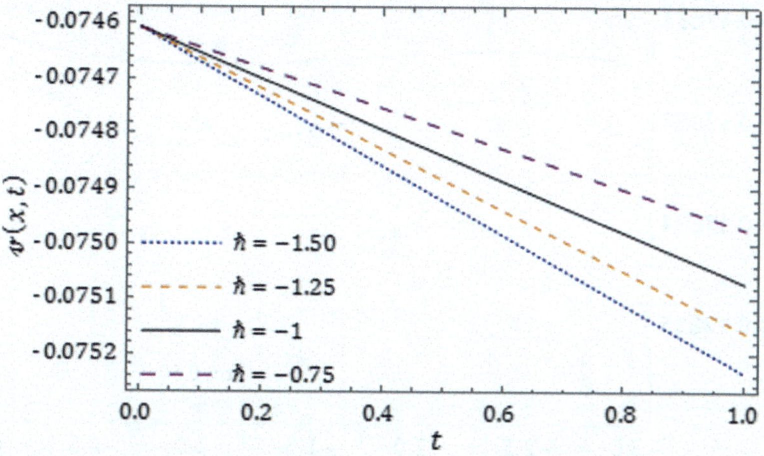

Fig. 6. Behaviour of q-HATM solution $v(x,t)$ with different values of $\hbar$ for Example 4.1 at $n = 5, \alpha = 1, \omega = 0.005, a = 1.5, b = 1.5, \ell = 0.1, c = 10$ and $x = 1$.

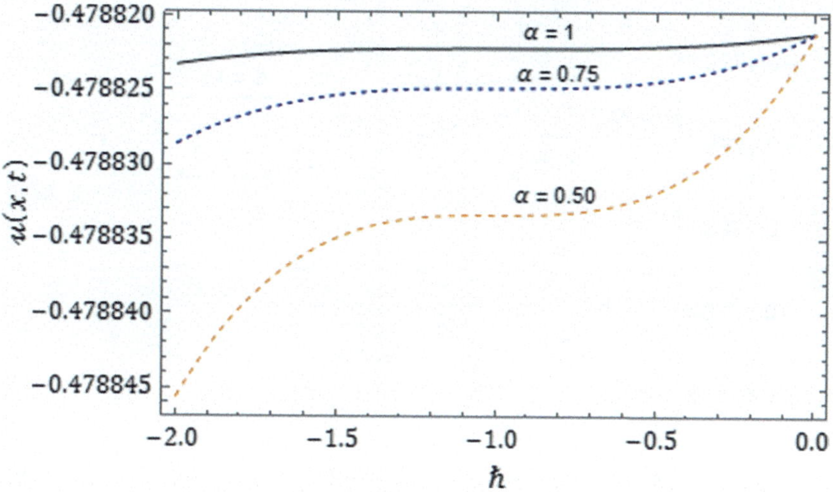

Fig. 7. $\hbar$-curves drown for the q-HATM solution $u(x,t)$ at $\omega = 0.005, a = 1.5, b = 1.5, \ell = 0.1$, $c = 10, x = 1, t = 0.01$ and $n = 1$ with diverse values of α for Example 4.1.

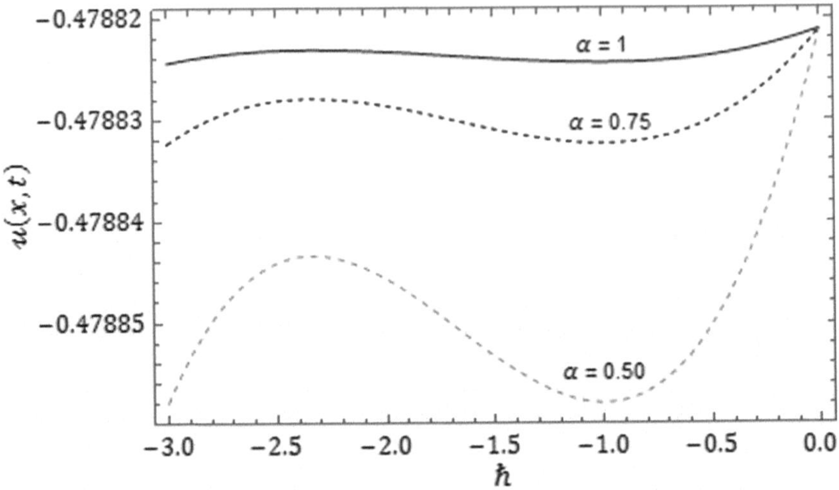

Fig. 8. $\hbar$-curves drown for the q-HATM solution $u(x,t)$ with different values of α for Example 4.1 at $\omega = 0.005, a = 1.5, b = 1.5, \ell = 0.1,\ c = 10, x = 1, t = 0.01$ and $n = 2$.

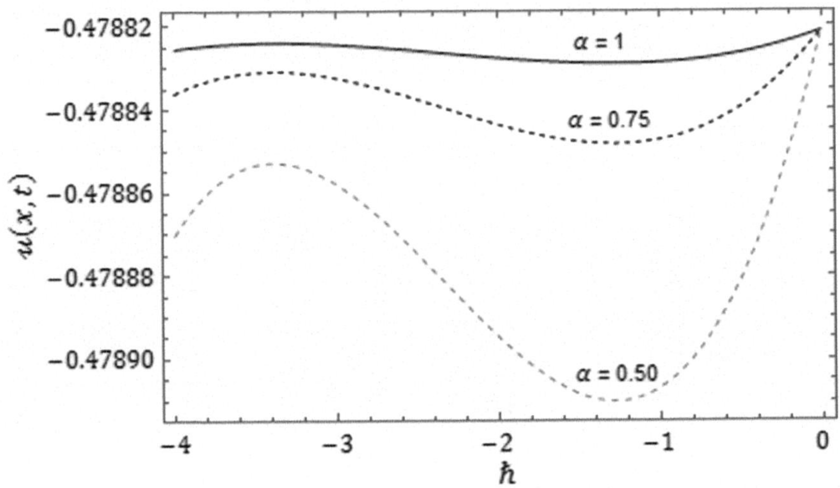

Fig. 9. $\hbar$-curves drown for the q-HATM solution $u(x,t)$ for Example 4.1 at $\omega = 0.005, a = 1.5, b = 1.5, \ell = 0.1,\ c = 10, x = 1, t = 0.01$ and $n = 3$ with distinct values of α.

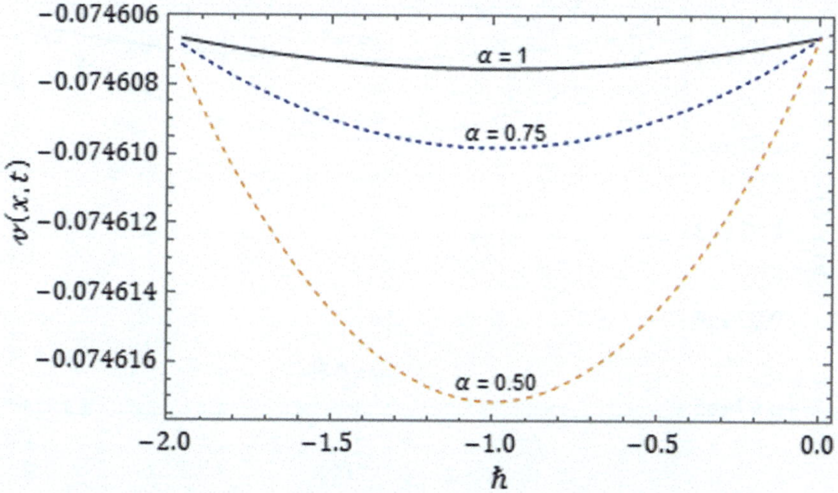

Fig. 10. $\hbar$-curves drown for the q-HATM solution $v(x,t)$ with different values of α for Example 4.1 at $\omega = 0.005, a = 1.5, b = 1.5, \ell = 0.1,\ c = 10, x = 1, t = 0.01$ and $n = 1$.

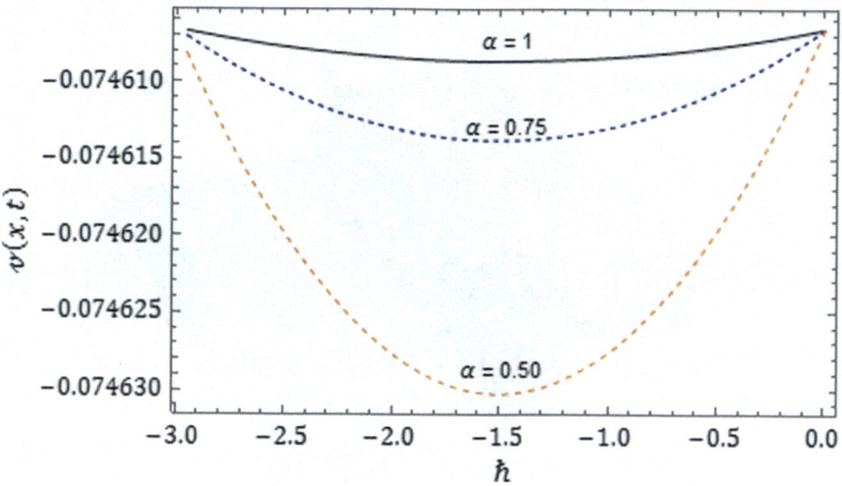

Fig. 11. $\hbar$-curves drown for the q-HATM solution $v(x,t)$ at $\omega = 0.005, a = 1.5, b = 1.5, \ell = 0.1,\ c = 10, x = 1, t = 0.01$ and $n = 2$ with diverse values of α for Example 4.1.

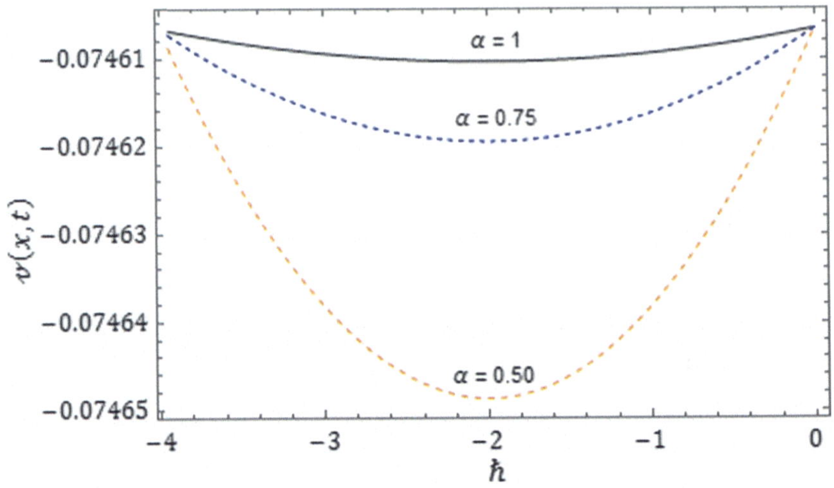

Fig. 12. $\hbar$-curves drown for the q-HATM solution $v(x,t)$ at $\omega = 0.005, a = 1.5, b = 1.5, \ell = 0.1, c = 10, x = 1, t = 0.01$ and $n = 3$ with distinct values of α for Example 4.1.

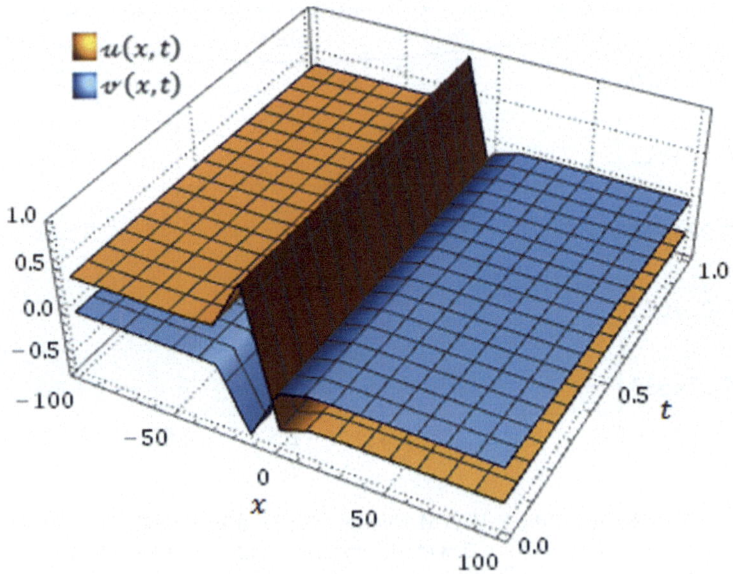

Fig. 13. Nature of coupled q-HATM solutions $u(x,t)$ and $v(x,t)$ at $n = 1, \hbar = -1, c = 10, \omega = 0.005, a = 1.5, b = 1.5, \ell = 0.1$ and $\alpha = 1$ for Example 4.1.

Table 1. Comparative study between ADM [22], VIM [23], CFRDTM [21] and q-HATM for the approximate solution $u(x,t)$ at $c = 10$, $\omega = 0.005, a = 1.5, b = 1.5, \ell = 0, n = 1, \hbar = -1$ and $\alpha = 1$ for Example 4.1

(x,t)	$\lvert u_{Exact} - u_{ADM}\rvert$	$\lvert u_{Exact} - u_{VIM}\rvert$	$\lvert u_{Exact} - u_{CFRDTM}\rvert$	$\lvert u_{Exact} - u^{(3)}_{q-HATM}\rvert$
$(0.1, 0.1)$	1.04892×10^{-4}	1.23033×10^{-4}	1.11022×10^{-16}	1.11022×10^{-16}
$(0.1, 0.3)$	9.64474×10^{-5}	3.69597×10^{-4}	1.11022×10^{-16}	1.11022×10^{-16}
$(0.1, 0.5)$	8.88312×10^{-5}	6.16873×10^{-4}	1.33227×10^{-15}	1.33227×10^{-15}
$(0.2, 0.1)$	4.25408×10^{-4}	1.19869×10^{-4}	2.22045×10^{-16}	2.22045×10^{-16}
$(0.2, 0.3)$	3.91098×10^{-4}	3.60098×10^{-4}	1.66533×10^{-16}	1.66533×10^{-16}
$(0.2, 0.5)$	3.60161×10^{-4}	6.01006×10^{-4}	1.49880×10^{-15}	1.49880×10^{-15}
$(0.3, 0.1)$	9.71922×10^{-4}	1.16789×10^{-4}	0	0
$(0.3, 0.3)$	8.93309×10^{-4}	3.50866×10^{-4}	2.77556×10^{-16}	2.77556×10^{-16}
$(0.3, 0.5)$	8.22452×10^{-4}	5.85610×10^{-4}	1.27676×10^{-15}	1.27676×10^{-15}
$(0.4, 0.1)$	1.75596×10^{-3}	1.13829×10^{-4}	5.55112×10^{-17}	5.55112×10^{-17}
$(0.4, 0.3)$	1.61430×10^{-3}	3.41948×10^{-4}	1.66533×10^{-16}	1.66533×10^{-16}
$(0.4, 0.5)$	1.48578×10^{-3}	5.70710×10^{-4}	1.27676×10^{-15}	1.27676×10^{-15}
$(0.5, 0.1)$	2.79519×10^{-3}	1.10936×10^{-4}	0	0
$(0.5, 0.3)$	2.56714×10^{-3}	3.33274×10^{-4}	2.22045×10^{-16}	2.22045×10^{-16}
$(0.5, 0.5)$	2.36184×10^{-3}	5.56235×10^{-4}	1.22125×10^{-15}	1.22125×10^{-15}

Table 2. Comparative study between ADM [22], VIM [23], CFRDTM [21] and q-HATM for the approximate solution $v(x,t)$ at $c = 10, \omega = 0.005, a = 1.5, b = 1.5, \ell = 0.1, n = 1, \hbar = -1$ and $\alpha = 1$ for Example 4.1

(x,t)	$\lvert v_{Exact} - v_{ADM}\rvert$	$\lvert v_{Exact} - v_{VIM}\rvert$	$\lvert v_{Exact} - v_{CFRDTM}\rvert$	$\lvert user2 v_{Exact} - v^{(3)}_{q-user2HATM}\rvert$
$(0.1, 0.1)$	6.41419×10^{-3}	1.10430×10^{-4}	2.77556×10^{-17}	2.77556×10^{-17}
$(0.1, 0.3)$	5.99783×10^{-3}	3.31865×10^{-4}	3.60822×10^{-16}	3.60822×10^{-16}
$(0.1, 0.5)$	5.61507×10^{-3}	5.54071×10^{-4}	2.40086×10^{-15}	2.40086×10^{-15}
$(0.2, 0.1)$	1.33181×10^{-2}	1.07016×10^{-4}	4.16334×10^{-17}	4.16334×10^{-17}
$(0.2, 0.3)$	1.24441×10^{-2}	3.21601×10^{-4}	3.05311×10^{-16}	3.05311×10^{-16}
$(0.2, 0.5)$	1.16416×10^{-2}	5.36927×10^{-4}	2.31759×10^{-15}	2.31759×10^{-15}
$(0.3, 0.1)$	2.07641×10^{-2}	1.03737×10^{-4}	5.55112×10^{-17}	5.55112×10^{-17}
$(0.3, 0.3)$	1.93852×10^{-2}	3.11737×10^{-4}	2.63678×10^{-16}	2.63678×10^{-16}
$(0.3, 0.5)$	1.81209×10^{-2}	5.20447×10^{-4}	2.15106×10^{-15}	2.15106×10^{-15}
$(0.4, 0.1)$	2.88100×10^{-2}	1.00579×10^{-4}	2.77556×10^{-17}	2.77556×10^{-17}
$(0.4, 0.3)$	2.68724×10^{-2}	3.02245×10^{-4}	2.49800×10^{-16}	2.49800×10^{-16}
$(0.4, 0.5)$	2.50985×10^{-2}	5.04593×10^{-4}	2.04003×10^{-15}	2.04003×10^{-15}
$(0.5, 0.1)$	3.75193×10^{-2}	9.75385×10^{-4}	0	0
$(0.5, 0.3)$	3.49617×10^{-2}	2.93107×10^{-4}	2.63678×10^{-16}	2.63678×10^{-16}
$(0.5, 0.5)$	3.26239×10^{-2}	4.89335×10^{-4}	1.09126×10^{-15}	1.09126×10^{-15}

$$u(x,0) = \frac{1}{2} - 8\,tanh(-2x),\ \ v(x,0) = 16 - 16\,tanh^2(-2x). \tag{28}$$

By performing LT on both sides of Eq. (27) and make use of conditions provided in Eq. (28), we have

$$\begin{aligned} &L[u(x,t)] - \tfrac{1}{s}\left(\tfrac{1}{2} - 8\,tanh(-2x)\right) + \tfrac{1}{s^\alpha}L\left\{u\tfrac{\partial u}{\partial x} + \tfrac{\partial v}{\partial x} + \tfrac{\partial^2 u}{\partial x^2}\right\} = 0,\\ &L[v(x,t)] - \tfrac{16-16\,tanh^2(-2x)}{s} + \tfrac{1}{s^\alpha}L\left\{u\tfrac{\partial v}{\partial x} + v\tfrac{\partial u}{\partial x} - \tfrac{\partial^2 v}{\partial x^2} + 3\tfrac{\partial^3 u}{\partial x^3}\right\} = 0. \end{aligned} \tag{29}$$

Define the non-linear operator as

$$\begin{aligned} N^1[\varphi_1(x,t;q),\varphi_2(x,t;q)] =& L[\varphi_1(x,t;q)] - \frac{1}{s}\left(\frac{1}{2} - 8tanh(-2x)\right)\\ &+ \frac{1}{s^\alpha}L\left\{\varphi_1(x,t;q)\frac{\partial\varphi_1(x,t;q)}{\partial x} + \frac{\partial\varphi_2(x,t;q)}{\partial y} + \frac{\partial^2\varphi_1(x,t;q)}{\partial x^2}\right\},\\ N^2[\varphi_1(x,t;q),\varphi_2(x,t;q)] =& L[\varphi_2(x,t;q)] - \frac{16-16\tanh^2(-2x)}{s} + \frac{1}{s^\alpha}L\{\varphi_1(x,t;q)\frac{\partial\varphi_2(x,t;q)}{\partial x}\\ &+ \varphi_2(x,t;q)\frac{\partial\varphi_1(x,t;q)}{\partial y} - \frac{\partial^2\varphi_2(x,t;q)}{\partial x^2} + 3\frac{\partial^3\varphi_1(x,t;q)}{\partial x^3}\}. \end{aligned} \tag{30}$$

The *m-th* order deformation equation for $H(x,t) = 1$, is given as

$$\begin{aligned} L[u_m(x,t) - \text{k}_m u_{m-1}(x,t)] &= h\Re_{1,m}[\vec{u}_{m-1},\vec{v}_{m-1}],\\ L[v_m(x,t) - \text{k}_m v_{m-1}(x,t)] &= h\Re_{2,m}[\vec{u}_{m-1},\vec{v}_{m-1}], \end{aligned} \tag{31}$$

where

$$\begin{aligned} \Re_{1,m}[\vec{u}_{m-1},\vec{v}_{m-1}] =& L[u_{m-1}(x,t)] - \left(1 - \frac{\text{k}_m}{n}\right)\frac{1}{s}\left(\frac{1}{2} - 8\,tanh(-2x)\right)\\ &+ \frac{1}{s^\alpha}L\left\{\sum\nolimits_{i=0}^{m-1} u_i\frac{\partial u_{m-1-i}}{\partial x} + \frac{\partial v_{m-1}}{\partial x} + \frac{\partial^2 u_{m-1}}{\partial x^2}\right\},\\ \Re_{2,m}[\vec{u}_{m-1},\vec{v}_{m-1}] =& L[v_{m-1}(x,t)] - \left(1 - \frac{\text{k}_m}{n}\right)\frac{16-16\,tanh^2(-2x)}{s}\\ &+ \frac{1}{s^\alpha}L\left\{\sum\nolimits_{i=0}^{m-1} u_i\frac{\partial v_{m-1-i}}{\partial x} + \sum\nolimits_{i=0}^{m-1} v_i\frac{\partial u_{m-1-i}}{\partial x} - \frac{\partial^2 v_{m-1}}{\partial x^2} + 3\frac{\partial^3 u_{m-1}}{\partial x^3}\right\}. \end{aligned} \tag{32}$$

By applying inverse LT on Eq. (31), we get

$$\begin{aligned} u_m(x,t) &= \mathsf{k}_m u_{m-1}(x,t) + \hbar L^{-1}\{\Re_{1,m}[\vec{u}_{m-1}, \vec{v}_{m-1}]\}, \\ v_m(x,t) &= \mathsf{k}_m v_{m-1}(x,t) + \hbar L^{-1}\{\Re_{2,m}[\vec{u}_{m-1}, \vec{v}_{m-1}]\}. \end{aligned} \tag{33}$$

On solving above equation, we have

$$\begin{aligned}
u_0(x,t) &= \frac{1}{2} - 8\,tanh(-2x),\ v_0(x,t) = 16 - 16\,tanh^2(-2x),\\
u_1(x,t) &= \frac{8\hbar\, sech^2(-2x)t^\alpha}{\Gamma[\alpha+1]},\ v_1(x,t) = \frac{32\hbar sech^2(-2x)tanh(-2x)t^\alpha}{\Gamma[\alpha+1]},\\
u_2(x,t) &= \frac{8(n+\hbar)\hbar\, sech^2(-2x)t^\alpha}{\Gamma[\alpha+1]} + \frac{16\hbar^2 sech^2(-2x)tanh(-2x)t^{2\alpha}}{\Gamma[2\alpha+1]},\\
v_2(x,t) &= \frac{32(n+\hbar)\hbar\, sech^2(-2x)tanh(-2x)t^\alpha}{\Gamma[\alpha+1]} + \frac{32\hbar^2(-2+cosh(4x))sech^4(-2x)t^{2\alpha}}{\Gamma[2\alpha+1]},\\
u_3(x,t) &= \frac{8(n+\hbar)^2\hbar\, sech^2(-2x)t^\alpha}{\Gamma[\alpha+1]} + \frac{32(n+\hbar)\hbar^2 sech^2(-2x)tanh(-2x)t^{2\alpha}}{\Gamma[2\alpha+1]}\\
&\quad + \frac{8\hbar^3 sech^5(2x)\left(-32\Gamma[2\alpha+1]sinh(2x) + \Gamma[\alpha+1]^2(-3cosh(2x)+cosh(6x)+64sinh(2x))\right)t^{3\alpha}}{\Gamma[\alpha+1]^2\Gamma[3\alpha+1]},\\
v_3(x,t) &= \frac{32(n+\hbar)^2\hbar\, sech^2(-2x)tanh(-2x)t^\alpha}{\Gamma[\alpha+1]} + \frac{64\hbar^2(n+\hbar)(-2+\cosh(4x))sech^4(2x)t^{2\alpha}}{\Gamma[2\alpha+1]}\\
&\quad - \frac{16\hbar^3 sech^6(-2x)\left(32(3-2\,cosh(4x))\Gamma[2\alpha+1] + \Gamma[\alpha+1]^2(-192+128\cosh(4x)-10\sinh(4x)+\sinh(8x))\right)t^{3\alpha}}{\Gamma[\alpha+1]^2\Gamma[3\alpha+1]},\\
&\ \vdots
\end{aligned}$$

Following in the same procedure, the rest of the iterative components can be easily obtained. Eventually, the family of ***q***-HATM series solution of the system of Eq. (27) is given by

$$\begin{aligned} u(x,t) &= u_0(x,t) + \sum_{m=1}^{\infty} u_m(x,t)\left(\frac{1}{n}\right)^m, \\ v(x,t) &= v_0(x,t) + \sum_{m=1}^{\infty} v_m(x,t)\left(\frac{1}{n}\right)^m. \end{aligned} \tag{34}$$

If we set $\alpha = 1, h = -1$, and $n = 1$, then the obtained solutions $\sum_{m=1}^{N} u_m(x,t)\left(\frac{1}{n}\right)^m$ and $\sum_{m=1}^{N} v_m(x,t)\left(\frac{1}{n}\right)^m$ when $N \to \infty$, converges to the exact solutions of classical order coupled WBK equations

$$u(x,t) = \frac{1}{2} - 8\,\tanh\left[-2\left(x - \frac{t}{2}\right)\right],\ v(x,t) = 16 - 16\,tanh^2\left[-2\left(x - \frac{t}{2}\right)\right].$$

5 Numerical Results and Discussion

In order to verify whether the proposed algorithm lead to greater accuracy, the numerical solutions have been evaluated. From results we can certainly conclude that, the proposed technique provides remarkable exactness in comparison to the method available in the literature [21–26, 36]. It can be observed from Tables 1 and 2, the absolute error is very tiny and less numbers of homotopy polynomial are needed. Further, the accuracy of proposed scheme is drowned in terms of numerical simulations for Example 4.2 is showed in Tables 3. The Mathematica code is used for numerical computations.

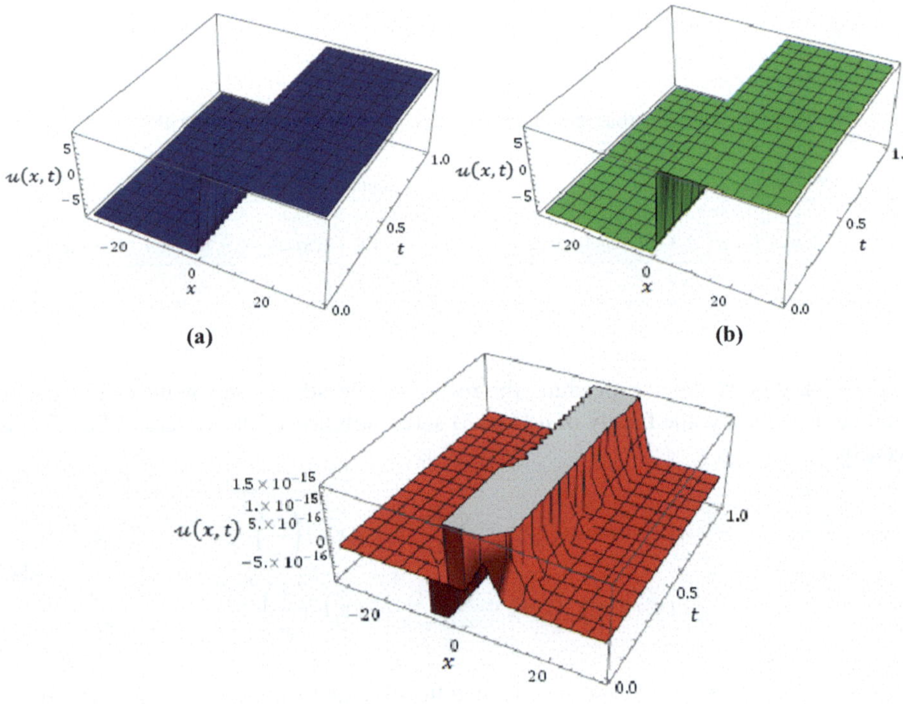

Fig. 14. **(a)** Surface of approximate solution **(b)** Surface of exact solution **(c)** Surface of absolute errore error $= |u_{exa.} - u_{app.}|$ at $n = 1, \alpha = 1$ and $\hbar = -1$ for Example 4.2.

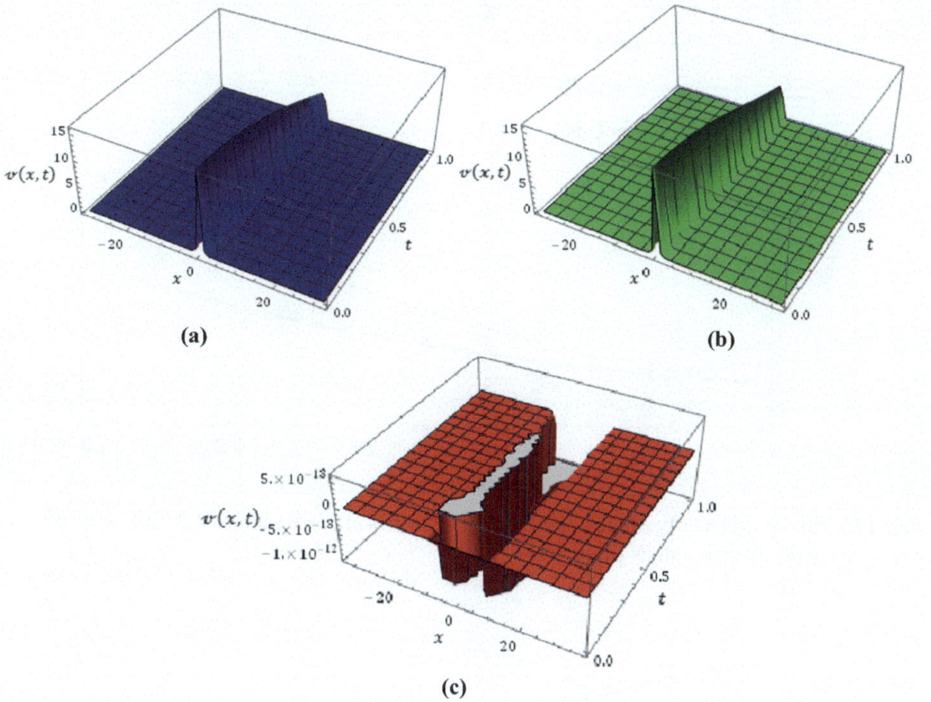

Fig. 15. **(a)** Surface of approximate solution **(b)** Surface of exact solution **(c)** Surface of absolute errore error $= \left| v_{exa.} - v_{app.} \right|$ at $n = 1, \alpha = 1$ and $\hbar = -1$ for Example 4.2.

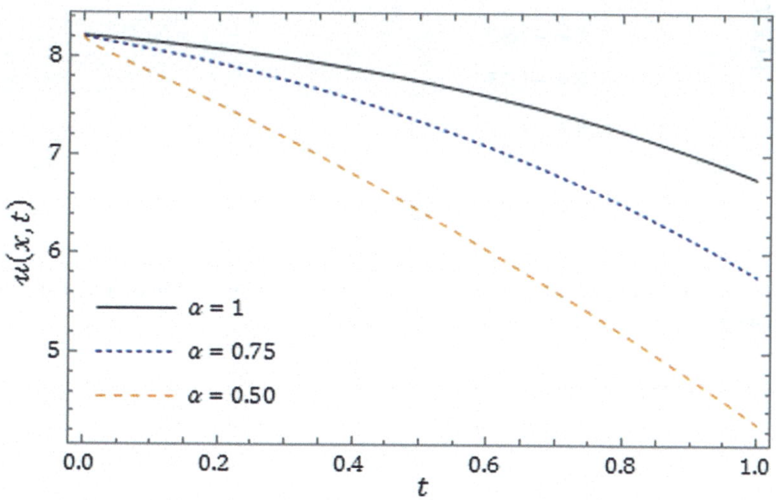

Fig. 16. Nature of q-HATM solution $u(x, t)$ with respect to t through distinct values of α for Example 4.2 when $n = 1, \hbar = -1$ and $x = 1$.

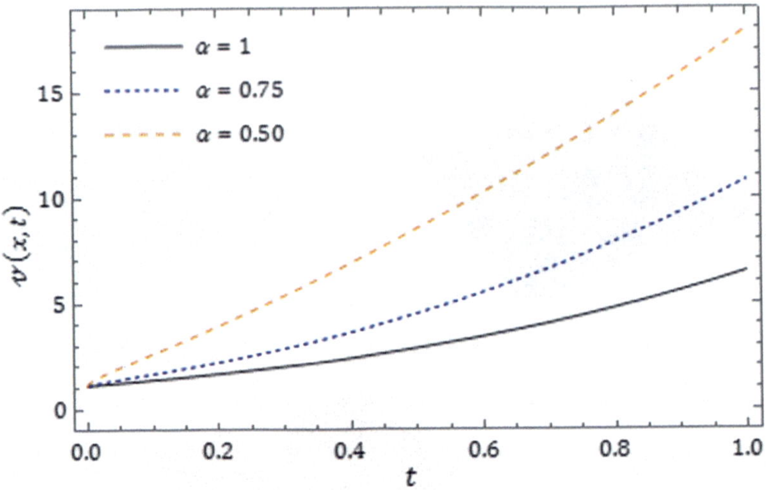

Fig. 17. Plot of q-HATM solution $v(x,t)$ with respect to t for Example 4.2. at $n = 1, \hbar = -1$ and $x = 1$ with diverse values of α.

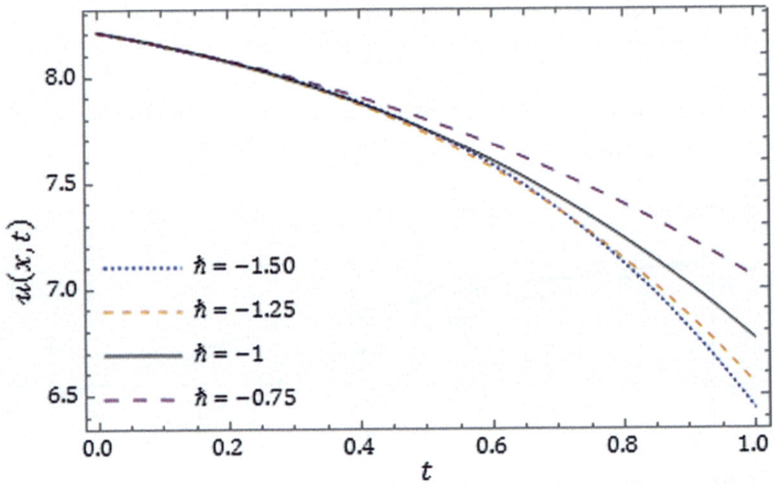

Fig. 18. Nature of q-HATM solution $u(x,t)$ at $n = 1, \alpha = 1$ and $x = 1$ with different values of $\hbar$ for Example 4.2.

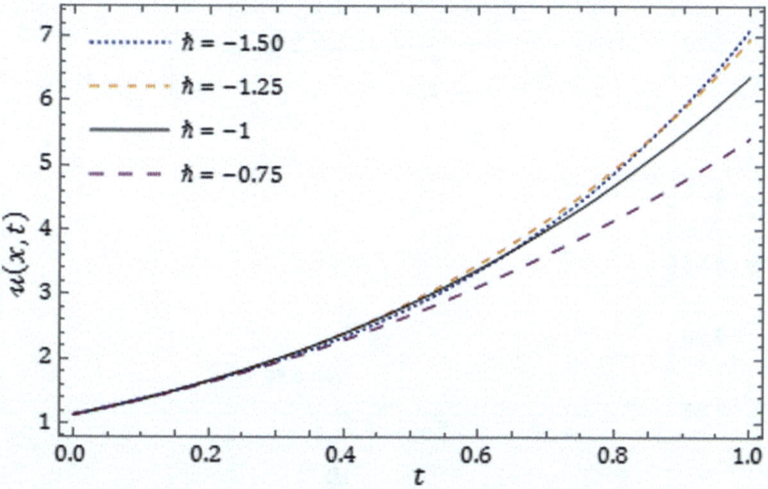

Fig. 19. Plot of q-HATM solution $v(x,t)$ for Example 4.2 with distinct values of $\hbar$ at $n = 1, \alpha = 1$ and $x = 1$.

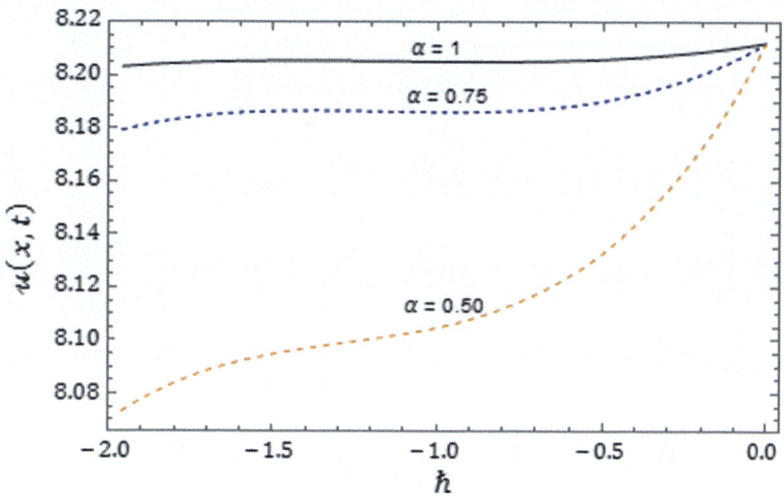

Fig. 20. $\hbar$-curve drown for the q-HATM solution $u(x,t)$ with different values of α for Example 4.2 at $x = 1, t = 0.01$ and $n = 1$.

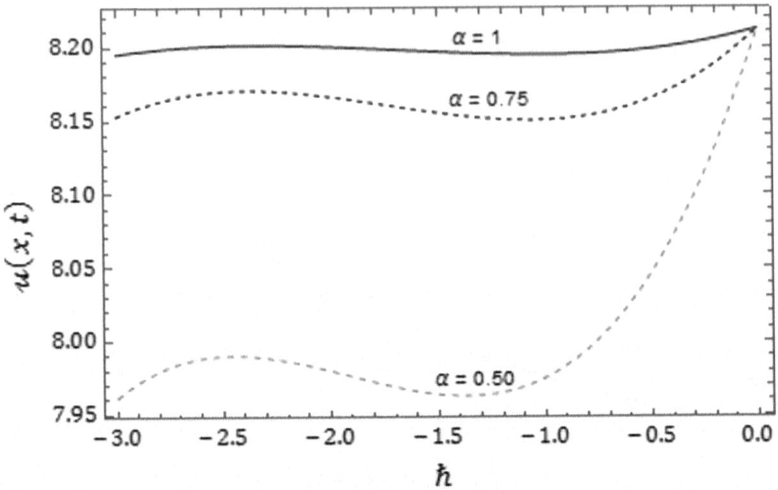

Fig. 21. $\hbar$-curve drown for the q-HATM solution $u(x,t)$ with diverse values of α for Example 4.2 at $x = 1, t = 0.01$ and $n = 2$.

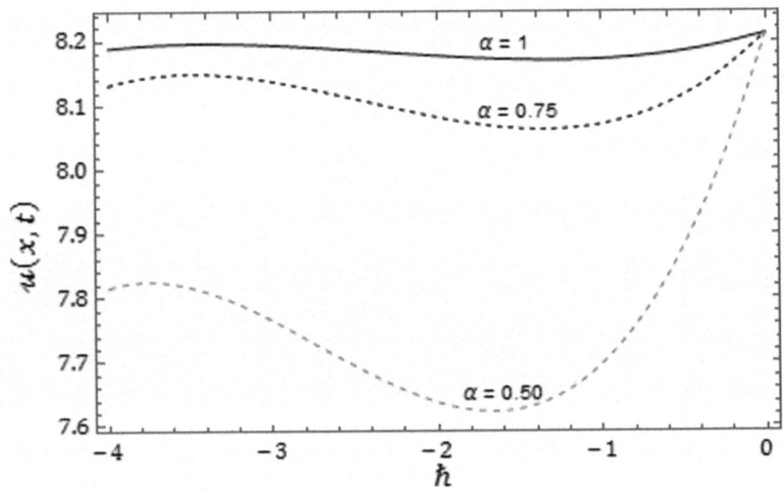

Fig. 22. $\hbar$-curve drown for the q-HATM solution $u(x,t)$ at $x = 1, t = 0.01$ and $n = 3$ with distinct values of α for Example 4.2.

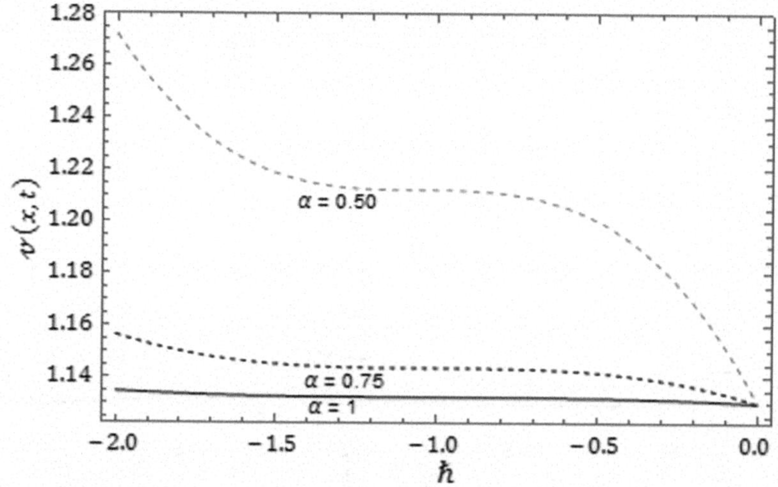

Fig. 23. $\hbar$-curve drown for the q-HATM solution $v(x, t)$ at $x = 1, t = 0.001$ and $n = 1$ with diverse values of α for Example 4.2.

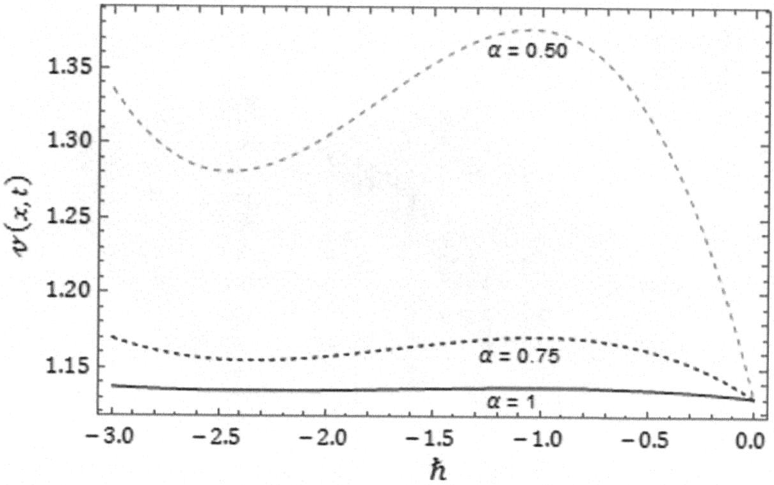

Fig. 24. $\hbar$-curve drown for the q-HATM solution $v(x, t)$ with different values of α for Example 4.2 at $x = 1, t = 0.001$ and $n = 2$.

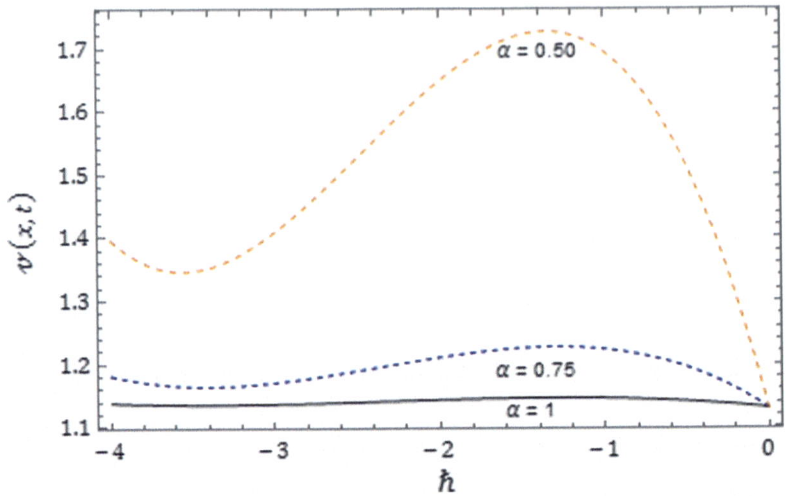

Fig. 25. $\hbar$-curve drown for the q-HATM solution $v(x,t)$ with distinct values of α for Example 4.2 at $x = 1, t = 0.001$ and $n = 3$.

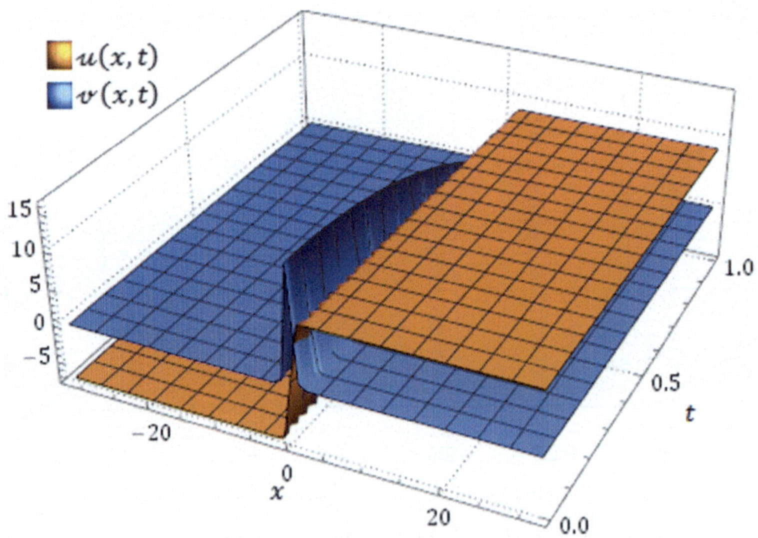

Fig. 26. Surface of q-HATM coupled solutions $u(x,t)$ and $v(x,t)$ when $n = 1, \alpha = 1$ and $\hbar = -1$ for Example 4.2.

Table 3. Comparison between exact and q-HATM solutions $u(x,t)$ and $v(x,t)$ in differing values of x and t at $n = 1, \hbar = -1$ and $\alpha = 1$ for Example 4.2

(x,t)	$u(x,t)$			$v(x,t)$		
	$q-HATM$	*Exact*	*Absolute error*	$q-HATM$	*Exact*	*Absolute error*
$(0.1, 0.01)$	2.0019696	2.0019696	9.45121×10^{-9}	15.4360256	15.4360217	3.89208×10^{-6}
$(0.1, 0.03)$	1.8470476	1.8470484	7.53287×10^{-7}	15.5464661	15.5463651	1.00978×10^{-4}
$(0.1, 0.05)$	1.6910745	1.6910802	5.71541×10^{-6}	15.6457798	15.6453319	4.47907×10^{-4}
$(0.2, 0.01)$	3.4708818	3.4708818	1.35768×10^{-8}	13.7934663	13.7934652	1.07888×10^{-6}
$(0.2, 0.03)$	3.3319326	3.3319336	1.09828×10^{-6}	13.9950665	13.9950379	2.86269×10^{-5}
$(0.2, 0.05)$	3.1909958	3.1910043	8.46058×10^{-6}	14.1897534	14.1896239	1.29544×10^{-4}
$(0.3, 0.01)$	4.7391648	4.7391648	1.16045×10^{-8}	11.5073691	11.5073703	1.15327×10^{-6}
$(0.3, 0.03)$	4.6228733	4.6228742	9.46436×10^{-7}	11.7504480	11.7504770	2.89910×10^{-5}
$(0.3, 0.05)$	4.5041543	4.5041616	7.35142×10^{-6}	11.9915477	11.9916723	1.24588×10^{-4}
$(0.4, 0.01)$	5.7672722	5.7672722	6.75430×10^{-9}	9.0639587	9.0639607	1.95748×10^{-6}
$(0.4, 0.03)$	5.6754350	5.6754356	5.55326×10^{-7}	9.3036664	9.3037166	5.01077×10^{-5}
$(0.4, 0.05)$	5.5811872	5.5811916	4.34857×10^{-6}	9.5451536	9.5453729	2.19238×10^{-4}
$(0.5, 0.01)$	6.5588985	6.5588985	2.25417×10^{-9}	6.8224352	6.8224369	1.75019×10^{-6}
$(0.5, 0.03)$	6.4896341	6.4896342	1.88687×10^{-7}	7.0310250	7.0310702	4.52158×10^{-5}
$(0.5, 0.05)$	6.4182629	6.4182644	1.50372×10^{-6}	7.2433368	7.2435366	1.99711×10^{-4}

Figures 1 and 2 explore the comparison of q-HATM solutions with exact solutions and absolute error for Example 4.1. Figures 3 and 4 cite the action of solutions obtained for Eq. (19) with distinct Brownian motions and standard motion ($\alpha = 1$). Figures 5 and 6 depict the q-HATM solutions for different values of auxiliary parameter $\hbar$, which helps us to control and adjust the convergence region. The Figs. 7, 8, 9, 10, 11 and 12 explore the role of asymptotic parameter n with respect to $\hbar$ in q-HATM solution.

Moreover, Figs. 14 and 15 cite the nature of q-HATM solutions in comparison with exact solutions for Example 4.2, in particular Figs. 14(c) and 15(c) revel the efficiency of proposed technique in terms of absolute error. Figures 16 and 17 explore the validity of Brownian motion and standard motion (i.e.,$\alpha = 1, 0.75, 0.50$). Figs. 18 and 19 depicts the q-HATM solutions for different values of auxiliary parameter $\hbar$ which helps us to control and adjust the convergence region. Figures 20, 21, 22, 23, 24 and 25 represent $\hbar$-curves and the horizontal line illustrate the range of convergence for Eq. (27). The surface of the coupled WBK equations consider in Examples 4.1 and 4.2 are respectively shown in Figs. 13 and 26, which helps us to understand the nature of the coupled equations.

6 Conclusion

In the present frame work, the q-homotopy analysis transform method is employed lucratively to find the numerical solution for nonlinear coupled time-fractional Whitham-Broer-Kaup equations. The obtained results expose that, the proposed technique is very accurate and it can be employed to explore wide class of fractional nonlinear differential equations interpreting complex phenomena. The suggested algorithm manipulates and controls the series solution, which rapidly converges to the exact solution in a short admissible domain. The novelty of the proposed technique is it provides nonlocal effect, promising large convergence region, free from any assumption and moreover it has straight forward solution procedure.

References

1. Caputo, M.: Elasticita e Dissipazione. Zanichelli, Bologna (1969)
2. Miller, K.S., Ross, B.: An Introduction to Fractional Calculus and Fractional Differential Equations. Wiley, New York (1993)
3. Podlubny, I.: Fractional Differential Equations. Academic Press, New York (1999)
4. Liao, S.J.: Homotopy analysis method: a new analytic method for nonlinear problems. Appl. Math. Mech. **19**, 957–962 (1998)
5. Toledo-Hernandez, R., Rico-Ramirez, V., Iglesias-Silva, G.A., Diwekar, U.M.: A fractional calculus approach to the dynamic optimization of biological reactive systems. Part I: fractional models for biological reactions. Chem. Eng. Sci. **117**, 217–228 (2014)
6. Drapaca, C.S., Sivaloganathan, S.: A fractional model of continuum mechanics. J. Elast. **107**, 105–123 (2012)
7. Nasrolahpour, H.: A note on fractional electrodynamics. Commun. Nonlinear Sci. Numer. Simul. **18**, 2589–2593 (2013)
8. West, B.J., Turalska1, M., Grigolini, P.: Fractional calculus ties the microscopic and macroscopic scales of complex network dynamics. New J. Phys. **17** (2015). https://doi.org/10.1088/1367-2630/17/4/045009
9. Tasbozan, O., Senol, M., Kurt, A., Ozkan, O.: New solutions of fractional Drinfeld-Sokolov-Wilson system in shallow water waves. Ocean Eng. **161**, 62–68 (2018)
10. Shchigolev, V.K.: Cosmological models with fractional derivatives and fractional action functional. Commun. Theor. Phys. **56**(2), 389–396 (2011)
11. Baskonus, H.M., Bulut, H.: On the numerical solutions of some fractional ordinary differential equations by fractional Adams-Bashforth-Moulton method. Open Math. **13**, 547–556 (2015)
12. Veeresha, P., Prakasha, D.G., Baleanu, D.: An efficient numerical technique for the nonlinear fractional Kolmogorov–Petrovskii–Piskunov equation. Mathematics **7**(3) (2019). https://doi.org/10.3390/math7030265
13. Baskonusa, H.M., Bulut, H.: Regarding on the prototype solutions for the nonlinear fractional-order biological population model. AIP Conf. Proc. **1738**, 290004 (2016). https://doi.org/10.1063/1.4952076
14. Prakasha, D.G., Veeresha, P., Rawashdeh, M.S.: Numerical solution for (2 + 1)-dimensional time-fractional coupled Burger equations using fractional natural decomposition method. Math. Meth. Appl. Sci. **42**(10), 3409–3427 (2019)
15. Veeresha, P., Prakasha, D.G.: Solution for fractional Zakharov-Kuznetsov equations by using two reliable techniques. Chin. J. Phys. **60**, 313–330 (2019)
16. Prakasha, D.G., Veeresha, P., Baskonus, H.M.: Residual power series method for fractional Swift–Hohenberg equation. Fractal Fract. **3**(1) (2019). https://doi.org/10.3390/fractalfract3010009
17. Singh, J., Kumar, D., Swroop, R.: Numerical solution of time- and space-fractional coupled Burgers' equations via homotopy algorithm. Alexandria Eng. J. **55**(2), 1753–1763 (2016)

18. Whitham, G.B.: Variational methods and applications to water waves. Proc. R. Soc. Lond. A **299**, 6–25 (1967)
19. Broer, L.J.F.: Approximate equations for long water waves. Appl. Sci. Res. **31**, 377–395 (1975)
20. Kaup, D.J.: A higher-order water-wave equation and the method for solving it. Prog. Theor. Phys. **54**, 396–408 (1975)
21. Saha Ray, S.: A novel method for travelling wave solutions of fractional Whitham-Broer-Kaup fractional modified Boussinesq and fractional approximate long wave equations in shallow water. Math. Meth. Appl. Sci. **38**, 1352–1368 (2015)
22. El-Sayed, S.M., Kaya, D.: Exact and numerical travelling wave solutions of Whitham–Broer–Kaup equations. Appl. Math. Comput. **167**, 1339–1349 (2005)
23. Rafei, M., Daniali, H.: Application of the variational iteration method to the Whitham-Broer-Kaup equations. Comput. Math Appl. **54**, 1079–1085 (2007)
24. Haq, S., Ishaq, M.: Solution of coupled Whitham-Broer-Kaup equations using optimal homotopy asymptotic method. Ocean Eng. **84**, 81–88 (2014)
25. Wang, L., Chen, X.: Approximate analytical solutions of time fractional Whitham-Broer-Kaup equations by a residual power series method. Entropy **17**, 6519–6533 (2015)
26. Ali, A., Shah, K., Khan, R.A.: Numerical treatment for travelling wave solutions of fractional Whitham-Broer-Kaup equations. Alexandria Eng. J. (2017). http://dx.doi.org/10.1016/j.aej.2017.04.012
27. Srivastava, H.M., Kumar, D., Singh, J.: An efficient analytical technique for fractional model of vibration equation. Appl. Math. Model. **45**, 192–204 (2017)
28. Veeresha, P., Prakasha, D.G., Baskonus, H.M.: Novel simulations to the time-fractional Fisher's equation. Math. Sci. **13**(1), 33–42 (2019)
29. Prakash, A., Veeresha, P., Prakasha, D.G., Goyal, M.: A homotopy technique for fractional order multi-dimensional telegraph equation via Laplace transform. Eur. Phys. J. Plus **134** (19), 1–18 (2019). https://doi.org/10.1140/epjp/i2019-12411-y
30. Prakasha, D.G., Veeresha, P., Baskonus, H.M.: Analysis of the dynamics of hepatitis E virus using the Atangana-Baleanu fractional derivative. Eur. Phys. J. Plus **134**(241), 1–11 (2019). https://doi.org/10.1140/epjp/i2019-12590-5
31. Prakash, A., Prakasha, D.G., Veeresha, P.: A reliable algorithm for time-fractional Navier-Stokes equations via Laplace transform. Nonlinear Eng. **8**(1), 695–701 (2019)
32. Veeresha, P., Prakasha, D.G., Baskonus, H.M.: New numerical surfaces to the mathematical model of cancer chemotherapy effect in Caputo fractional derivatives. Chaos **29**, 013119 (2019). https://doi.org/10.1063/1.5074099
33. Prakasha, D.G., Veeresha, P., Baskonus, H.M.: Two novel computational techniques for fractional Gardner and Cahn-Hilliard equations. Comp and Math Methods. **1**, 1–19 (2019). https://doi.org/10.1002/cmm4.1021
34. Veeresha, P., Prakasha, D.G., Kumar, D.: An efficient technique for nonlinear time-fractional Klein–Fock–Gordon equation. Appl. Math. Comput. **364** (2019). https://doi.org/10.1016/j.amc.2019.124637
35. Prakash, A., Veeresha, P., Prakasha, D.G., Goyal, M.: A new efficient technique for solving fractional coupled Navier-Stokes equations using q-homotopy analysis transform method. Pramana J. Phys. **93**(6), 1–10 (2019)
36. Ahmad, J., Mushtaq, M., Sajjad, N.: Exact solution of Whitham-Broer-Kaup shallow water wave equations. J. Sci. Arts **1**(30), 5–12 (2015)

An Efficient Computational Technique for Nonlinear Emden-Fowler Equations Arising in Astrophysics and Space Science

Sumit Gupta[1], Devendra Kumar[2(✉)], Jagdev Singh[3], and Sushila[4]

[1] Department of Mathematics, Swami Keshvanand Institute of Technology, Management and Gramothan, Ramnagaria, Jaipur 302017, Rajasthan, India
guptasumit.edu@gmail.com

[2] Department of Mathematics, University of Rajasthan, Jaipur 302004, Rajasthan, India
devendra.maths@gmail.com

[3] Department of Mathematics, JECRC University, Jaipur 303905, Rajasthan, India
jagdevsinghrathore@gmail.com

[4] Department of Physics, Vivekananda Global University, Jaipur 303012, Rajasthan, India
sushila.jag@gmail.com

Abstract. In the present article, we suggest an efficient computational scheme to examine nonlinear Emden-Fowler equations arising in astrophysics and space science. The suggested scheme is based on a modified theory of the Adomian polynomials, and the two steps Adomian decomposition technique mixed with the padé approximant. Moreover, a maple software package ADMP is used to apply the suggested computational scheme, which is very simple to perform and well organized. The input of the system requires initial or boundary conditions and many desired parameters to find the analytic approximate solutions within a very short time. The following algorithm does not require linearization, perturbations, guessing the initial terms and any restrictive supposition, which may leads the solutions in closed form. Several examples are discussed to illustrate the reliability of the algorithm.

Keywords: Emden Fowler equations · Lane Emden type equations · Astrophysics · ADM · TSADM · Adomian polynomials

1 Introduction

Analytical techniques have made a comeback in research methodology after proceeds a backseat to the numerical schemes for the end of the preceding century. The superiority of analytical schemes are manifolds, the main being that they give a much effective intuition than the numbers crunched by a computer using a purely numerical algorithm. Many such physical phenomena are represented in the form of nonlinear differential equations. A wide class of analytic techniques and computational techniques have been applied to examine nonlinear mathematical models [1–10].

The ADM has been proven to be one of the most powerful techniques to solve nonlinear differential equations. Numerous researchers have concentrated on the ADM

H. Dutta et al. (Eds.): CMES 2019, AISC 1111, pp. 76–98, 2020.
https://doi.org/10.1007/978-3-030-39112-6_5

[11–20]. In 1999, Wazwaz suggested a modification of the ADM [21], and shown the ability of the suggested moderation in many numerical experiments. Subsequently, Luo proposed the two-step ADM (TSADM) [22], which gives the exact solution without employing to the Adomian polynomials (AP). Zhang et al. [23] analysed the experimentation with TSADM. Babolian and Javadi [24] discussed one more new scheme to calculate the AP quickly. Gu and Li [25] enlarged the this operator and employed it to examine the nonlinear problems. Furthermore, Rach [26] again defined and amalgamated the class of AP in another from. Moreover, he has shown that the novel methods are much more effective.

Here, we investigate the Emden-Fowler type equation in the following manner

$$y'' + \frac{a}{x}y' + a\, f(x)g(y) = Q(x), \tag{1}$$

with initial conditions (ICs) $y(0) = A, \quad y'(0) = 0$.

In Eq. (1) a represents a constant, $f(x)$ and $g(y)$ indicate functions of x and y respectively. For $Q(x) = 0, f(x) = 1$ and $g(y) = y^m$, Eq. (1) is the classical Lane –Emden equation, which is employed to describe the thermal behaviour of a spherical cloud of gas acting under the mutual attraction of the molecules [27] and under the well known laws of thermodynamics. The Emden-Fowler type equation has notable uses in several disciplines of science and technology world. Different of forms of $g(y)$ have been studied by several mathematicians and scientists with the aid of numerous techniques involving numerical and perturbation schemes have been employed to handle the Emden-Fowler equations [28–33]. The Emden-Fowler equations play a key role to describe many facts in physical sciences and astrophysics such as thermal explosions, stellar structure, the thermal behaviour of a spherical cloud of a gas, isothermal gas spheres and thermionic currents [34, 35]. The detailed plan of the padé approach was given by Pozzi and Bassano [36].

The principal aim of this work is to implement a systematic numerical approach based on the theory of Lin et al. [1], where the mathematicians have transformed the solution of the model with Dirichlet and Robin boundary conditions from the developed computer algorithm to the approximate solution, which converges to exact solution.

This article is developed as follows. In part 2, we discuss the key features of the ADM, the novel concept of the AP as well as the TSADM. In portion 3, a method is considered to develop analytical and approximate solution of nonlinear differential equation having ICs. The part 4 presents the MAPLE package ADMP and the stability and convergence of ADM. Finally, in Sect. 5 various kinds of numerical experiments are discussed to demonstrate the usefulness the software package ADMP.

2 Basic Ideas

In this portion, the fundamental plans of the ADM, the novel theory of the AP and the TSADM have been discussed.

2.1 The Classical ADM

Let us take a nonlinear differential equation (NDE) of the form

$$Lv + Rv + Nv = \Gamma. \tag{2}$$

In Eq. (2) L denotes the higher order linear derivative and R represents the extra part of the linear operator, Nv indicates the nonlinear operator and Γ represents the function due to source. The Eq. (2) can be expressed as

$$Lv = \Gamma - Rv - Nv, \tag{3}$$

Employing the inverse operator L^{-1} on the Eq. (3), it gives

$$v = \zeta - L^{-1}Rv - L^{-1}Nv, \tag{4}$$

where ζ denotes the terms occurring from integration the source function Γ and the associated ICs with Eq. (2).

Using the ADM, the solution v can be presented in the subsequent manner

$$v = \sum_{m=0}^{\infty} \lambda^m v_m, \tag{5}$$

The term Nv is analytic in nature and can be expressed into a particular series modified to the specific nonlinearity Nv, which is given as

$$Nv = \sum_{m=0}^{\infty} \lambda^m A_m, \tag{6}$$

where λ is a grouping parameter of convenience, and A_m, $m = 0, 1, 2\ldots$ indicate the AP.

On replacing, we arrive at the below result

$$\sum_{m=0}^{\infty} \lambda^m v_m = \zeta - L^{-1}R\sum_{m=0}^{\infty} \lambda^m v_m - L^{-1}\sum_{m=0}^{\infty} \lambda^m v_m, \tag{7}$$

For the easiness of evaluation, we put $\lambda = 1$ the iterates v_m $(m \geq 0)$ can be simply obtained from the subsequent classical Adomian iterative technique

$$\begin{aligned} v_0 &= \zeta, \\ v_{m+1} &= -L^{-1}Rv_m - L^{-1}A_m, \quad m \geq 0. \end{aligned} \tag{8}$$

In the above procedure, the r-term partial sum is given by

$$v = \sum_{m=0}^{r-1} v_m(x). \tag{9}$$

2.2 The Modified Theory of the AP

Rach [26] suggested a new concept to obtain the uniform result for the family of the classes of the AP [26].

The introduced method for all classes of the AP can be presented in the subsequent from

$$A_0 \equiv T_1\left(\sum A_n\right).$$

For $m \geq 1: \quad A_m \equiv T_{m+1}(\sum A_n) - T_m(\sum A_n)$,
where $T_m(\sum A_n) = T_m(\sum A_n;\, q_1, q_2, q_3) = \sum_{n=0}^{m-1} A_n = \zeta_m[N(u)]$ or similar form is

$$\left[\sum_{n=0}^{\infty} A_n\right]_{n\geq m} = \left[\left[\sum_{n=0}^{\infty}\left[\frac{1}{n!}\left(\sum_{r=0}^{\infty} v_r - v_0\right)^n \times \frac{\partial^n}{\partial v_0^n} N(v_0)\right]_{r\geq q_1(m)}\right]_{r\geq q2(m)}\right]_{\sum jrj\geq q_3(m)}$$

For more details of this scheme one can see the work of Rach [26].

2.3 Analysis of TSADM

Recently, TSADM was suggested which is a systematic improvement of the ADM [23], it yields a encouraging plan for several utilizations [24]. The plan of the TSADM is demonstrated in the subsequent way.

Step 1. Employing the inverse operator L^{-1} to Γ, and then applying the associated condition, we have

$$\phi = \xi + L^{-1}\Gamma$$

By setting

$$\phi = \xi_0 + \xi_1 + \ldots + \xi_l,$$

where $\xi_0, \xi_1, \ldots \xi_l$ are the terms arising from integrating the source term Γ and from the given conditions.

In view of the above, we suppose that

$$u_0 = \xi_{k_1} + \ldots + \xi_{k_1+k_2},$$

In which $k_1 = 0, 1 \ldots l,\ k_2 = 0, 1 \ldots l - k_1$. then we prove that u_0 fulfils the given problem and the associated conditions by substitutions, once we proceed to the Step 2.

Step 2. We substitute $v_0 = \phi$ and use the classical Adomian recursive algorithm to compute the succeeding solutions terms.

3 Main Algorithms

3.1 A Novel Algorithm for Handling Nonlinear IVP

In view of the techniques discussed in the previous portion, an effective scheme is studied in the present part to derive the solution for NDE having the ICs.

To illustrates the procedure of the suggested technique, we study the system of NDE

$$\begin{cases} L_1 v_1 + R_1(v_1, v_2, \ldots v_s) + N_1(v_1, v_2, \ldots v_s) = \Gamma_1, \\ L_2 v_2 + R_2(v_1, v_2, \ldots v_s) + N_2(v_1, v_2, \ldots v_s) = \Gamma_2, \\ \vdots \\ L_s v_s + R_s(v_1, v_2, \ldots v_s) + N_s(v_1, v_2, \ldots v_s) = \Gamma_s. \end{cases} \tag{10}$$

Subject to the ICs

$$\begin{aligned} &v_i(t = 0, x, y, \ldots) = q_i(x, y, \ldots), \quad i = 1, 2, \ldots s, \\ &\left.\frac{\partial v_i(t, x, y, \ldots)}{\partial t}\right|_{t=0} = p_i(x, y, \ldots), \quad i = 1, 2, \ldots s, \end{aligned} \tag{11}$$

Where Γ_i are the sources terms and L_i are linear derivative of the highest orde. R_i denotes rest part of the linear operators and N_i stands for the nonlinear operators. For the IVP, we normally describe L_i^{-1} for $L_i = d^n / dt^n$ as the n-fold definite integration operator lies from 0 to t. Hence, the inversion of the operators L_i^{-1} are defined as

$$L_i^{-1} = \underbrace{\int_0^t \cdots \int_0^t (\bullet) dt \ldots dt}_{n-times} = F_i(t) - \sum_{j=0}^{n-1} \frac{t^j}{j!} \left.\frac{d^j F_i(t)}{dt^j}\right|_{t=0}, \tag{12}$$

where $F_i(t)$ stand for a n- fold purely integrations of the integrand.

Employing L_i^{-1} on Eq. (10), it gives

$$\begin{cases} v_1 = \zeta_1 - L_1^{-1} R_1(v_1, v_2, \ldots v_s) - L_1^{-1} N_1(v_1, v_2, \ldots v_s), \\ v_2 = \zeta_2 - L_2^{-1} R_2(v_1, v_2, \ldots v_s) - L_2^{-1} N_2(v_1, v_2, \ldots v_s), \\ \vdots \\ v_s = \zeta_s - L_s^{-1} R_s(v_1, v_2, \ldots v_s) - L_s^{-1} N_s(v_1, v_2, \ldots v_s), \end{cases} \tag{13}$$

where

$$\zeta_i = \xi_i + L_i^{-1} \Gamma_i, \tag{14}$$

Following the TSADM, we put

$$\zeta_i = \xi_{i,0} + \xi_{i,1} + \ldots + \xi_{i,l_i} \tag{15}$$

where $\xi_{i,0}, \xi_{i,1}, \ldots + \xi_{i,l_i}$ are the function occurring from integration of the source parts Γ_i and the given ICs. In view of the above consideration, we take

$$v_{i,0} = \xi_{i,k_{i_1}} + \xi_{i,k_{i_2}} + \ldots + \xi_{i,k_{i_1}+k_{i2},} \tag{16}$$

in which $k_{i1} = 0, 1, \ldots, l_i, \quad k_{i2} = 0, 1, \ldots, l_i - k_{i1}$. Then we prove whether $v_{i,0}$ fulfils the given problem and the ICs or not, one time when the exact solution is attained, we terminate the procedure.

In the other case, by employing the classical ADM, the solution v_i is decomposed into following series

$$v_i = \sum_{m=0}^{\infty} v_{i,m}\lambda^m, \quad i = 1, 2, \ldots s. \tag{17}$$

The terms $N_i(v_1, v_2, \ldots v_s)$ are analytic in nature and can be expressed in the form of infinite series of their corresponding AP, which are given in the following manner

$$N_i(v_1, v_2, \ldots v_s) = \sum A_{i,m}(v_{1,0}, \ldots, v_{1,m}, v_{2,0}, \ldots, v_{2,m}, v_{s,0,\ldots,} v_{s,m})\lambda^m, \tag{18}$$

where the $A_{i,m}$ can be computed by making use of any one of Rach's novel theories.

4 Convergence Analysis

The present section offers a scheme to demonstrate the examination of convergence of the used technique. Many mathematicians [37, 38] have studied the convergence of ADM for handling NDE of various kinds.

Consider that X $= C[0, 1]$ be a banach space having the norm $\|v\| = \max_{0\le t\le 1} |v(t), v \in X|$ then the succeeding theorems will show the convergence of the suggested scheme.

Theorem 1. Let $N(v)$ be the nonlinear operator describe by $N(y) = \xi^\alpha f(\xi)$ that hold the Lipschitz condition having the Lipschitz constant $\rho < 1$ then there holds $\|v_{k+1}\| \le \rho\|v_k\|, k = 0, 1, 2, \ldots$ and sequence $\{\psi_n\}$ defined by $\psi_n = \sum_{j=0}^{n} v_j(t)$ indicate the n-term approximate series, converges to the exact solution v.

Proof. Please see, [39].

Theorem 2. Suppose that $v(t)$ be the exact solution of the NDE (2). Assume that $\psi_m(t)$ be the sequence of the series solution expressed by $\psi_n = \sum_{j=0}^{n} v_j(t)$. Then there holds

$$\max_{0\le t\le 1}\left| v(t) - \sum_{j=0}^{m} v_j(t)\right| \le \frac{\rho^{m+1}}{1-\rho}\|v_0\|. \tag{19}$$

Proof. Please see, [39].

5 Numerical Experiments

In this portion, some numerical experiments are discussed to represent the technique and show the successfulness of the package.

Example 1. Here we examine the following nonlinear singular initial value problem (NSIVP)

$$y'' + \frac{3}{x}y' + 2x^2y^2 = 0$$

along with ICs $y(0) = 1,\ y'(0) = 0$.

Taking *class* = *4* and *index* = *15*, the package automatically provides the ADM-padé approximations and outputs the Figs. 1, 2, 3 and 4 in 0.187 s.

The expression of the Padé [10/10]:

$$\text{Pade}_4[10/10] = \frac{1.0000 - 0.00021x^8 - 0.01736x^4}{1.0000 + 0.00112x^8 + 0.06597x^4}.$$

To authenticate the package, in the following we made comparison of the approximate solutions of different orders as shown by the outstanding agreement. The ADMP provides outputs Figs. 1 and 2, in which the approximate solutions are closely the identical in different order of approximations. Also Figs. 3 and 4 represent the error and absolute error among the approximate solution of 15-order and exact solution, which are showing that the error is very less and the approximate solution yields effective approximation by suggested scheme.

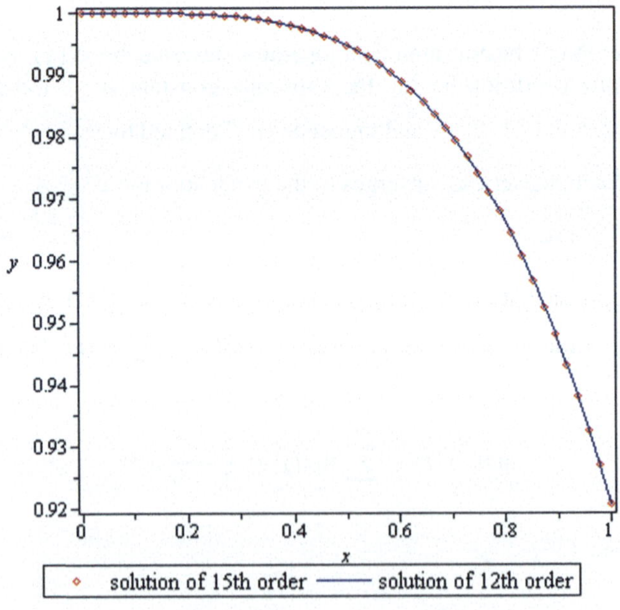

Fig. 1. Plots of the solution of different order for $0 \leq x \leq 1$.

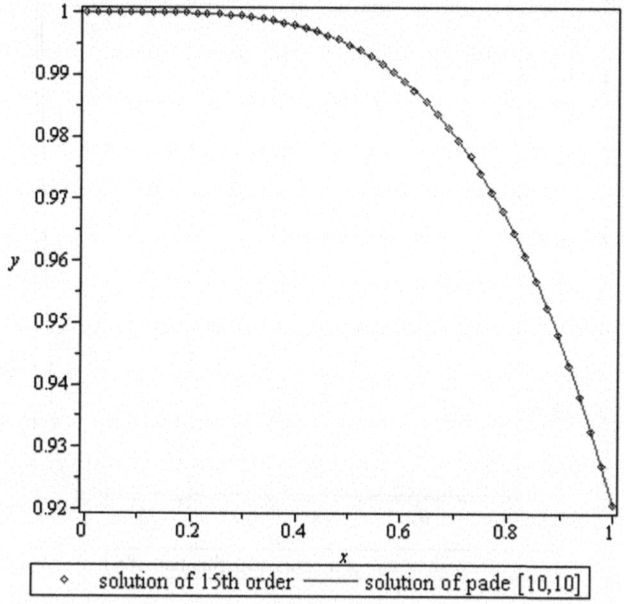

Fig. 2. Plots of TSADM and combination of TSADM-pade solutions.

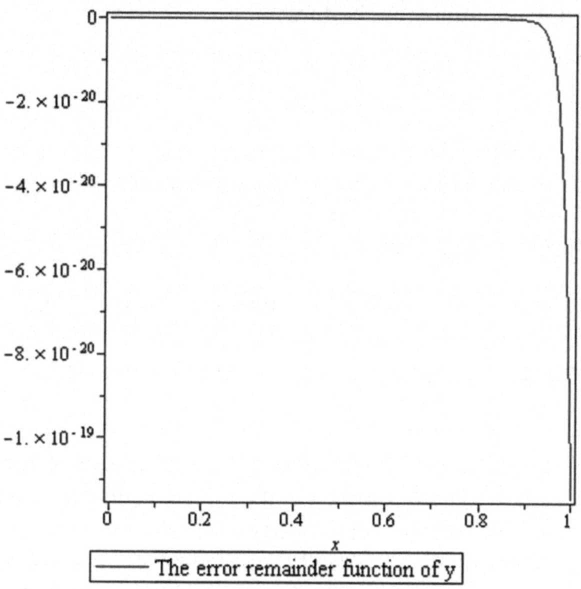

Fig. 3. Plot of the error function.

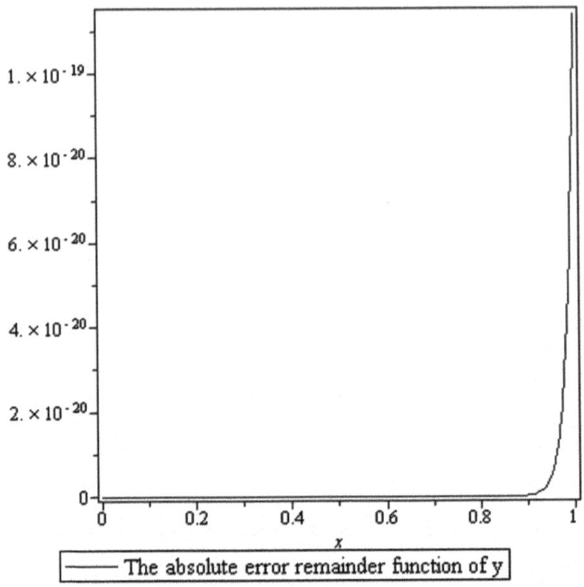

Fig. 4. Plot of the absolute error function.

Example 2. We study the following NSIVP

$$y'' + \frac{8}{x}y' + xy^2 = x^5 + x^4$$

with ICs $y(0) = 1,\ y'(0) = 0$.

Taking *class* = *4* and *index* = *15*, the package automatically provides the ADM-Padé approximations and outputs the Fig. 4 in 1.435 s.

The expression of the Padé [10/10]:

$$\begin{aligned} &\text{Pade}_4[10/10] = [0.00217x^2 - 0.00395x - 0.02371x^3 + 0.01321x^6 + 0.01015x^7 \\ &+ 6.8963910^{-9}x^7 + 0.0000x^{10} + 0.00016x^4 - 0.00009x^5 - 0.00001x^8 + 1.0000]/ \\ &[0.00217x^2 - 0.00395x + 0.00962x^3 - 0.00015x^6 - 4.7892310^{-7}x^7 + 0.00004x^9 \\ &0.00002x^{10} + 0.00002x^4 - 0.00002x^5 + 3.8809910^{-9}x^8 + 1.0000. \end{aligned}$$

To authenticate the package, in the following we made comparison of the approximate solutions of distinct orders as demonstrated by the great concurrence. The applied scheme delivers outputs Figs. 5 and 6, where the approximate solutions are closely the identical in different order of approximations. Also Figs. 7 and 8 represent the error and absolute error among the exact and approximate solutions, which are showing that the error is very less and the approximate solution yields good approximation with aid of TSADM.

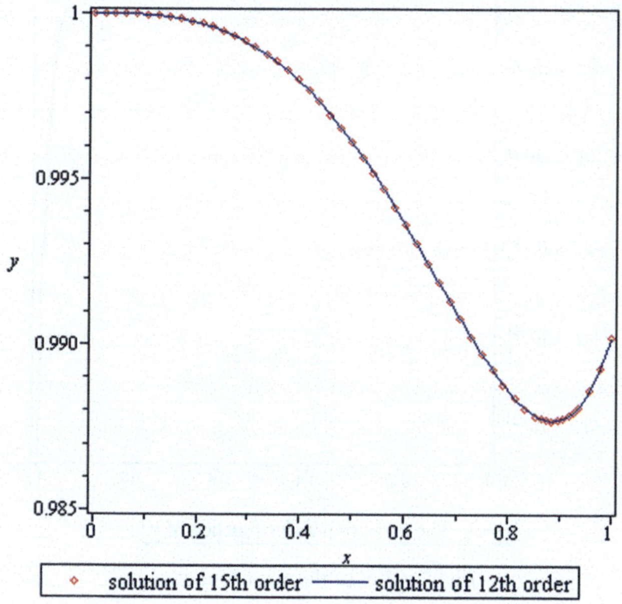

Fig. 5. Plots of the solution of different order for $0 \leq x \leq 1$.

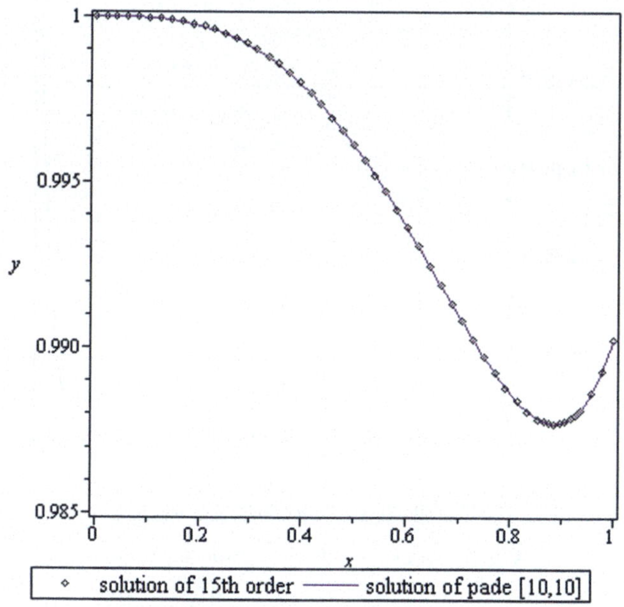

Fig. 6. Plots of 15th order TSADM and combination of TSADM-pade approximants [10] solutions.

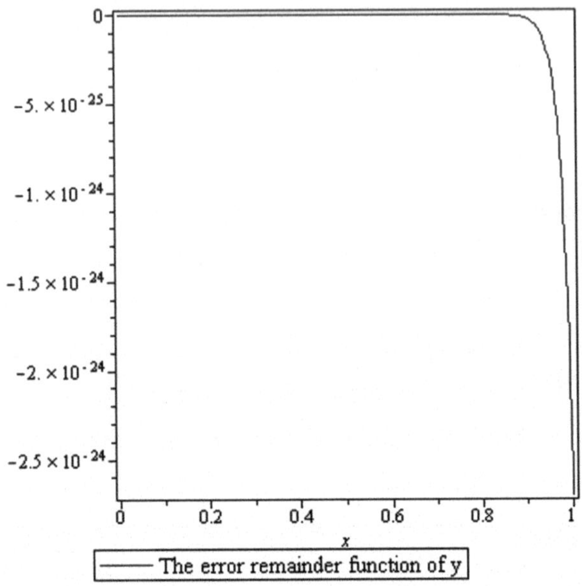

Fig. 7. Plot of the error function.

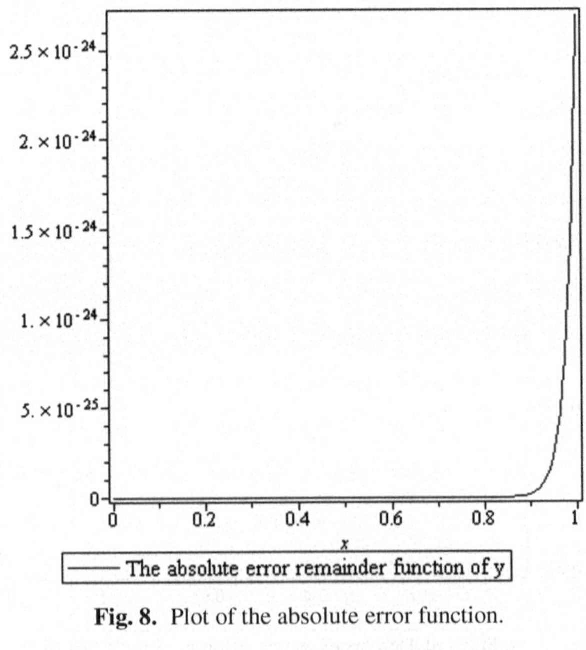

Fig. 8. Plot of the absolute error function.

Example 3. We study the following NSIVP

$$y'' + \frac{2}{x}y' + e^y = 0$$

with ICs $y(0) = 0,\ y'(0) = 0$.

Taking *class* = *4* and *index* = *15*, the package automatically provides the ADM-Padé approximations in 0.187 s.

The presentation of the Padé [10/10] is as:

$$\begin{aligned}&\mathrm{Pade}_4[10/10] = \\ &[-0.16667x^2 - 0.03114x^4 - 0.00190x^6 - 0.00004x^8 - 2.31254 10^{-7}x^{10}]/[1.0000 \\ &+ 0.23684x^2 + 0.02007x^4 + 0.00073x^6 + 0.00001x^8 + 3.60058 10^{-8}x^{10}].\end{aligned}$$

To authenticate the package, in the following we made comparison of the approximate solutions of distinct orders as demonstrated by the great concurrence. The used scheme delivers outputs Figs. 9 and 10, in which the approximate solutions are closely the identical in different order of approximations. Also Figs. 11 and 12 show the error and absolute error among the approximate solution of 15-order and the exact solution, which are showing that the error is very less and the approximate solution yields effective approximation by the suggested scheme.

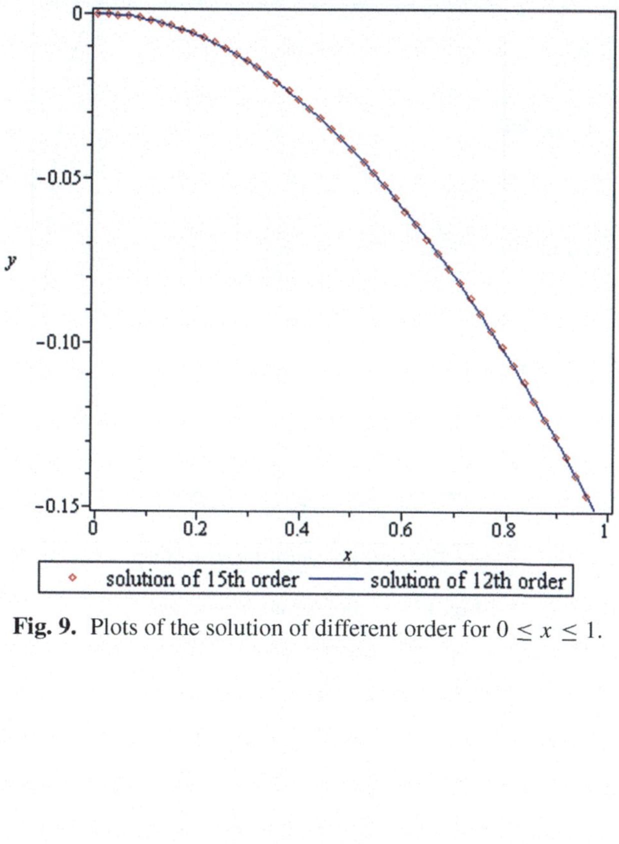

Fig. 9. Plots of the solution of different order for $0 \le x \le 1$.

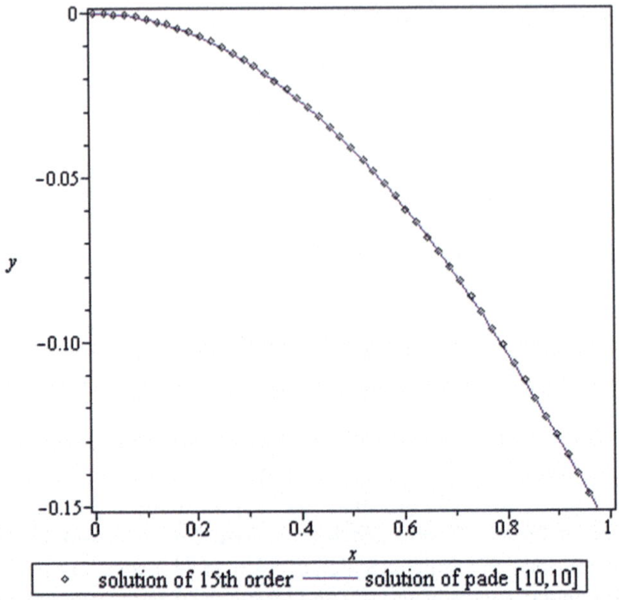

Fig. 10. Plots of 15th order TSADM and combination of TSADM-pade solutions.

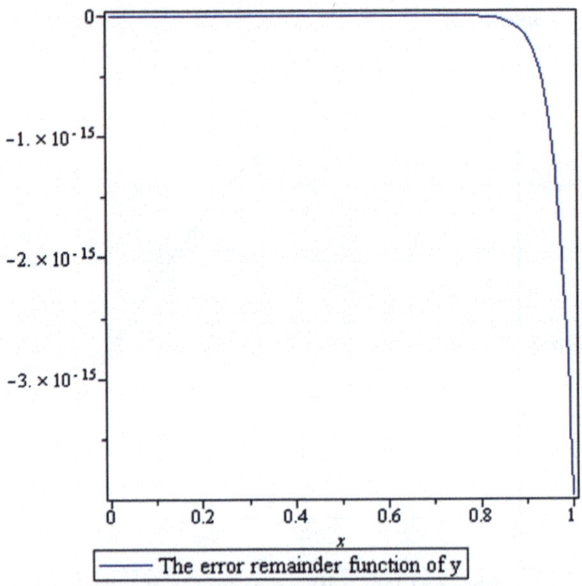

Fig. 11. Plot of the error function.

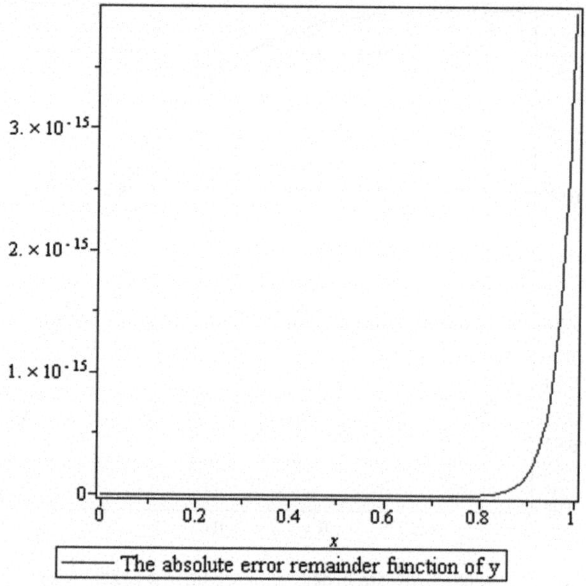

Fig. 12. Plot of the absolute error function.

Example 4. Consider the following NSIVP

$$y'' + \frac{2}{x}y' + x^2 \ln y = 0,$$

with ICs $y(0) = 1, \; y'(0) = 0$.

Taking *class* = *4* and *index* = *15*, the package automatically provides the ADM-Padé approximations in 1.202 s.

The presentations of the Padé [15/15] is as follows:

$$\begin{aligned}&\text{Pade}_4[15/15] = \\ &1.0000 - 9.03954 \times 10^{-10}x^{14} - 7.33607 \times 10^{-8}x^{12} + 0.0000x^{10} - 0.00132x^6 - 0.01638x^4 \\ &+ 0.8771x^2]/[1.0000 - 1.47463 \times 10^{-9}x^{14} - 5.551399 \times 10^{-8}x^{12} - 0.0000x^{10} - 0.00005x^8 \\ &+ 0.00020x^6 + 0.02405x^4 + 0.28358x^2].\end{aligned}$$

To authenticate the package, in the following we made comparison of the approximate solutions of distinct orders as demonstrated by the superb concurrence. The ADMP gives outputs Figs. 13 and 14, in which the approximate solutions are closely the identical in different order of approximations. Also Figs. 15 and 16 represent the error and absolute error between approximate solution of 15-order and the exact solution, which are showing that the error is very less and the approximate solution yields efficient approximation by the suggested scheme.

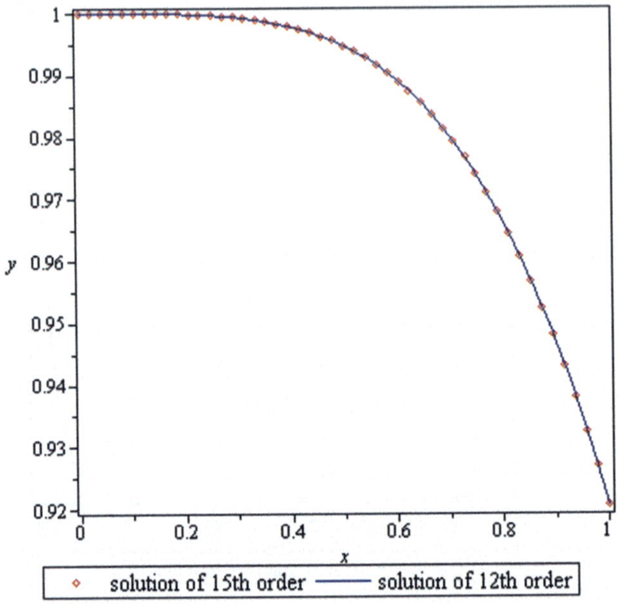

Fig. 13. Plots of the solution of different order for $0 \leq x \leq 1$.

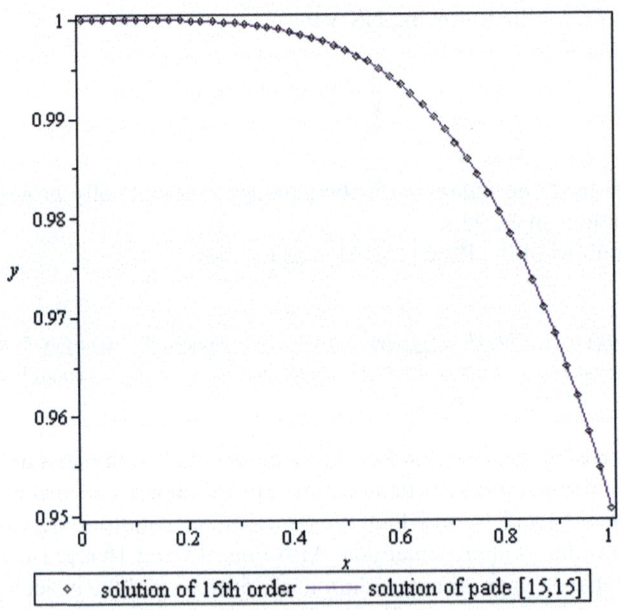

Fig. 14. Plots of 15th order TSADM and combination of TSADM-pade solutions.

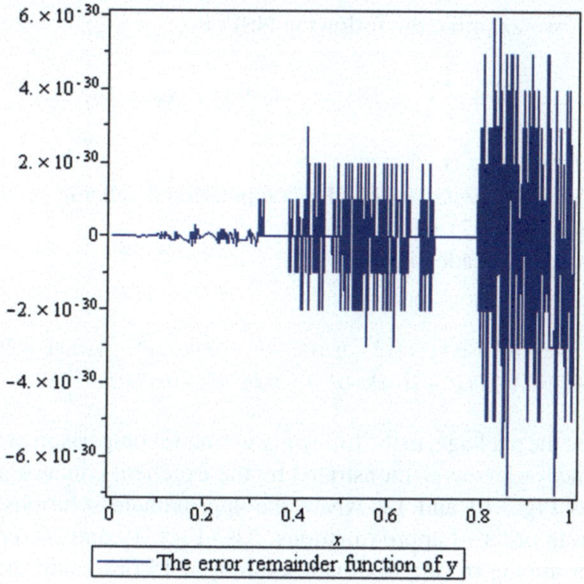

Fig. 15. Plot of the error function.

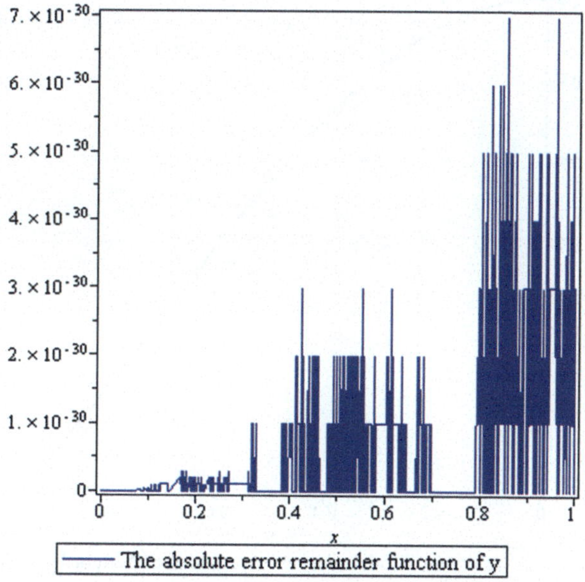

Fig. 16. Plot of the absolute error function.

Example 5. Next, we examine the following NSIVP

$$y'' + \frac{2}{x}y' + 18y = -4y\ln(y) \quad (x > 0),$$

with ICs $y(0) = 1,\ y'(0) = 0$.

Taking *class* = *4* and *index* = *15*, the computational scheme provides the ADM-Padé approximations in 0.821 s.

The expression of the Padé [10/10]:

$$\begin{aligned}&\text{Pade}_4[10/10] = \\ &[1.0000 - 0.50000x^2 + 0.11111x^4 - 0.01389x^6 - 0.00099x^8 - 0.00003x^{10}]/[1.0000 \\ &+ 0.50000x^2 + 0.11111x^4 + 0.01389x^6 + 0.00099x^8 + 0.00003x^{10}\Big].\end{aligned}$$

To authenticate the package, in the following we made comparison of the approximate solutions of distinct orders as demonstrated by the excellent concurrence. The ADMP yields the outputs Figs. 17 and 18, where the approximate solutions are closely the identical in different order of approximations. Also Figs. 19 and 20 represent the error and absolute error among the approximate solution of 15-order and the exact solution, which are showing that the error is very less and the approximate solution yields better approximation by the suggested technique.

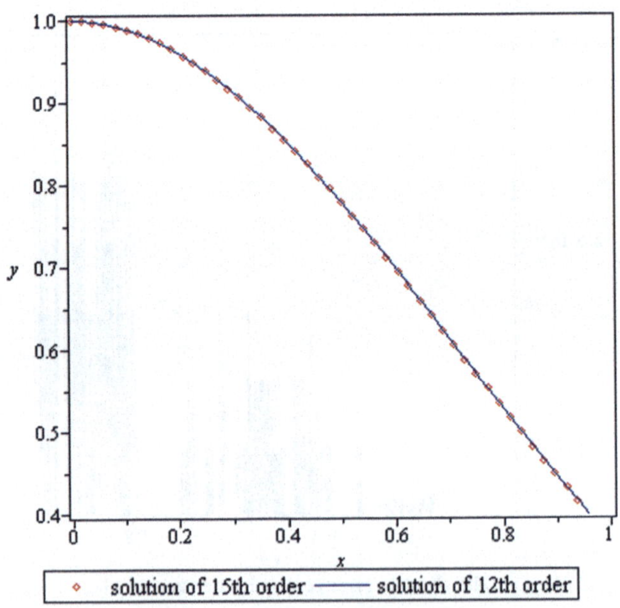

Fig. 17. Plots of the solution of different order for $0 \leq x \leq 1$.

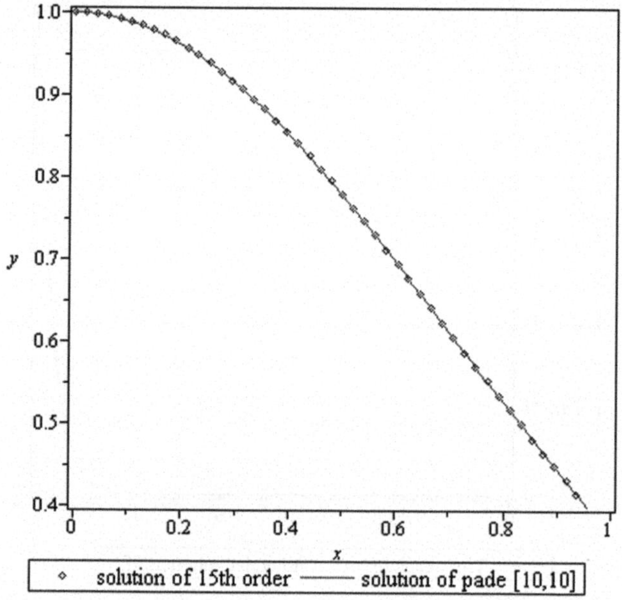

Fig. 18. Plots of 15th order TSADM and combination of TSADM-pade solutions.

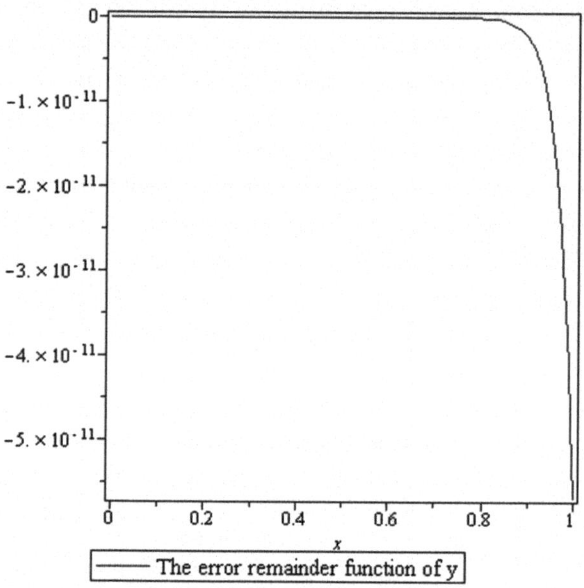

Fig. 19. Plot of the error function.

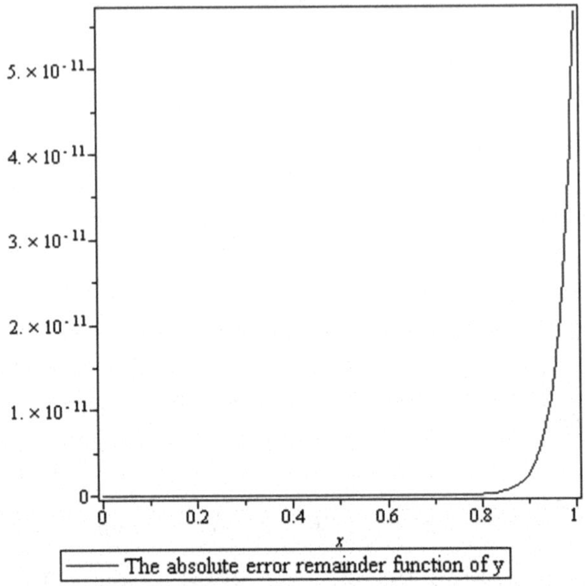

Fig. 20. Plot of the absolute error function.

Example 6. Consider the following NSIVP

$$y'' + \frac{2}{x}y' + 8\left(e^{y} + e^{y/2}\right) = 0, \quad (x > 0),$$

with the initial conditions $y(0) = 1, \ y'(0) = 0$.

Taking *class* = *4* and *index* = *15*, the computational scheme yields the ADM- Padé approximations in 31.419 s.

The expression of the Padé [10/10]:

$$Pade_4[10, 10] = \Big[(1.00000 + 1.31927x^2 - 3.37473x^4 - 6.57331x^6 - 3.15754x^8 - 0.35145x^{10})/ \\ (1.00000 + 4.23061x^2 + 6.36354x^4 + 4.0314x^6 + 0.96516x^8 + 0.05151x^{10})\Big].$$

To authenticate the package, in the following we made the comparison of the approximate solutions of distinct orders as displayed by the outstanding concurrence. The ADMP provides outputs Figs. 21 and 22, in which the approximate solutions are closely the identical in different order of approximations. Also Figs. 23 and 24 represent the error and absolute error among the approximate solution of 15-order and the exact solution, which are showing that the error is very less and the approximate solution yields efficient approximation by the proposed scheme.

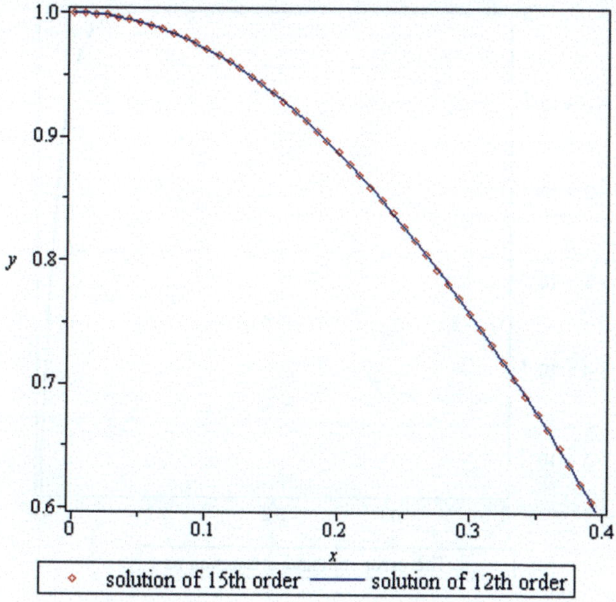

Fig. 21. Plots of the solution of different order for $0 \leq x \leq 0.4$.

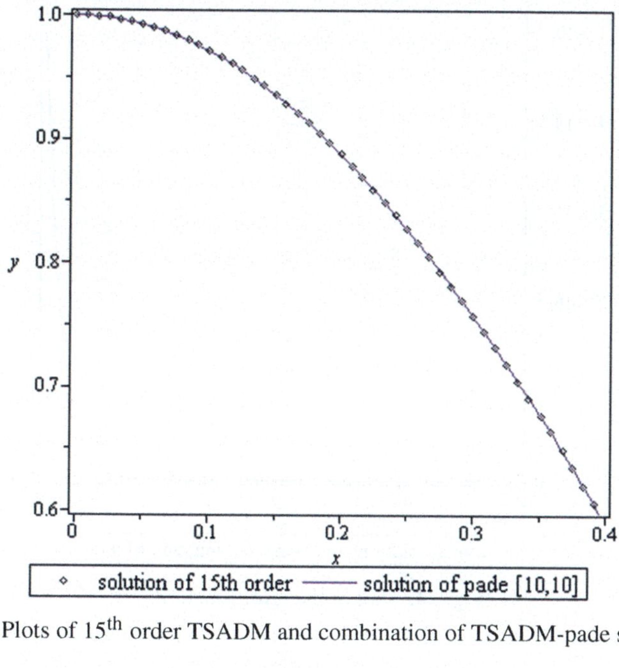

Fig. 22. Plots of 15th order TSADM and combination of TSADM-pade solutions.

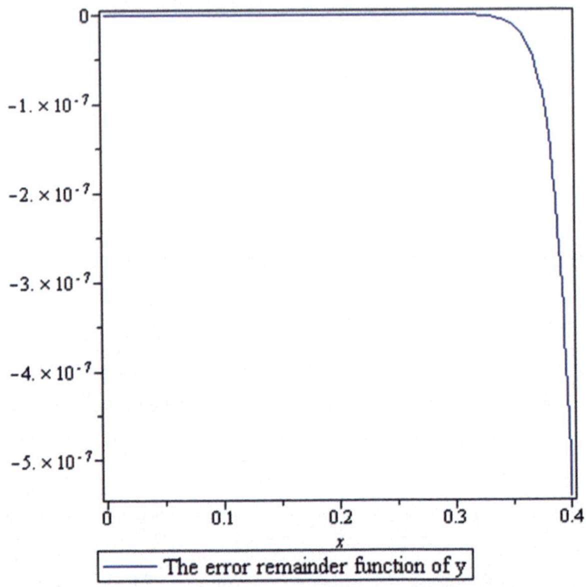

Fig. 23. Plot of the error function.

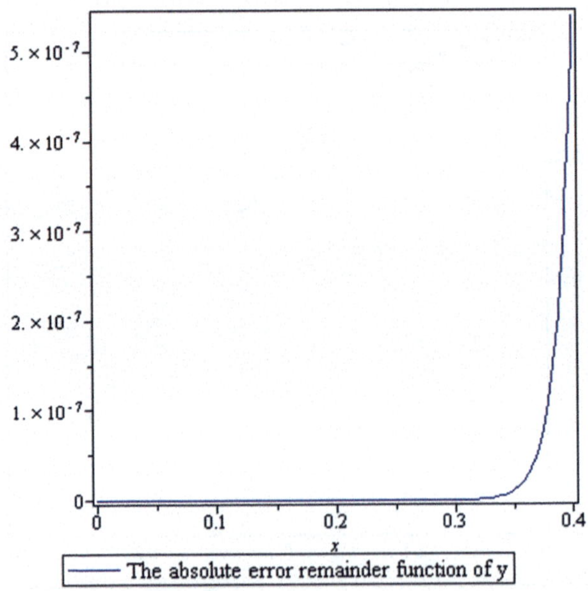

Fig. 24. Plot of the error function.

6 Concluding Remarks and Observations

In the present work, a software package ADMP based on maple has been applied to obtain numerical solutions of Emden-Fowler equations. Several working problems are taken to demonstrate the productiveness and accuracy of the suggested numerical algorithm. The suggested computational scheme has been employed with a great success to examine six distinct kinds of nonlinear differential equations. At the same time, it also depicts graph of comparative study of distinct order of approximations, their corresponding pade approximants and graphs for error examinations. Thus it can be concluded that the used approach is user friendly and easy to use instrument in scientific and technological areas.

Acknowledgement. Authors are sincerely thankful to Dr. Yezhi Lin (Department of Computer Science and Technology, East China Normal University, Shanghai 200241, PR China), for his kind help and support to complete the manuscript.

References

1. Lin, Y., Liu, Y., Li, Z.: Symbolic computation of analytic approximate solution of nonlinear fractional differential equations. Comput. Phys. Commun. **184**, 130–141 (2013)
2. He, J.H.: Homotopy perturbation technique. Comput. Methods Appl. Mech. Eng. **178**, 257–262 (1999)
3. Kumar, D., Singh, J., Baleanu, D.: A new numerical algorithm for fractional Fitzhugh–Nagumo equation arising in transmission of nerve impulses. Nonlinear Dyn. **91**, 307–317 (2018)
4. He, J.H.: Variational iteration method-a kind of nonlinear analytical technique: Some examples. Int. J. Nonlinear Mech. **34**(3), 699–708 (1999)
5. Liao, S.J.: Beyond Perturbation: Introduction to the Homotopy Analysis Method. Chapman & Hall/CRC Press, Boca Raton (2003)
6. Kumar, D., Agarwal, R.P., Singh, J.: A modified numerical scheme and convergence analysis for fractional model of Lienard's equation. J. Comput. Appl. Math. **339**, 405–413 (2018)
7. Zhou, J.K.: Differential Transformation and Its Applications for Electrical Circuits. Huazhong University Press, Wuhan (1986)
8. Kumar, D., Singh, J., Baleanu, D., Rathore, S.: Analysis of a fractional model of Ambartsumian equation. Eur. J. Phys. Plus **133**, 259 (2018)
9. Gupta, S., Kumar, D., Singh, J.: MHD mixed convective stagnation point flow and heat transfer of an incompressible nanofluid over an inclined stretching sheet with chemical reaction and radiation. Int. J. Heat Mass Transf. **118**, 378–387 (2018)
10. Singh, J., Kumar, D., Baleanu, D., Rathore, S.: An efficient numerical algorithm for the fractional Drinfeld-Sokolov-Wilson equation. Appl. Math. Comput. **335**, 12–24 (2018)
11. Adomian, G.: Solving Frontier Problems of Physics: The Decomposition Method. Kluwer Academic Publishers, Dordrecht (1994)
12. El-Sayed, S.M., Kaya, D.: Exact and numerical travelling wave solution of Whitham-Broer-Kaup equations. Appl. Math. Comput. **167**, 1339–1349 (2005)
13. Bougoffa, L., Bougoffa, S.: Adomian method for solving some coupled systems of equations. Appl. Math. Comput. **177**, 553–560 (2006)
14. Srivastava, H.M., Kumar, D., Singh, J.: An efficient analytical technique for fractional model of vibration equation. Appl. Math. Model. **45**, 192–204 (2017)

15. Singh, J., Kumar, D., Qurashi, M.A., Baleanu, D.: A novel numerical approach for a nonlinear fractional dynamical model of interpersonal and romantic relationships. Entropy **19**(7), 375 (2017)
16. Wazwaz, A.M.: A comparison between Adomian-decomposition method and Taylor's series method in series solution. Appl. Math. Comput. **197**, 37–44 (1998)
17. Saravanan, A., Magesh, N.: A comparison between reduced differential transform method and Adomian decomposition method for the Newell-Whitehead-Segel equation. J. Egypt. Math. Soc. **21**(3), 259–265 (2013)
18. Noor, M.A., Noor, K.I., Waseem, M.: A new decomposition method for solving a system of linear equation. J. Assoc. Arab Univ. Basic Appl. Sci. **16**, 27–33 (2014)
19. Gejji, V.D., Jafari, H.: Adomian decomposition method: a tool for solving a system of fractional differential equations. J. Math. Anal. Appl. **301**, 508–518 (2005)
20. Singh, R., Nelakanti, G., Kumar, J.: Approximate solution of Urysohn integral equation using the Adomian decomposition method. World Sci. J. **2014**, 6 (2014). Article ID 150483
21. Wazwaz, A.M.: A reliable modification of Adomian decomposition method. Appl. Math. Comput. **102**, 77–86 (1999)
22. Luo, X.: A two step Adomian decomposition method. Appl. Math. Comput. **170**(1), 570–583 (2005)
23. Zhang, B., Wu, Q., Luo, X.: Experimentation with two step Adomian decomposition method. Appl. Math. Comput. **175**(2), 1495–1502 (2009)
24. Babolian, E., Javidi, S.: New method for calculating Adomian polynomials. Appl. Math. Comput. **153**(1), 253–259 (2004)
25. Gu, H., Li, Z.: A modified Adomian method for system of nonlinear differential equations. Appl. Math. Comput. **187**(2), 748–755 (2007)
26. Rach, R.: A new definition of the Adomian polynomials. Kybernetes **37**, 910–955 (2008)
27. Davis, H.T.: Introduction to Nonlinear Differential and Integral Equations. Dover, New York (1962)
28. Demir, H., Sungrii, I.C.: Numerical solution of a class of nonlinear Emden-Fowler equation by differential transform method. J. Arts Sci. **12**, 75–81 (2009)
29. Wazwaz, A.M.: Adomian decomposition method for a reliable treatment of the Emden-Fowler equation. Appl. Math. Comput. **161**(2), 543–560 (2005)
30. Chowdhury, M.S.H., Hashim, I.: Solutions of Emden-Fowler equation by homotopy perturbation method. Nonlinear Anal. RWA **10**(1), 104–115 (2009)
31. Khan, J.A., Raja, M.A., Qureshi, I.M.: Numerical treatment of nonlinear Emden-Fowler equation using stochastic technique. Ann. Math. Artif. Intell. **63**(2), 185–207 (2011)
32. Wazwaz, A.M.: The variational iteration method for solution of system of equations of Emden-Fowler type. Int. J. Comput. Math. **88**(16), 3406–3415 (2011)
33. Lima, P.M.: Numerical method and asymptotic error expansion for Emden-Fowler equation. J. Comput. Appl. Math. **70**(2), 245–266 (1996)
34. Chandrasekhar, S.: Introduction to the Study of Stellar Structure. Dover Publications, New York (1967)
35. Richardson, O.U.: The Emission of Electricity from Hot Bodies. Longsman's Green and Company, London (1921)
36. Pozzi, A., Bassano, E.: Application of Pade approximants to the study of plane jets. Comput. Math Appl. **30**(11), 107–123 (1995)
37. Abbaoui, K., Cherrault, Y.: Convergence of Adomian's method applied to differential equations. Comput. Math Appl. **28**(5), 103–109 (1994)
38. Benabidallah, M., Cherrault, Y.: Application of Adomian method for solving a class of boundary problems. Kybernetes **33**(1), 118–132 (2004)
39. Gupta, S., Kumar, D., Singh, J.: ADMP: a maple package for symbolic computation and error estimating to singular two-point boundary value problems with initial conditions. Proc. Nat. Acad. Sci. Phys. Sci. A (2018). https://doi.org/10.1007/s40010-018-0540-4

Using Genetic Algorithms for Parameter Estimation of a Two-Component Circular Mixture Model

Muhammet Burak Kılıç([✉])

Department of Business Administration,
Burdur Mehmet Akif Ersoy University, Burdur, Turkey
mburak@mehmetakif.edu.tr

Abstract. In many engineering and biological applications, the data sets such as wave direction, orientations of animal are circular. This type of data refers as circular data and cannot be analyzed using linear statistical methods. The most common distributions for analyzing circular data are the von Mises (vM) and wrapped Cauchy (wC) distributions. In the present chapter, we consider a two-component circular mixture model of the vM and wC distributions. In order to obtain the maximum likelihood estimators of the parameters of the circular mixture model, we consider four optimization methods as the Newton-Raphson, Nelder-Mead, simulated annealing and the proposed genetic algorithm (GA). Here, GAs are a class of evolutionary algorithms and based on the principles of biological systems. The search space in GA addresses for the circular mixture model. To compare the performance of four optimization methods, we present the simulation study and phase data examples. The results indicate that the proposed GA seems to perform well in terms of parameter estimations as seen in simulated and phase data examples.

Keywords: Circular data · A mixture of circular distributions · Genetic algorithms

1 Introduction

In many types of research including environmental, engineering and biological applications, circular data are commonly used and have the property of bimodality. Modeling the bimodal circular data plays an important role in the biological and engineering sciences and cannot be analyzed using linear statistical methods.

In the literature, bimodal circular data have been analyzed by different authors using mixtures of von Mises (vM) distributions (Mardia 1975; Spurr and Koutbeiy 1991; Mooney *et al.* 2003), extensions of vM distribution (Yfantis and Borgman 1982; Gatto and Jammalamadaka 2007; Kim and SenGupta 2013), sine-skewed circular distributions (Abe and Pewsey 2011), wrapped flexible generalized skew-normal distribution (Hernández-Sánchez and Scarpa 2012) and the general projected normal distribution (Wang and Gelfand 2013). In this study, we focus on a mixture of two different circular

H. Dutta et al. (Eds.): CMES 2019, AISC 1111, pp. 99–110, 2020.
https://doi.org/10.1007/978-3-030-39112-6_6

distributions of vM and wrapped Cauchy (wC) and inference techniques have developed for estimating its parameters.

One of the popular approaches for estimating the parameters of circular distributions is maximum likelihood estimation which is an attractive method. However, maximum likelihood estimation possesses some difficult problems due to the Newton-type optimization which involves inequality constraints and gradient problems. Another problem for circular distributions is the normalizing constants which have a complex form and leads inferential and predictive problems. To overcome these problems, we provide a metaheuristic approach based on genetic algorithm (GA) to obtain the maximum likelihood estimators of the parameters of a two-component circular mixture model.

GAs are a class of stochastic search algorithms based on the principles of biological systems including the genetic process of selection, crossover, and mutation proposed by Holland (1992). GA is commonly used in many areas for solving optimization problems such as machine learning systems (Goldberg and Holland 1988), image processing (Booker 1982), pipeline control system (Goldberg and Kuo 1987) and estimating the parameters of Weibull (Thomas *et al.* 1995), the negative binomial gamma (Gençtürk and Yiğiter 2016) and skew-normal parameters (Yalçınkaya *et al.* 2018). In the context of circular probability distributions, the parameters of finite mixtures of vM distributions were estimated by using the GA technique (Heckenbergerová *et al.* 2013), but the search space in GA was not well addressed. The main advantage of GA uses a search space instead of random starting points over other iterative algorithms. Recently, an adaptive search space that is depending on the data has become an essential issue in the aspect of the performance of evolutionary algorithms (Acitas *et al.* 2019). Therefore, this paper is addressed an adaptive search space in GA for estimating the parameters of a two-component circular mixture model.

The contribution of this chapter is to present a novel metaheuristic approach based on GA with Nelder-Mead (NM) search space for maximum likelihood estimators of the parameters of the mixture of two different distributions of vM and wC. Then, we compare the performances of the maximum likelihood estimators by using the Newton-Raphson (NR), NM, simulated annealing (SA) and the proposed GA methods. The remainder of this chapter is structured as follows. In Sect. 2, we will be described the functional form and the basic properties including probability functions of a mixture of the circular distributions used in this study. In Sect. 3, we briefly describe the parameter estimation frameworks which are based on the maximization of the log-likelihood function by using NR, NM, SA, and the proposed GA methods and an adaptive search space in GA is addressed for estimating of the parameters of a two-component circular mixture model. Section 4 presents the simulation study and a phase data example to show the efficiency of the proposed GA approach. In Sect. 5, we provide some concluding remarks.

2 Circular Data

Circular (directional) data analysis is different than linear data analysis because of the restriction of the support on the unit circle and a geometric structure of the data. Standard methods cannot be appropriate to compute descriptive statistics. Let $\theta_1, \theta_2, \ldots, \theta_n$ be a set of directional observations. Consider the polar to rectangular transformation for each observation, then it can be obtained the resultant vector as $\boldsymbol{R} = (\sum_{i=1}^{n} \cos\theta_i, \sum_{i=1}^{n} \sin\theta_i) = (C, S)$. The length of the resultant vector is computed by $R = \|\boldsymbol{R}\| = \sqrt{C^2 + S^2}$. The direction of the resultant vector is named as the mean direction denoted by $\overline{\theta}$ which is computed by $\overline{\theta} = \arctan^*(S/C)$ ($\arctan^*(S/C)$ is defined as $\arctan(S/C)$ if $C > 0, S \geq 0$; $\pi/2$ if $C = 0, S > 0$; $\arctan(S/C) + \pi$ if $C < 0$; $\arctan(S/C) + 2\pi$ if $C \geq 0, S < 0$; undefined if $C = 0, S = 0$ given in Jammalamadaka and SenGupta 2001).

2.1 A Two-Component Circular Mixture Model

Most commonly used distribution in circular data analysis is the vM distribution with respective parameters μ and κ as shown below

$$f(\theta; \mu, \kappa) = \frac{1}{2\pi I_0(\kappa)} e^{\kappa \cos(\theta - \mu)}, \qquad 0 \leq \theta < 2\pi \tag{1}$$

where $0 \leq \mu < 2\pi$ and $\kappa > 0$ are the circular mean and concentration parameters, respectively. $I_0(\kappa)$ is modified Bessel function of first the kind of order zero which can be defined by: $I_0(\kappa) = \sum_{i=0}^{\infty} (i!)^{-2} \left(\frac{1}{2}\kappa\right)^{2i}$. Another commonly used circular distribution is the wC distribution with respective parameters μ and ρ as shown below

$$f(\theta; \mu, \rho) = \frac{1}{2\pi} \frac{1 - \rho^2}{1 + \rho^2 - 2\rho \cos(\theta - \mu)}, \qquad 0 \leq \theta < 2\pi \tag{2}$$

where $0 \leq \mu < 2\pi$ and $0 \leq \rho < 1$ are the circular mean and concentration parameters, respectively. The maximum likelihood estimator of the parameters of wC distribution is given in Kent and Tyler (1988).

The standard approach to constructing bimodal circular distribution is to use a mixture of unimodal circular distributions. Then, the probability density function of the mixture of two different distributions of vM and wC is defined by:

$$f(\theta; \varphi) = p\, f(\theta; \mu_1, \kappa) + (1 - p) f(\theta; \mu_2, \rho), \qquad 0 \leq \theta < 2\pi \tag{3}$$

where $0 \leq \mu_1, \mu_2 < 2\pi$, $\kappa > 0$, $0 \leq \rho < 1$ and $0 < p < 1$ are the circular mean, concentration and mixing parameters, respectively. $\varphi = (p, \mu_1, \kappa, \mu_2, \rho)$ is called as the parameter vector of the mixture of two different distributions of vM and wC. In the following section, we have investigated the parameter estimation frameworks of a two-component circular mixture model.

3 Estimation

Estimation procedures for the parameters of a mixture of circular distributions are performed by maximizing the log-likelihood function. The log-likelihood function for the mixture of two different distributions of vM and wC are given below

$$\ln \mathrm{L}(\varphi) = \sum_{i=1}^{n} \ln\left(p \frac{1}{2\pi I_0(\kappa)} e^{\kappa \cos(\theta_i - \mu_1)} + (1 - p) \frac{1}{2\pi} \frac{1 - \rho^2}{1 + \rho^2 - 2\rho \cos(\theta_i - \mu_2)}\right). \tag{4}$$

Since there are no explicit solutions for estimators, we use the NR, NM, SA and GA frameworks for estimating unknown parameters vector φ of the mixture of circular distributions.

3.1 Newton-Raphson Framework

The NR method can be used to estimate the parameters of a mixture model. Here, we summarize steps for NR framework (McLachan and Krishnan 2007) of a two-component circular mixture model as follows.

Step 1. Determine initial values $\varphi^{(0)}$ for φ.
Step 2. Compute the score equation denoted by $S(\varphi) = \frac{\partial \ln \mathrm{L}(\varphi)}{\partial \varphi} = 0$.
Step 3. Compute the matrix of the negative of second-order partial derivatives of the log-likelihood function in respect of the elements of φ as shown below

$$J(\varphi) = J(\varphi; \theta) = -\frac{\partial^2 \ln \mathrm{L}(\varphi; \theta)}{\partial \varphi \partial \varphi^T}.$$

Step 4. Compute the elements of φ at $(m+1)$ iteration by using Eq. (5)

$$\varphi^{(m+1)} = \varphi^{(m)} + J^{-1}\left(\varphi^{(m)}; \theta\right) S\left(\varphi^{(m)}\right). \tag{5}$$

Step 5. Continue the iterations until the convergence criteria are satisfied.

Here, the NR method for mixture distributions has some disadvantages. The first disadvantage is that the log-likelihood function is not concave. Hence, the NR method is not assured to converge from a random initial value (McLachlan and Krishnan 2007). Secondly, the matrix $J(\varphi)$ can be lead to an inverse problem. Therefore, we propose a metaheuristic approach to obtain the maximum likelihood estimators of the parameters of the mixture of circular distributions.

3.2 Nelder-Mead Framework

The Nelder-Mead method proposed by Nelder and Mead (1965). The steps of NM (Lagarias et al. 1998) are as follows.

Step 1. Identify the four scalar parameters of the NM method: $\beta > 0$ (reflection), $\chi > 1$ (expansion), $0 < \lambda < 1$ (contraction), $0 < \sigma < 1$ (shrinkage). These parameters are generally chosen as $\beta = 1$, $\chi = 2$, $\lambda = 1/2$ and $\sigma = 1/2$ in the literature.
Step 2. Order the $d+1$ vertices to satisfy $f(\varphi_1) \leq f(\varphi_2) \leq \ldots \leq f(\varphi_{d+1})$ where $f(\varphi) = -\ln \mathrm{L}(\varphi)$ is used to minimize and d is the number of optimization parameters.
Step 3. Compute the reflection point φ_r from $\varphi_r = \overline{\varphi} + \beta(\overline{\varphi} - \varphi_{d+1})$ and evaluate $f_r = f(\varphi_r)$.
Step 4. If $f(\varphi_r) < f(\varphi_1)$, compute the expansion point from $\varphi_e = \overline{\varphi} + \chi(\varphi_r - \varphi_{d+1})$ and evaluate $f_e = f(\varphi_e)$. If $f_e < f_r$, accept φ_e and terminate the iteration; otherwise, accept φ_r and terminate the iteration.

Step 5. If $f_r \geq f_d$, perform a contraction between $\overline{\varphi}$ and the better of φ_{d+1} and φ_r.
Step 6. Perform a shrink step. Evaluate f at d points, $\varphi_i = \varphi_1 + \sigma(\varphi_i - \varphi_1), i = 2, \ldots, (d+1)$, and apply the next iteration as $\varphi_1, \varphi_2, \ldots, \varphi_{d+1}$.
Step 7. Stop the algorithm when the point is the lowest value.

3.3 Simulated Annealing

Kirkpatrick et al. (1983) first introduced the SA algorithm. Heating of solid matters and cooling them until crystallization called as annealing process. The simulated annealing algorithm evaluates the energy function denoted by E, which shows the objective function and its parameters as $E = E(f, \varphi)$. Here, the objective function is to be maximized the log-likelihood given in Eq. (4). The steps of SA (Al-Ageelee et al. 2017) are below.

Step 1. Choose an initial point φ_0.
Step 2. Initialize T with a large value.
Step 3. Repeat.

a. Repeat.

   1. Apply random perturbations to the state $\varphi = \varphi + \Delta\varphi$.
   2. Evaluate $\Delta\text{E}(\varphi) = \text{E}(\Delta + \Delta\varphi) - \text{E}(\varphi)$ if $\Delta\text{E}(\varphi) \leq 0$, keep the new state; otherwise, accept the new state with probability $\text{P} = e^{-\frac{\Delta\text{E}}{T}}$ until the number of accepted transitions is lower than a threshold level.

b. Set $T = T - \Delta T$ until T is too small.

3.4 Genetic Algorithm Framework

GAs are well-known evolutionary algorithms inspired by the natural principles of biological systems. Here, we summarize the steps of the GA framework for estimating the parameters of the mixture of two different distributions of vM and wC as follows.

Step 1. Determine the fitness function, the search space and initial GA's parameters which contain population size, selection rate, crossover probability, and mutation probability. The fitness function is the log-likelihood function $\ln \text{L}(\varphi)$ given in Eq. (4).
Step 2. Start with randomly generating an initial population of N chromosomes from search space via random generation. The initial population is given by $\varphi_1^{(0)}, \varphi_2^{(0)}, \ldots, \varphi_N^{(0)}$ where $\varphi = (p, \mu_1, \kappa, \mu_2, \rho)$ represents the parameters of a two-component circular mixture model.
Step 3. The fitness value, $\ln \text{L}(\varphi_j^{(m)})$ at any iteration m, for each chromosome, is evaluated. The individuals, having the worst fitness value based on a selection rate, are replaced by new individuals. These individuals are named as elite individuals.

Step 4. Perform crossover and mutation operators to obtain best candidate individuals based on crossover probability and mutation probability. Cross over of parents, which are selected from two best individuals, is conducted to obtain new offspring individuals and then apply mutation for determining new individuals. Hence $(m + 1)$th iteration can be obtained as $\varphi_1^{(m+1)}, \varphi_2^{(m+1)}, \ldots, \varphi_N^{(m+1)}$.
Step 5. Repeat the steps 2–5 with the fitness evaluation step until convergence criteria are met the evolution stops. Finally, the best individual is accepted as the optimum denoted by $\varphi^* = \text{argmax}_{\varphi_j^{(m)}} \ln \mathrm{L}\left(\varphi_j^{(m)}\right)$.

3.4.1 Adaptive Search Space

The search space in GA is a most valuable issue in the aspect of the performance of GA. Here, we have addressed an adaptive search space in GA for estimating the parameters of a two-component mixture model. The proposed algorithm for identifying the search space in GA is illustrated below.

Step 1. From a random sample, $\Theta = (\theta_1, \theta_2, \ldots, \theta_n)$, compute $\hat{\varphi}$ using NM.
Step 2. Sample with replacement from the original sample to obtain $\Theta^{*b} = (\theta_1^{*b}, \theta_2^{*b}, \ldots, \theta_n^{*b})$, $b = 1, 2, \ldots B$.
Step 3. Compute $\hat{\varphi}^{*b}$ using the sample in step 2 from the bootstrap replicates.
Step 4. Use the percentile bootstrap method for constructing the confidence intervals for each element of φ.

Here, the developed search space is constructed by NM method which is both fast and robust estimation method. For circular mean parameters, the method of symmetric arc (Fisher and Hall 1989) is considered by $\gamma_1^b = \left|\hat{\mu}_1^{*b} - \hat{\mu}_1\right|$ and $\gamma_2^b = \left|\hat{\mu}_2^{*b} - \hat{\mu}_2\right|$. Here, order the γ_1^b and γ_2^b from the smallest to largest. The lower endpoint of the interval is in the integer part of $l = (1/2\,B\alpha + 1/2)$th and the upper endpoint is in the $m = (B - l)$th position of the same ordered list. A 100 $(1 - \alpha)$% confidence interval for the circular mean parameters is given by $\left(\hat{\mu}_1 - \gamma_1^l, \hat{\mu}_1 + \gamma_1^m\right)$ and $\left(\hat{\mu}_2 - \gamma_2^l, \hat{\mu}_2 + \gamma_2^m\right)$. The method of the minimum length is used to construct the confidence interval of vM and wC concentration parameters and mixing parameter. This confidence interval based on minimum length performs well and reduces bias (Buckland 1983). The procedure of a confidence interval 100 (1–2α)% is given by $(\hat{\kappa}^{*k}, \hat{\kappa}^{*(k+s-r)})$, $(\hat{\rho}^{*k}, \hat{\rho}^{*(k+s-r)})$ and $(\hat{p}^{*k}, \hat{p}^{*(k+s-r)})$ where k $(k = 0, 1, 2, \ldots, r)$ is identified as minimum value of $(\hat{\kappa}^{*k}, \hat{\kappa}^{*(k+s-r)})$, $(\hat{\rho}^{*k}, \hat{\rho}^{*(k+s-r)})$ and $\left(\hat{p}^{*k}, \hat{p}^{*(k+s-r)}\right)$, respectively. $r = (n + 1)\alpha$ and $s = (n + 1)(1 - \alpha)$ values are rounded to the nearest integer.

4 Applications

The objective of this section is to investigate the performance of the proposed GA approach under two different simulation scenarios and phase data example. Here, the initial GA parameters whose population size, elitism number, crossover probability, and mutation probability are chosen as 15, 1, 0.8 and 0.01, respectively. These parameters are similar to the other studies given in the literature.

4.1 Simulation Study

The purpose of the simulation study is to compare the efficiency of the maximum likelihood estimators of the parameters of the mixture of two different distributions of vM and wC densities using NR, NM, SA, and GA for different sample sizes 30, 50, 100 and 200 respectively. For each sample size, 1000 circular data sets are generated from the mixture of two different distributions of vM and wC as follows

$$\theta_i \sim p\,\text{vM}\,(\mu_1, \kappa) + (1-p)\text{wC}\,(\mu_2, \rho), \quad i = 1, 2, \ldots, n$$

where $\boldsymbol{\mu} = (\mu_1, \mu_2) = (\pi/3, 3\pi/2)$ and we here consider two different scenarios for concentration parameters $(\kappa, \rho) = (5, 0.7)$ with $p = 0.2$ and $(\kappa, \rho) = (15, 0.8)$ with $p = 0.5$ indicating circular observations with low and high concentrations, respectively. To evaluate the performance of the maximum likelihood estimators using the NR, NM, SA and GA frameworks, we compute bias, mean absolute error (MAE), and circular bias, mean absolute cosine error (MACE) (Jammalamadaka and SenGupta 2001) as follows

$$\begin{aligned}
\text{Bias}\left(\hat{\phi}\right) &= \frac{1}{1000}\sum_{i=1}^{1000}\left(\hat{\phi}_i - \phi\right),\\
\text{MAE}\left(\hat{\phi}\right) &= \frac{1}{1000}\sum_{i=1}^{1000}\left|\hat{\phi}_i - \phi\right|,\\
\text{Circular Bias}\left(\hat{\psi}\right) &= \frac{1}{1000}\sum_{i=1}^{1000}\sin\left(\hat{\psi}_i - \psi\right),\\
\text{MACE}\left(\hat{\psi}\right) &= \frac{1}{1000}\sum_{i=1}^{1000}\left|\cos\left(\hat{\psi}_i\right) - \cos(\psi)\right|, \qquad (6)
\end{aligned}$$

where ϕ shows one of the true parameters vector (p, κ, ρ) and $\hat{\phi}_i$ shows an estimation of true parameters ϕ in the ith run. Here, we compute circular bias and the MACE for μ_1 and μ_2 since these represent the circular mean parameters. Therefore, ψ shows one of the true parameter vector (μ_1, μ_2) and $\hat{\psi}_i$ shows an estimation of true parameters ψ in the ith run for each component. The smaller absolute value of the bias, MAE, absolute value of the circular bias and the MACE show the high efficiency of estimators. All computational codes for NR, NM, SA, and GA frameworks are made by maxLik (Henningsen and Toomet 2011) and GA packages (Scrucca 2013) of R software and these codes can be provided upon request from the authors. Table 1 demonstrates the bias, MAE, circular bias, and the MACE of maximum likelihood estimators using NR, NM, SA, and the proposed GA approach via Monte Carlo simulations. In this table, the best values of the bias, MAE, circular bias, and the MACE are highlighted in bold. From Table 1, we note that:

1. the biases of all estimators of the parameters tend to zero for large n except the circular mean parameters from 50 to 100 sizes. These results are also consistent with the simulation results of a mixture of two vM distributions (Spurr and Koutbeiy 1991),

2. NR and NM have almost identical biases for circular mean and mixing parameters,
3. NR has the largest biases for concentration parameters among the presented four optimization methods,
4. GA and SA estimators have the smallest biases for the concentration parameters,
5. GA has the smallest biases for the circular mean parameters,
6. the estimators of κ are positively biased, and the estimators of ρ, p, μ_1 and μ_2 are both positively and negatively biased,
7. MAE and MACE of the corresponding estimators of the parameters tend to zero for large *n*,
8. GA has the smallest MAE and MACE among the presented four-optimization methods.

Generally, the absolute value of bias, MAE, the absolute value of circular bias and the MACE are smaller for GA among the presented methods. Accordingly, we have also concluded that the GA method based on NM search space has also very good estimation performance than for NM from simulation results.

Table 1. Simulated bias, MAE and MACE values for the estimators of the parameters of a two-component circular mixture model using the NR, NM, SA and GA frameworks.

	n = *30*		*n* = *50*		*n* = *100*		*n* = *200*	
	Bias & Circ. Bias	MAE & MACE	Bias & Circ. Bias	MAE & MACE	Bias & Circ. Bias	MAE & MACE	Bias & Circ. Bias	MAE & MACE
NR								
$p = 0.2$	0.0263	0.0624	0.0141	0.0482	0.0084	0.0335	0.0035	0.0208
$\mu_1 = \pi/3$	**−0.4645**	0.1845	−0.3872	0.1498	−0.1461	0.1010	−0.0549	0.0688
$\mu_2 = 3\pi/2$	**0.0023**	0.0892	−0.2447	0.0662	−0.0674	0.0480	−0.1336	0.0337
$\kappa = 5$	97.8252	99.3926	29.7984	31.2713	2.9555	4.4006	0.7987	1.8919
$\rho = 0.7$	0.0213	0.0718	0.0118	0.0544	0.0056	0.0384	0.0020	0.0261
$p = 0.5$	0.0038	0.0229	0.0008	0.0159	0.0010	0.0116	0.0000	0.0078
$\mu_1 = \pi/3$	−0.0638	0.0518	−0.0118	0.0382	0.0263	0.0268	0.0745	0.0188
$\mu_2 = 3\pi/2$	−0.4534	0.0741	−0.1007	0.0541	0.0086	0.0366	0.0447	0.0248
$\kappa = 15$	4.6809	7.7503	2.5179	5.0876	1.0073	3.0200	0.4541	1.9437
$\rho = 0.8$	0.0063	0.0579	0.0039	0.0434	0.0003	0.0315	0.0007	0.0209
NM								
$p = 0.2$	0.0320	0.0635	0.0148	0.0463	0.0082	0.0328	0.0035	0.0208
$\mu_1 = \pi/3$	−0.6178	0.1921	−0.2943	0.1495	−0.0945	0.1012	−0.0549	0.0689
$\mu_2 = 3\pi/2$	−0.0863	0.0898	−0.2736	0.0666	−0.677	0.0479	**−0.1336**	0.0337
$\kappa = 5$	15.2812	16.9516	8.0604	9.5169	2.6584	4.0866	0.7987	1.8850
$\rho = 0.7$	0.0242	0.0723	0.0120	0.0540	0.0055	0.0382	0.0020	0.0262
$p = 0.5$	0.0041	0.0229	0.0009	0.0159	0.0009	0.0118	−0.0001	0.0079

(*continued*)

Table 1. (*continued*)

	n = *30*		*n* = *50*		*n* = *100*		*n* = *200*	
	Bias & Circ. Bias	MAE & MACE	Bias & Circ. Bias	MAE & MACE	Bias & Circ. Bias	MAE & MACE	Bias & Circ. Bias	MAE & MACE
$\mu_1 = \pi/3$	−0.0945	0.0517	−0.0098	0.0381	0.0291	0.0268	0.0773	0.0189
$\mu_2 = 3\pi/2$	−0.2336	0.0739	−0.0992	0.0543	0.0095	0.0366	0.0468	0.0248
$\kappa = 15$	4.3219	7.2529	2.3004	4.7676	0.9400	2.8806	0.4621	1.8925
$\rho = 0.8$	0.0065	0.0574	0.0038	0.0433	0.0001	0.0315	0.0007	0.0209
SA								
$p = 0.2$	0.0732	0.0953	0.0239	0.0514	0.0101	0.0327	0.0044	0.0199
$\mu_1 = \pi/3$	0.8902	0.2160	−0.9986	0.1569	**0.0408**	0.1020	−0.2089	0.0699
$\mu_2 = 3\pi/2$	0.0543	0.1275	−0.6014	0.0800	−0.0538	0.0497	−0.1555	0.0346
$\kappa = 5$	**1.8201**	**3.3847**	**1.5305**	**2.7946**	**0.6599**	**1.9194**	**0.2490**	**1.2200**
$\rho = 0.7$	0.0089	0.0926	0.0050	0.0597	0.0053	0.0387	0.0015	0.0265
$p = 0.5$	**0.0016**	0.0264	**−0.0001**	0.0209	**0.0002**	0.0164	−0.0014	0.0134
$\mu_1 = \pi/3$	−0.0991	0.0521	**−0.0029**	0.0388	0.0376	0.0277	**0.0433**	0.0198
$\mu_2 = 3\pi/2$	−0.2394	0.0752	−0.1315	0.0548	**0.0076**	0.0385	0.0407	0.0258
$\kappa = 15$	**1.4352**	**3.3674**	**1.0054**	**2.6394**	0.8087	**1.7519**	0.7221	**1.2076**
$\rho = 0.8$	0.0030	0.0582	0.0021	0.0440	−0.0013	0.0326	−0.0008	0.0223
GA								
$p = 0.2$	**0.0014**	**0.0390**	**0.0059**	**0.0309**	**0.0020**	**0.0222**	**0.0016**	**0.0152**
$\mu_1 = \pi/3$	−0.8586	**0.1605**	**−0.0821**	**0.1252**	0.1648	**0.0859**	**0.0228**	**0.0597**
$\mu_2 = 3\pi/2$	0.1733	**0.0767**	**−0.2229**	**0.0591**	**−0.0467**	**0.0424**	−0.1378	**0.0298**
$\kappa = 5$	14.2948	14.7359	6.2738	6.8418	1.4216	2.7308	0.6445	1.4672
$\rho = 0.7$	**−0.0008**	**0.0566**	**−0.0002**	**0.0428**	**0.0013**	**0.0316**	**0.0006**	**0.0221**
$p = 0.5$	0.0016	**0.0170**	0.0013	**0.0130**	0.0010	**0.0103**	**0.0000**	**0.0065**
$\mu_1 = \pi/3$	**−0.0062**	**0.0449**	0.0250	**0.0335**	**0.0115**	**0.0247**	0.0772	**0.0171**
$\mu_2 = 3\pi/2$	**−0.1091**	**0.0635**	**−0.0533**	**0.0478**	0.0132	**0.0338**	**0.0305**	**0.0219**
$\kappa = 15$	4.1145	6.4037	2.1451	4.2815	**0.7922**	2.6958	**0.3933**	1.7083
$\rho = 0.8$	**0.0011**	**0.0495**	**0.0020**	**0.0380**	**0.0000**	**0.0292**	**0.0001**	**0.0185**

4.2 Phase Data Example

We consider phase differences in hand flexion-extension movements. Data are obtained from a study by Puglisi *et al.* (2017) on the role of attention in human motor resonance. For analysis phase data, we here apply the NR, NM, SA and the proposed GA approaches to obtain maximum likelihood estimators of the parameters of the mixture of vM and wC distributions. Table 2 presents the parameter estimates, maximize log-likelihood (lnL), Akaike information criterion (AIC), Bayesian information criterion (BIC). We also provide the value of the Watson U^2 goodness of fit test which is useful for circular data. The proposed GA has the highest lnL and the lowest AIC, BIC, Watson U^2 values which denote the best fit. Accordingly, the proposed GA shows superiority in terms of parameter estimations among other optimization methods. A linear histogram of phase data and the fitted densities obtained from four different methods are shown in Fig. 1. Accordingly, the proposed GA approach seems to capture the modes of data well.

Table 2. The parameter estimations using the NR, NM, SA and GA frameworks, lnL, AIC, BIC, Watson U^2 values for phase data example.

	NR	NM	SA	Proposed GA
$\hat{p}$	0.2118	0.2422	0.1979	0.0658
$\hat{\mu}_1$	0.3109	0.3434	1.1605	1.5036
$\hat{\mu}_2$	0.7884	0.7724	0.3745	0.5010
$\hat{\kappa}$	15.6262	12.4908	3.6156	157.2482
$\hat{\rho}$	0.5141	0.5161	0.6269	0.6148
ln L	−58.4510	−58.5042	−58.6501	**−57.2336**
AIC	126.9020	127.0084	127.3002	**124.4672**
BIC	135.5903	135.6967	135.9885	**133.1555**
U^2	0.0125	0.0141	0.0158	**0.0123**

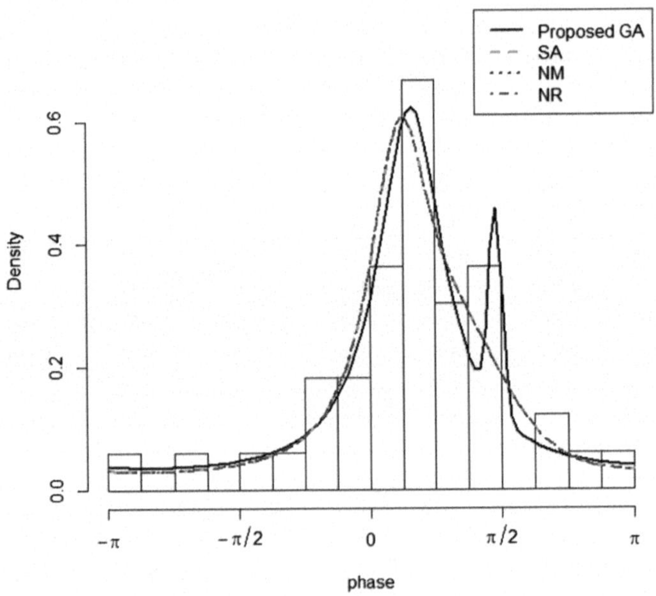

Fig. 1. A histogram of phase data and the fitted densities obtained from four different parameter estimation methods.

5 Concluding Remarks

In this chapter, we have provided a novel metaheuristic GA method to obtain the maximum likelihood estimators of the parameters of a two-component mixture circular model based on adaptive search space. The proposed GA based on NM search space seems to perform well in terms of parameter estimations as seen in the simulation study and phase data example and it has also improved the performance of maximum likelihood estimation. In addition to analysis phase data, we would like to emphasize that the mixture of two different distributions of vM and wC can be used as an alternative circular distribution for modeling bimodal circular data.

References

Abe, T., Pewsey, A.: Sine-skewed circular distributions. Stat. Pap. **52**(3), 683–707 (2011)

Acitas, S., Aladag, C.H., Senoglu, B.: A new approach for estimating the parameters of Weibull distribution via particle swarm optimization: an application to the strengths of glass fibre data. Reliab. Eng. Syst. Saf. **183**, 116–127 (2019)

Al-Ageelee, A.A., Salim, N.J., Kadhum, R.: Cryptanalysis of nonlinear stream cipher cryptosystem based on improved particle swarm optimization. Int. J. Appl. Inf. Syst. **11**(11), 43–53 (2017)

Booker, L.B.: Intelligent behavior as an adaptation to the task environment. Dissertation presented to the University of Michigan, at Ann Arbor, Mich., in partial fulfillment of the requirements for the degree of Doctor of Philosophy (1982)

Buckland, S.T.: Monte Carlo methods for confidence interval estimation using the bootstrap technique. J. Appl. Stat. **10**(2), 194–212 (1983)

Fisher, N.I., Hall, P.: Bootstrap confidence regions for directional data. J. Am. Stat. Assoc. **84**(408), 996–1002 (1989)

Gatto, R., Jammalamadaka, S.R.: The generalized von Mises distribution. Stat. Methodol. **4**(3), 341–353 (2007)

Gençtürk, Y., Yiğiter, A.: Modeling claim number using a new mixture model: negative binomial gamma distribution. J. Stat. Comput. Simul. **86**(10), 1829–1839 (2016)

Goldberg, D.E., Kuo, C.H.: Genetic algorithms in pipeline optimization. J. Comput. Civ. Eng. **1**(2), 128–141 (1987)

Goldberg, D.E., Holland, J.H.: Genetic algorithms and machine learning. Mach. Learn. **3**(2), 95–99 (1988)

Heckenbergerová, J., Musilek, P., Mejznar, J., Vančura, M.: Estimation of wind direction distribution with genetic algorithms. In: 26th Annual IEEE Canadian Conference on Electrical and Computer Engineering (CCECE), pp. 1–4. IEEE (2013)

Henningsen, A., Toomet, O.: maxLik: a package for maximum likelihood estimation in R. Comput. Stat. **26**(3), 443–458 (2011)

Hernández-Sánchez, E., Scarpa, B.: A wrapped flexible generalized skew-normal model for a bimodal circular distribution of wind directions. Chil. J. Stat. **3**(2), 129–141 (2012)

Holland, J.H.: Adaptation in Natural and Artificial Systems: An Introductory Analysis with Applications to Biology, Control, and Artificial Intelligence. MIT Press, Cambridge (1992)

Jammalamadaka, S.R., SenGupta, A.: Topics in Circular Statistics. World Scientific, Singapore (2001)

Kent, J.T., Tyler, D.E.: Maximum likelihood estimation for the wrapped Cauchy distribution. J. Appl. Stat. **15**(2), 247–254 (1988)

Kim, S., SenGupta, A.: A three-parameter generalized von Mises distribution. Stat. Pap. **54**, 685–693 (2013)

Kirkpatrick, S., Gelatt, C.D., Vecchi, M.P.: Optimization by simulated annealing. Science **220**(4598), 671–680 (1983)

Lagarias, J.C., Reeds, J.A., Wright, M.H., Wright, P.E.: Convergence properties of the Nelder–Mead simplex method in low dimensions. SIAM J. Optim. **9**(1), 112–147 (1998)

Mardia, K.V.: Statistics of directional data. J. R. Stat. Soc. Ser. B (Methodol.) **37**(3), 349–371 (1975)

McLachlan, G.J., Krishnan, T.: The EM Algorithm and Extensions, vol. 382. Wiley, Hoboken (2007)

Mooney, J.A., Helms, P.J., Jolliffe, I.T.: Fitting mixtures of von Mises distributions: a case study involving sudden infant death syndrome. Comput. Stat. Data Anal. **41**, 505–513 (2003)

Nelder, J.A., Mead, R.: A simplex method for function minimization. Comput. J. **7**(4), 308–313 (1965)

Puglisi, G., Leonetti, A., Landau, A., Fornia, L., Cerri, G., Borroni, P.: The role of attention in human motor resonance. PLoS ONE **12**(5), 1–21 (2017)

Scrucca, L.: GA: a package for genetic algorithms in R. J. Stat. Softw. **53**(4), 1–37 (2013)

Spurr, B.D., Koutbeiy, M.A.: A comparison of various methods for estimating the parameters in mixtures of von Mises distributions. Commun. Stat. Simul. Comput. **20**(2–3), 725–741 (1991)

Thomas, G.M., Gerth, R., Velasco, T., Rabelo, L.C.: Using real-coded genetic algorithms for Weibull parameter estimation. Comput. Ind. Eng. **29**(1–4), 377–381 (1995)

Wang, F., Gelfand, A.E.: Directional data analysis under the general projected normal distribution. Stat. Methodol. **10**(1), 113–127 (2013)

Yalçınkaya, A., Şenoğlu, B., Yolcu, U.: Maximum likelihood estimation for the parameters of skew-normal distribution using a genetic algorithm. Swarm Evol. Comput. **38**, 127–138 (2018)

Yfantis, E.A., Borgman, L.E.: An extension of the von Mises distribution. Commun. Stat. Theory Methods **11**(15), 1695–1706 (1982)

An Early Detection Model for a Brain Tumor-Is (Immune System) Interaction with Fuzzy Initial Values

Fatma Berna Benli and Onur Alp İlhan(✉)

Department of Mathematics, Faculty of Education, Erciyes University, 38039 Kayseri, Turkey
{akpinarb,oailhan}@erciyes.edu.tr

Abstract. In this paper, we adopt a model by including fuzzy initial values to study the interaction of a monoclonal brain tumor and the macrophages for an early detection treatment. Numerical simulations will give detailed information on the behavior of the model at the end of the paper. We perform all the computations in this study with the help of the Maple software.

Keywords: Fuzzy number · Fuzzy derivative · Fuzzy differential equations (FDE) · Fuzzy initial values

2000 AMS Classification: 05C38 · 15A15

1 Introduction

In 1965, Zadeh [1] introduced fuzzy set theory. Kandel and Byatt [2] were among the first to study fuzzy differential equation in 1978. Within the past ten years, Hüllermeier [3], Bede and Gal [4], Bede et al. [5], Khastan, Bahrami and Ivaz [6] studied fuzzy differential equations and have also explained the concept of strongly generalized derivative of higher order fuzzy differential equation.

Many terms we use randomly daily life have a fuzziness. There is often a vague structure in life the language and numerical expression when we use describing something, explain an event, command and many other situations contain ambiguity. As for the fuzzy set, it has two basic features. The first is the modeling of systems whose mathematical model is uncertain or whose behavior can be estimated approximately. The second can determine when there is incomplete and uncertain information. Because among these features, new mathematical concepts have emerged, and new research problems and engineering practices have emerged. Particularly interesting applications in the field of medicine and artificial intelligence are beginning to emerge. Nowadays, diseases can now be expressed and solved mathematically. This leads to a better way to treat the disease. Fuzzy numbers can get closer results than classical mathematics. Therefore, in our research, by using these numbers, we can obtain more accurate results and be more effective in the diagnosis and treatment of diseases. One of the diseases is cancer, which is one of the biggest killers in the world. Controlling tumor growth requires special attention

H. Dutta et al. (Eds.): CMES 2019, AISC 1111, pp. 111–120, 2020.
https://doi.org/10.1007/978-3-030-39112-6_7

[7] and interdisciplinary research, such as biology, medicine, and mathematics, many of which are attracted by the spread of the disease. Typical methods for treating GBM (Glioblastoma Multiforme) include surgical resection followed by radiation therapy and chemotherapy [8]. Work on multi-subgroup modeling can be shown in [5, 9–11].

Differential equations have high importance in biological modeling. In the past few years, by using different types of models, Bozkurt uses the differential equation system as a model of brain tumors and its interaction with the human immune system (IS) [12, 13]. In addition, another interesting model is the GBM model constructed by Bozkurt [14], which explains the interaction between cancer and the human body. Also, at [15], Akın and Oruç studied a predator-prey model with fuzzy initial values. Then, by considering the second order initial value Akın, Khaniyev, Oruç and Turksen [16] generalized the model. Studies about the prey-predator model of fuzzy numbers can be seen in [17–19]. Finally, Benli and Keskin [20] considered a model with a predator-prey structure between monoclonal tumor and macrophages. In this study, a new model for the work of Bozkurt [14] was established, which used fuzzy initial values. In addition, the Allee threshold functions are embedded in the system to study the threshold effect on the system and to interpret the extinction conditions.

The paper has four sections and constructed as follows: The model and the preliminary definitions are in Sect. 2, nonlinear fuzzy differential equations are given in Sect. 3, a numerical study of the model are given in Sect. 4. The last part we have the conclusion, to summarize the study in the paper.

2 The Model and Preliminary Definitions

The model is constructed as follows:

$$\begin{cases} \frac{d_x}{d_t} = \left(\frac{x}{E+x}\right)(px + r_1x(K_1 - \alpha_1 x) - d_1x - \tau_1 xy) \\ \frac{d_y}{d_t} = r_2y(K_2 - \beta_1 y) - d_2y - \tau_2 xy \end{cases} \tag{2.1}$$

where $t \geq 0$ denotes the time, the parameters α_1, β_1, τ_1, τ_2, p, d_1, d_2, K_1, K_2, r_1 and r_2 are positive numbers (see Table 1). $x(t)$ is used for the GB which is used to represent the monoclonal brain tumor. On the other hand, $y(t)$ represents the activated macrophages in the system [21].

Differential equations are indispensable for modeling the real world phenomena unfortunately, whenever uncertainty can interfere with real-world problems, the uncertainty can come from insufficient data, measurement errors, or when determining initial conditions. Fuzzy set theory is a powerful tool to overcome these problems. The following are some of the definitions needed to have a basic idea of the work.

We referee the reader to [4, 5, 15] to more information about the definitions that have been used in current section.

Hence, we recall some basic definition about the subject.

Definition 1. A fuzzy set A in a universe set X is a mapping $A(x) : X \to [0, 1]$. We think of A as assigning to each element $x \in X$ a degree of membership, $0 \leq A(x) \leq 1$. Let us denote by $\mathcal{F}$ the class of fuzzy subsets of the real axis, $A(x) : X \to [0, 1]$ satisfying the following properties:

Table 1. Values of the parameter of system (2.1)

p	Division rate of the sensitive cells	0,192
K_1	Carrying capacity of the tumor cells	4,704
K_2	Carrying capacity of the macrophages	1,232
r_1	The growth rate of the macrophage	0,55
r_2	The growth rate of the macrophage	0,5
τ_i	Destroying rate caused from the interaction	0,01
d_1	Causes of drug treatment to the tumor cells	0,6
d_2	Causes of drug treatment to the macrophages	0,06
β_1	Logistic population rate of macrophages	$\beta_2 \in [0{,}05;\ 0{,}25]$
α_1	Logistic population rate of tumor cell population	$\alpha_1 \in [0{,}5;\ 0{,}95]$

i. A is a convex fuzzy set, i.e. $A(r\lambda + (1-\lambda)s) \geq \min[A(r), A(s)]$, $\lambda \in [0, 1]$ and $r, s \in X$
ii. A is normal, i.e. $\exists x_0 \in X$ with $A(x_0) = 1$;
iii. A is upper semicontinuous, i.e. $A(x_0) \geq \lim_{x \to x_0^{\mp}} A(x)$;
iv. $[A]^0 = \overline{sup\, p(A)} = \overline{\{x \in R | A(x) \geq 0\}}$ is compact, where $\overline{A}$ denotes the closure of A.

Then $\mathcal{F}$ is called the space of fuzzy numbers.

If A is a fuzzy set, we define $[A]^\alpha = \{x \in X | \mu_A \geq \alpha\}$, the α-level (cut) of A, with $0 < \alpha \leq 1$. For $u, v \in \mathcal{F}$ and $\lambda \in R$ the sum $u \oplus v$ and the product $\lambda \odot u$ are defined by $[u \oplus v]^\alpha = [u]^\alpha + [v]^\alpha$, $\lambda \odot u = \lambda[u]^\alpha$, $\forall \alpha \in [0, 1]$. Additionally, $u \oplus v = v \oplus u$, $\lambda \odot u = u \odot \lambda$. Also, if $u \in \mathcal{F}$ the α-cut of u, denoted by $[u]^\alpha = \left[\underline{u^\alpha}, \overline{u^\alpha}\right]$, $\forall\, \alpha \in [0, 1]$.

Definition 2. Let $D : \mathcal{F}x\mathcal{F} \to R_+ \cup \{0\}$, $D(u, v) = sup_{\alpha \in [0,1]} max\{|\underline{u^\alpha}, \underline{v^\alpha}|, |\overline{u^\alpha}, \overline{v^\alpha}|\}$ be a Hausdorff distance between fuzzy numbers, where $[u]^\alpha = \left[\underline{u^\alpha}, \overline{u^\alpha}\right]$ and $[v]^\alpha = \left[\underline{v^\alpha}, \overline{v^\alpha}\right]$. The following properties are well-known [21, 22].

$$D(u \oplus \omega, v \oplus \omega) = D(u, v), \quad \forall u, v, \omega \in \mathcal{F},$$
$$D(k \odot u, k \odot v) = |k| D(u, v), \ \forall k \in R, \quad u, v \in \mathcal{F},$$
$$D(u \oplus v, \omega \oplus e) \leq D(u, \omega) + D(v, e), \ \forall u, v, \omega, e \in \mathcal{F},$$

a $(\mathcal{F}, D)$ is a complete metric space.

Definition 3 (H-Difference). Let $\forall\, u, v \in \mathcal{F}$. If there exists $\omega \in \mathcal{F}$ such that $u = v \oplus \omega$, then ω is called the H-difference of u and v and is denoted by $u \ominus v$.

Definition 4 (Hukuhara Derivative) [23]. Consider a fuzzy mapping $F : (a, b) \to \mathcal{F}$ and $t_0 \in (a, b)$. We say that F is differentiable at $t_0 \in (a, b)$ if there exists an element $F'(t_0) \in \mathcal{F}$ such that for all $h > 0$ sufficiently small $F(t_0+h) \ominus F(t_0)$, $F(t_0) \ominus F(t_0-h)$, and the limits (in the metric D)

$$\lim_{x \to 0^+} \frac{F(t_0+h) \ominus F(t_0)}{h} = \lim_{x \to 0^-} \frac{F(t_0) \ominus F(t_0-h)}{h}$$

exist and are equal to $F'(t_0)$.

Note that this definition of the derivative is very restrictive; for instance in [4, 5] the authors showed that if $F(t) = c.g(t)$ where c is a fuzzy number and $g : [a, b] \to R^+$ is a function with $g'(t) < 0$, then F is not differentiable. To avoid this difficulty, the authors of [4, 5] introduce a more general definition of the derivative for fuzzy mappings.

Definition 5 (Generalized Fuzzy Derivative) [4, 5]. Let $F : (a, b) \to \mathcal{F}$ and $t_0 \in (a, b)$. We say that F is strongly generalized differentiable at t_0 if there exists an element $F'(t_0) \in \mathcal{F}$ such that

i. for $h > 0$ sufficiently small $\exists F(t_0+h) \ominus F(t_0)$, $F(t_0) \ominus F(t_0-h)$ and the limits satisfy

$$\lim_{h \to 0} \frac{F(t_0+h) \ominus F(t_0)}{h} = \lim_{h \to 0} \frac{F(t_0) \ominus F(t_0-h)}{h} = F'(t_0)$$

ii. for $h > 0$ sufficiently small $\exists F(t_0) \ominus F(t_0+h)$, $F(t_0-h) \ominus F(t_0)$ and the limits satisfy

$$\lim_{h \to 0} \frac{F(t_0) \ominus F(t_0+h)}{(-h)} = \lim_{h \to 0} \frac{F(t_0-h) \ominus F(t_0)}{(-h)} = F'(t_0)$$

or

iii. for $h > 0$ sufficiently small $\exists F(t_0+h) \ominus F(t_0)$, $F(t_0-h) \ominus F(t_0)$, and the limits satisfy

$$\lim_{h \to 0} \frac{F(t_0+h) \ominus F(t_0)}{h} = \lim_{h \to 0} \frac{F(t_0-h) \ominus F(t_0)}{(-h)} = F'(t_0)$$

or

iv. for $h > 0$ sufficiently small $\exists F(t_0) \ominus F(t_0+h)$, $F(t_0) \ominus F(t_0-h)$ and the limits satisfy

$$\lim_{h \to 0} \frac{F(t_0) \ominus F(t_0+h)}{(-h)} = \lim_{h \to 0} \frac{F(t_0-h) \ominus F(t_0)}{h} = F'(t_0)$$

Definition 5 is equivalent to Definition 6 that we will use in this paper.

Definition 6. Let $F : (a, b) \rightarrow \mathcal{F}$ and $t_0 \in (a, b)$.

(1) for $h > 0$ sufficiently small, $\exists F(t_0 + h) \ominus F(t_0)$, $F(t_0) \ominus F(t_0 - h)$ and

$$\lim_{h \to 0^+} \frac{F(t_0 + h) \ominus F(t_0)}{h} = \lim_{h \to 0^+} \frac{F(t_0) \ominus F(t_0 - h)}{h} = F'(t_0)$$

or

(2) for $h > 0$ sufficiently small $\exists F(t_0 + h) \ominus F(t_0)$, $F(t_0) \ominus F(t_0 - h)$ and

$$\lim_{h \to 0^-} \frac{F(t_0 + h) \ominus F(t_0)}{h} = \lim_{h \to 0^-} \frac{F(t_0) \ominus F(t_0 - h)}{h} = F'(t_0).$$

The following theorem is very important to solve fuzzy differential equations.

Theorem 1 [24, 25]. Let $F : T \rightarrow \mathcal{F}$ be a function and set $[F(t)]^\alpha = [f_\alpha(t), g_\alpha(t)]$ for each $\alpha \in [0, 1]$. Then

i. If F is differentiable following the form (1) in **Definition** 6, then $f_\alpha(t)$ and $g_\alpha(t)$ are differentiable functions and $\left[F'(t)\right]^\alpha = \left[f'_\alpha(t), g'_\alpha(t)\right]$.
ii. If F is differentiable following the form (2) in **Definition** 6, then $f_\alpha(t)$ and $g_\alpha(t)$ are differentiable functions and $\left[F'(t)\right]^\alpha = \left[g'_\alpha(t), f'_\alpha(t)\right]$.

3 Solving Fuzzy Differential Equations with Fuzzy Initial Values

Consider the following equation with fuzzy initial values

$$x'(t) = F(t, x(t)), \quad x(0) = x_0 \tag{3.1}$$

$F : [0, \alpha] \times \mathcal{F} \rightarrow \mathcal{F}$ and x_0 is a fuzzy number $[x(t)]^\alpha = [u_\alpha(t), v_\alpha(t)]$, $[x_0]^\alpha = \left[u^0_\alpha, v^0_\alpha\right]$ and

$$[F(t, x(t))]^\alpha = \left[f_\alpha(t, u_\alpha(t), v_\alpha(t)), g_\alpha(t, u_\alpha(t), v_\alpha(t))\right].$$

Then, we get the following alternatives for solving the initial value problem (3.1):

1. If we consider $x'(t)$ by using the derivative in the first form (1), then from Theorem 1

$$\left[x'(t)\right]^\alpha = \left[u'_\alpha(t), v'_\alpha(t)\right]$$

So we have the following equalities:

$$u'_\alpha(t) = f_\alpha(t, u_\alpha(t), v_\alpha(t)), \quad u_\alpha(0) = u^0_\alpha$$
$$v'_\alpha(t) = g_\alpha(t, u_\alpha(t), v_\alpha(t)), \quad v_\alpha(0) = v^0_\alpha.$$

By solving the above system for u_α and v_α; we get the fuzzy solution $[x(t)]^\alpha = [u_\alpha(t), v_\alpha(t)]$. Finally, we ensure that $[x(t)]^\alpha = [u_\alpha(t), v_\alpha(t)]$ and $\left[x'(t)\right]^\alpha = \left[u'_\alpha(t), v'_\alpha(t)\right]$ are valid level sets.

2. If we consider $x'(t)$ by using the derivative in the second form (2), then from Theorem 1 $\left[x'(t)\right]^{\alpha} = \left[u'_{\alpha}(t), v'_{\alpha}(t)\right]$ So we get following:

$$u'_{\alpha}(t) = g_{\alpha}(t, u_{\alpha}(t), v_{\alpha}(t)), \quad u_{\alpha}(0) = u^{0}_{\alpha}$$
$$v'_{\alpha}(t) = f_{\alpha}(t, u_{\alpha}(t), v_{\alpha}(t)), \quad v_{\alpha}(0) = v^{0}_{\alpha}.$$

Solving the above system for u_{α} and v_{α}, we get the fuzzy solution $[x(t)]^{\alpha} = [u_{\alpha}(t), v_{\alpha}(t)]$ Finally we ensure that $[x(t)]^{\alpha} = [u_{\alpha}(t), v_{\alpha}(t)]$ and $\left[x'(t)\right]^{\alpha} = \left[u'_{\alpha}(t), v'_{\alpha}(t)\right]$ are valid level sets.

4 Numerical Study for Early Brain Tumor Growth Model with Fuzzy Initial Values

Now we consider the following numerical study for the following model, which represents the early diagnosis tumor model with fuzzy initial values.

$$\frac{dx}{dt} = \left(\frac{x}{0,1+x}\right)[0,192x + 0,65x(4,704 - 0,55x) - 0,6x - 0,01xy]$$
$$\frac{dy}{dt} = 0,5y(1,232 - 0,25y) - 0,06y - 0,01xy$$
$$x(0) = 0,35ml, \quad y(0) = 0,2ml. \tag{4.1}$$

where $x(t)$ and $y(t)$ are the density of the tumor and macrophages at time t, respectively. Table 1 shows the values of the parameters in Eq. (1), which gives us (4.1)

Now, we consider the following model given as:

$$\tfrac{dx}{dt} = \left(\tfrac{x}{0,1+x}\right)\left[2,6496x - 0,3575x^2 - 0,01xy\right]$$
$$\tfrac{dy}{dt} = 0,556y - 0,125y^2 - 0,01xy$$
$$x(0) = 0,35\,ml, \quad y(0) = 0,ml.$$

Linearization of the equations, we get a system of two linear equations which can be solved to get a solution closer to the real situation,

$$J(7,303392; 3,863729) = \begin{pmatrix} -2,5757 & -0,07205 \\ -0,03864 & -0,48297 \end{pmatrix}$$
$$\tfrac{dx}{dt} = -2,5757x - 0,07205y$$
$$\tfrac{dy}{dt} = -0,03864x - 0,48297y \tag{4.2}$$
$$x(0) = 0,35\,ml, \quad y(0) = 0,2\,ml.$$

The crisp solutions for the problem (4.2) are shown in Fig. 1.

Let the initial values be fuzzy, that is, $x(0) = \widetilde{0,35}$, $y(0) = \widetilde{0,2}$ and let their α-level sets be as follows;

$$x(0) = \left[\widetilde{0,35}\right] = [0,20 + 0,15\alpha, \quad 0,50 - 0,15\alpha]$$

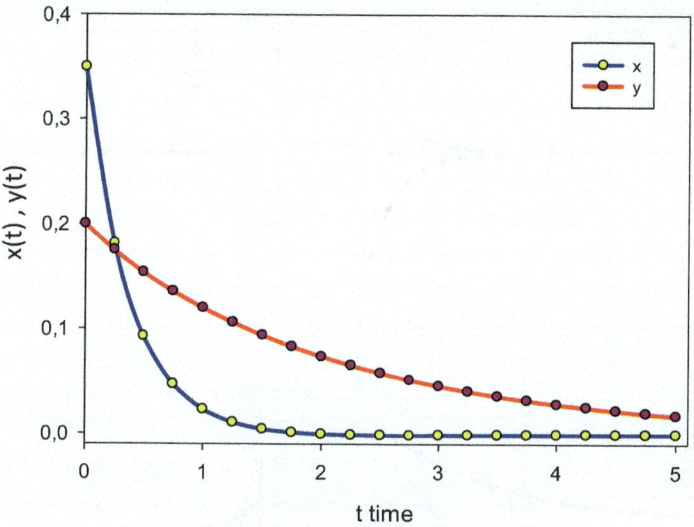

Fig. 1. Crisp solution for problem (4.2)

$$y(0) = \left[\widetilde{0,2}\right] = [0{,}10 + 0,\ 10\alpha, \quad 0{,}30 - 0{,}10\alpha]$$

Let the α-level sets of x(t, α) be $[x(t,\alpha)]^{\alpha} = [u(t,\alpha), v(t,\alpha)]$ and for simplicity denote them as [u; v]; similarly $\left[y(t,\alpha)\right]^{\alpha} = [r(t,\alpha), s(t,\alpha)] = [r; s]$.

If x(t, α) and y(t, α) are (1) differentiable according to Definition 6, system (4.2) becomes

$$\left[u', v'\right] = -2{,}5757[u, v] - 0{,}07205[r, s],$$
$$\left[r', s'\right] = -0{,}03864[u, v] - 0{,}48297[r, s].$$

Hence for $\alpha = 0$ the following initial value problem derives from (4.2):

$$u' = -2,5757v - 0{,}07205s,$$
$$v' = -2{,}5757u - 0{,}07205r,$$
$$r' = -0{,}03864v - 0{,}48297s,$$
$$s' = -0{,}03864u - 0{,}48297r.$$

$$u(0) = 0{,}20\,ml, \quad v(0) = 0{,}50\,ml, \quad r(0) = 0{,}10\,ml, \quad s(0) = 0{,}30\,ml.$$

$u(0)$, $v(0)$, $r(0)$ and $s(0)$ are fuzzy initial conditions for the system. Now if x(t, α) and y(t, α) are (2) differentiable according to Definition 6, system (4.2) becomes

$$u' = -2{,}5757u - 0{,}07205r,$$
$$v' = -2{,}5757v - 0{,}07205s,$$
$$r' = -0{,}03864u - 0{,}48297r,$$
$$s' - 0{,}03864v - 0{,}48297s$$

$$u(0) = 0{,}20\,ml, \quad v(0) = 0{,}50\,ml, \quad r(0) = 0{,}10\,ml, \quad s(0) = 0{,}30\,ml.$$

In Fig. 2, we can see the graphical solution of all cases for $\alpha = 0$.

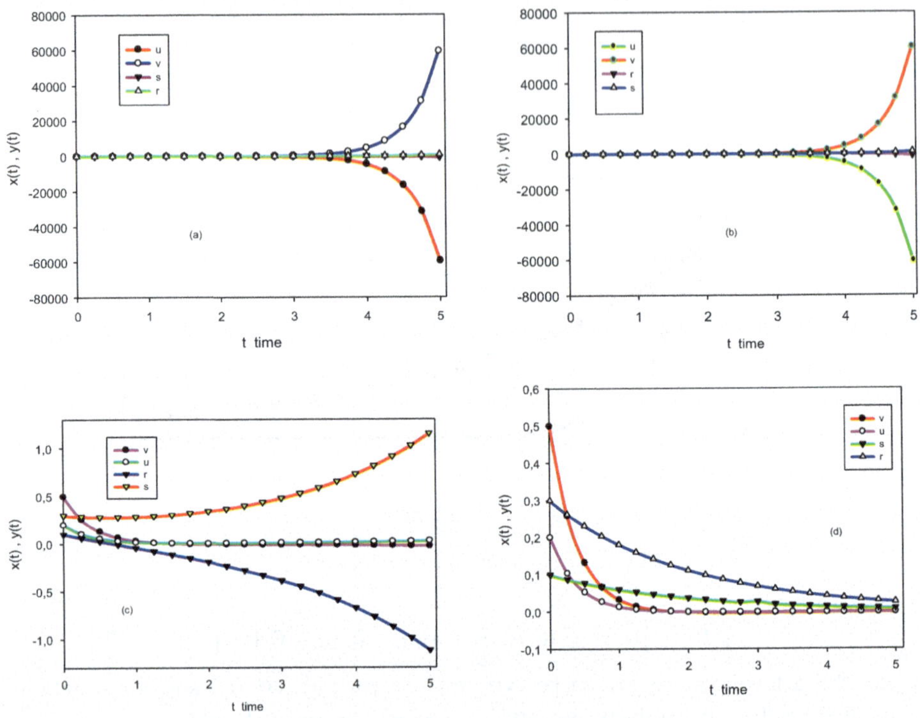

Fig. 2. Fuzzy solution of problem (4.2) for $\alpha = 0$

In Fig. 2; according to Definition 6, (a) means that $x(t, \alpha)$ and $y(t, \alpha)$ are (1) differentiable, (b) means that $x(t, \alpha)$ is (1) differentiable and $y(t, \alpha)$ is (2) differentiable, (c) means that $x(t, \alpha)$ is (2) differentiable and $y(t, \alpha)$ is (1) differentiable, (d) means that $x(t, \alpha)$ and $y(t, \alpha)$ are (2) differentiable. Now, if Fig. 2 is analyzed, we see that when $x(t, \alpha)$ and $y(t, \alpha)$ are (2) differentiable graphical solution (Fig. 2(d)) is biologically meaningful. In addition, the graphical solution is compatible with a crisp solution. In contrast, when $x(t, \alpha)$ and $y(t, \alpha)$ are differentiable as in (a), (b) and (c), the graphical solutions are incompatible with biological facts. $x(t)$ and $y(t)$ are the density of tumors and macrophages at time t respectively. Being compatible with biological facts means that the tumor is increasing or decreasing and the macrophages are increasing or decreasing. It does not give us information about tumors and macrophages when it is differentiated as in (a), (b) and (c), so these conditions are biologically meaningless.

Now, we will focus on the situation when $x(t, \alpha)$ and $y(t, \alpha)$ are (2) differentiable. When the crisp graphical solution and the fuzzy graphical solution $x(t, \alpha)$ and $y(t, \alpha)$ are (2) differentiable, we will plot their graphs on the same graph for $\alpha = 0$ and $\alpha = [0, 1]$. The fuzzy solution for $\alpha = 0$ and the crisp solution are given in Table 1 and Fig. 3.

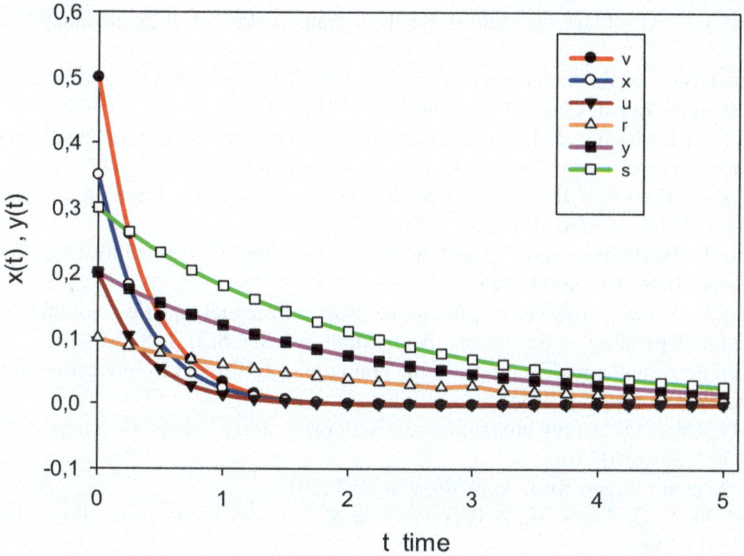

Fig. 3. Crisp solution and fuzzy solution for $\alpha = 0$

5 Conclusion

In this paper, we consider a model which has a predator-prey structure between the monoclonal tumor and the macrophages. Building upon the work of Bozkurt [14], we include fuzzy initial values to study the interaction of a monoclonal brain tumor and the macrophages to see the extinction conditions for the tumor population. From one hand and as a result of using fuzzy initial values, the uniqueness of the solution is lost. On the other hand, and using the strongly generalized derivative, biologically we obtain more realistic behavior that explains the interaction phenomena.

References

1. Zadeh, L.A.: Fuzzy sets. Inf. Control **8**, 338–353 (1965)
2. Kandel, A., Byatt, W.J.: Fuzzy differential equations. In: Proceedings of the International Conference on Cybernetics and Society, Tokyo, Japan, pp. 1213–1216 (1978)
3. Hüllerheimer, E.: An approach to modeling and simulation of uncertain dynamical systems. Int. J. Uncertainty Fuzziness Knowl.-Based Syst. **5**(2), 117–137 (1997)
4. Bede, B., Gal, S.: Generalizations of differentiability of fuzzy number valued functions with applications to fuzzy differential equation. Fuzzy Sets Syst. **151**, 581–599 (2005)
5. Bede, B., et al.: First order linear fuzzy differential equations under generalized differentiability. Inform. Sci. **177**, 1648–1662 (2007)
6. Khastan, A., Bahrami, F., Ivaz, K.: New results on multiple solutions for nth-order fuzzy differential equations under generalized differentiability, boundary value problems, Article ID 395714 (2009)
7. El-Gohary, A.: Chaos and optimal control of cancer self-remission and tumor system steady states. Chaos Solutions Fractals **37**, 1305–1316 (2008)

8. Holland, E.C.: Glioblastoma multiforme, the terminator. Proc. Nat. Acad. Sci. **97**,6242–6244 (2000)
9. Birkhuad, B.G., et al.: A mathematical model of the development of drug resistance to cancer chemotherapy. Eur. J. Cancer Clin. Oncol. **23**, 1421–1427 (1987)
10. Coldman, A.J., Goldiec, J.H.: A mathematical model for relating the drug sensitivity of tumors to their spontaneous mutation rate. Cancer Treat. Rep. **63**, 1727–1731 (1979)
11. Schmitz, E., Kansal, A.R., Torquato, S.: A cellular automaton of brain tumor treatment and resistance. J. Theor. Med. **4**(4), 223–239 (2002)
12. Bozkurt, F.: Modeling a tumor growth with piecewise constant arguments. Discret. Dyn. Nat. Soc. 1–8 (2013). Article ID 841764
13. Bozkurt, F.: Stability analysis of a fractional-order differential equation system of a GBM-IS interaction depending on the density. Appl. Math. Inf. Sci. **8**(3), 1–8 (2014)
14. Bozkurt, F.: A mathematical model of the brain tumor GBM and IS interaction. Int. J. Math. Comput. **22**(1), 58–65 (2014)
15. Akın, Ö., Oruç, Ö.: A prey predator model with fuzzy initial values. Hacettepe J. Math. Stat. **41**(3), 387–395 (2012)
16. Akın, Ö., et al.: Expert Syst. Appl. **40**, 953–957 (2013)
17. Ahmad, M.Z., De Baets, B.: A Prey-Predator Model with Fuzzy Initial Populations. IFSA EUSFLAT (2009)
18. Liu, X., Lou, Y.: Global dynamics of a predator-prey model. J. Math. Anal. Appl. **371**(1), 323–340 (2010)
19. Peixoto, M.S., Barros, L.C., Baasanezi, R.C.: Predator-prey model. Ecol. Model. **214**, 39–44 (2008)
20. Benli, F.B., Keskin, Ö.: A tumor-macrophage interaction model with fuzzy initial values. AIP Conf. Proc. **1648** (2015). Article ID 370007
21. Gal, S.G.: Approximation theory in fuzzy setting. In: Anastassiou, G.A. (ed.) Handbook of Analytic-Computational Methods in Applied Mathematics, pp. 617–666. Chapman & Hall/CRC Press, Boca Raton (2000)
22. Wu, C., Gong, Z.: On Henstock integral of fuzzy-number-valued functions (I). Fuzzy Sets Syst. **120**, 523–532 (2001)
23. Puri, M., Ralescu, D.: Differential and fuzzy functions. J. Math. Anal. Appl. **91**, 552–558 (1983)
24. Chalco-Cano, Y., Roman-Flores, H.: On the new solution of fuzzy differential equations. Chaos Solitons Fractals **38**, 112–119 (2006)
25. Kaleva, O.: Fuzzy differential equations. Fuzzy Sets Syst. **24**, 301–317 (1987)

Laguerre Matrix-Collocation Method to Solve Systems of Pantograph Type Delay Differential Equations

Burcu Gürbüz[1,2,3](✉) and Mehmet Sezer[4]

[1] Institute of Mathematics, Johannes Gutenberg-University Mainz, Mainz, Germany
burcu.gurbuz@uni-mainz.de, burcu.gurbuz@uskudar.edu.tr
[2] Department of Computer Engineering, Üsküdar University, İstanbul, Turkey
[3] Jean Leray Mathematics Lab, University of Nantes, Nantes, France
[4] Department of Mathematics, Manisa Celal Bayar University, Manisa, Turkey
mehmet.sezer@cbu.edu.tr

Abstract. In this study, an improved matrix method based on collocation points is developed to obtain the approximate solutions of systems of high-order pantograph type delay differential equations with variable coefficients. These kinds of systems described by the existence of linear functional argument play a critical role in defining many different phenomena and particularly, arise in industrial applications and in studies based on biology, economy, electrodynamics, physics and chemistry. The technique we have used reduces the mentioned delay system solution with the initial conditions to the solution of a matrix equation with the unknown Laguerre coefficients. Thereby, the approximate solution is obtained in terms of Laguerre polynomials. In addition, several examples along with error analysis are given to illustrate the efficiency of the method; the obtained results are scrutinized and interpreted.

Keywords: Laguerre polynomials and series · Matrix method · Pantograph equations · System of delay differential equations · Collocation method

1 Introduction

Mathematical models have an importance in many areas such as engineering, biology, physics, and social science. Especially in biology, we define the systems of high-order pantograph type delay differential equations to represent many biological phenomena. For instance, epidemiological models are the subject of study in biology. This is the study of disease dynamics which describe the mechanism of disease transmission. Various biological models can be explained by systems of high-order pantograph type delay differential equations. A delay model of predator-prey interaction is an example of these types of systems [1].

A specific example for the applications of these type of models can be given as the glucose-insulin regulatory system and ultradian insulin secretory oscillations which include two explicit time delays (Fig. 1).

H. Dutta et al. (Eds.): CMES 2019, AISC 1111, pp. 121–132, 2020.
https://doi.org/10.1007/978-3-030-39112-6_8

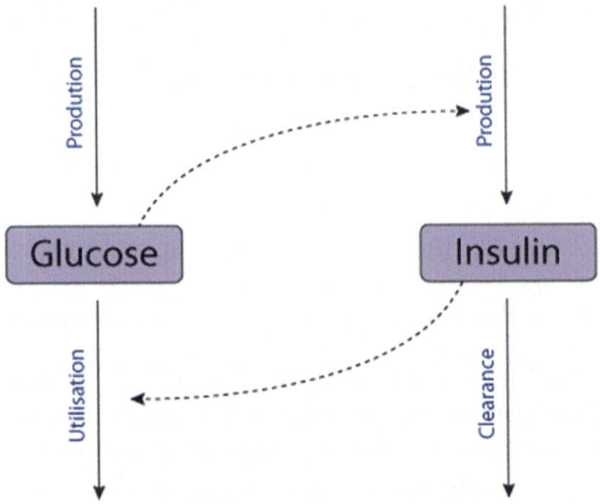

Fig. 1. Schematic diagram of the glucose-insulin regulatory system model [2].

The diagram illustrates the modeled schematic diagram of the glucose-insulin regulatory system. The lines with the dots indicate exalted glucose concentration which stimulates insulin synthesis excretion by the pancreatic beta cells. Moreover, insulin propagates utilisation in muscle, fat and other tissues [2–4].

These type of models are difficult to solve and often arises in the study of numerical approaches which have been investigated by many authors. So that, Runge-Kutta method, collocation methods, spectral method, Adomian decomposition method have been studied to obtain approximate solutions of system of high-order pantograph type delay differential equations with variable coefficients [5–12]. Also, the model's asymptotic behaviours, harmonic balance analysis, stability, and control have been reached over the last decades [13–15].

2 The Model

In this study, we consider system of high-order pantograph type delay differential equations which includes variable coefficients as

$$\sum_{r=0}^{m}\sum_{i=1}^{k}\sum_{s=0}^{S} P_{ji}^{r,s}(t)y_i^{(r)}(t-\beta_s) + Q_{ji}^{r}y_i^{(r)}(t) = g_j(t), \;\; j = 1, 2, \ldots, k, \;\; 0 \le t \le 1, \tag{1}$$

with the initial conditions are given as

$$\sum_{i=0}^{m-1}\left(a_{ji}^{n}y_n^{(i)}(0)\right) = \lambda_{ni}, \;\; j = 0, 1, \ldots, m-1, \;\; n = 1, 2, \ldots, k. \tag{2}$$

where a_{ji}^{n}, λ_{ni} and β_s are real constants. Also, $P_{ji}^{r,s}(t)$ and Q_{ji}^{r} are continuous functions defined in $0 \le t \le 1$. Our aim is to find approximate solution of the problem in (1)–(2)

in the truncated Laguerre series form

$$y_i(t) \cong y_{i,N}(t) = \sum_{n=0}^{N} a_{i,n} L_n(t), \quad i = 1, 2, \ldots, k, \ 0 \le t \le 1 \tag{3}$$

where a_n, $n = 0, 1, \ldots, N$ are unknown coefficients; and $L_n(t)$, $n = 0, 1, \ldots, N$, are the Laguerre polynomials which are defined as

$$L_n(t) = \sum_{r=0}^{n} \frac{(-1)^r}{r!} \binom{n}{r} t^r. \tag{4}$$

3 Fundamental Relations

In this section, we compose the matrix forms of $y_i(t)$ and their derivatives in the matrix forms:

$$[y_i(t)] = \mathbf{L}(t)\mathbf{A}_i = \mathbf{X}(t)\mathbf{H}\mathbf{A}_i, \quad i = 1, 2, \ldots, k \tag{5}$$

$$\begin{aligned}
\left[y_i^{(1)}(t)\right] &= \mathbf{L}^{(1)}(t)\mathbf{A}_i = \mathbf{X}(t)\mathbf{B}\mathbf{H}\mathbf{A}_i, \\
\left[y_i^{(2)}(t)\right] &= \mathbf{L}^{(2)}(t)\mathbf{A}_i = \mathbf{X}(t)\mathbf{B}^2\mathbf{H}\mathbf{A}_i, \\
&\vdots \\
\left[y_i^{(r)}(t)\right] &= \mathbf{L}^{(r)}(t)\mathbf{A}_i = \mathbf{X}(t)\mathbf{B}^r\mathbf{H}\mathbf{A}_i, \quad r = 1, 2, \ldots, m
\end{aligned} \tag{6}$$

where

$$\mathbf{L}(t) = \left[L_0(t) \ \ L_1(t) \ \ L_2(t) \ldots L_N(t) \right], \ \mathbf{X}(t) = [1 \ \ t \ \ldots \ t^N],$$

$$\mathbf{H} = \begin{bmatrix}
\frac{(-1)^0}{0!}\binom{0}{0} & \frac{(-1)^0}{0!}\binom{1}{0} & \frac{(-1)^0}{0!}\binom{2}{0} & \cdots & \frac{(-1)^0}{0!}\binom{N}{0} \\
0 & \frac{(-1)^1}{1!}\binom{1}{1} & \frac{(-1)^1}{1!}\binom{2}{1} & \cdots & \frac{(-1)^1}{1!}\binom{N}{1} \\
0 & 0 & \frac{(-1)^2}{2!}\binom{2}{2} & \cdots & \frac{(-1)^2}{2!}\binom{N}{2} \\
\vdots & \vdots & \vdots & \ddots & \vdots \\
0 & 0 & 0 & \cdots & \frac{(-1)^N}{N!}\binom{N}{N}
\end{bmatrix}, \ \mathbf{B} = \begin{bmatrix}
0 & 1 & 0 & \ldots & 0 \\
0 & 0 & 2 & \ldots & 0 \\
\vdots & \vdots & \vdots & \ddots & \vdots \\
0 & 0 & 0 & N & 0 \\
0 & 0 & 0 & 0 & 0
\end{bmatrix},$$

and $\mathbf{A}_i = [a_{i,0} \ a_{i,1} \ \cdots \ a_{i,N}]^T$, $i = 1, \ldots, k$. By replacing $t \to (t - \beta_s)$ in Eq. (6) then the matrix form is the following

$$\left[y_i^{(r)}(t - \beta_s)\right] = \mathbf{L}^{(r)}(t - \beta_s)\mathbf{A}_i = \mathbf{X}(t - \beta_s)\mathbf{B}^r\mathbf{H}\mathbf{A}_i, \quad r = 1, 2, \ldots, m, \ \ i = 1, 2, \ldots, k. \tag{7}$$

We have the relation between $\mathbf{X}(t-\beta_s)$ and $\mathbf{X}(t)$ as

$$\mathbf{X}(t-\beta_s)=\mathbf{X}(t)\mathbf{B}_{(-\beta_s)}. \tag{8}$$

where

$$\mathbf{B}(-\beta_s)=\begin{bmatrix} \binom{0}{0}(-\beta_s)^0 & \binom{1}{0}(-\beta_s)^1 & \binom{2}{0}(-\beta_s)^2 & \dots & \binom{N}{0}(-\beta_s)^N \\ 0 & \binom{1}{1}(-\beta_s)^0 & \binom{2}{1}(-\beta_s)^1 & \dots & \binom{N}{1}(-\beta_s)^{N-1} \\ 0 & 0 & \binom{2}{2}(-\beta_s)^0 & \dots & \binom{N}{2}(-\beta_s)^{N-2} \\ \vdots & \vdots & \vdots & \ddots & \vdots \\ 0 & 0 & 0 & \dots & \binom{N}{N}(-\beta_s)^0 \end{bmatrix}.$$

By substituting Eq. (8) into Eq. (7), we have

$$\left[y_i^{(r)}(t-\beta_s)\right]=\mathbf{X}(t)\mathbf{B}(-\beta_s)\mathbf{B}^r\mathbf{H}\mathbf{A}_i,\quad r=1,2,\dots,m,\quad i=1,2,\dots,k \tag{9}$$

Then, from (6), (8) and (9), the following matrix forms are found

$$\mathbf{y}^{(r)}(t)=\bar{\mathbf{X}}(t)\bar{\mathbf{B}}^r\bar{\mathbf{H}}\mathbf{A}_i,\quad r=1,2,\dots,m, \tag{10}$$

and

$$\mathbf{y}^{(r)}(t-\beta_s)=\bar{\mathbf{X}}(t)\bar{\mathbf{B}}(-\beta_s)\bar{\mathbf{B}}^r\bar{\mathbf{H}}\mathbf{A}_i,\quad r=1,2,\dots,m, \tag{11}$$

where

$$\mathbf{y}^{(r)}(t)=\begin{bmatrix} y_1^{(r)}(t) \\ y_2^{(r)}(t) \\ \vdots \\ y_k^{(r)}(t) \end{bmatrix},\ \mathbf{y}^{(r)}(t-\beta_s)=\begin{bmatrix} y_1^{(r)}(t-\beta_s) \\ y_2^{(r)}(t-\beta_s) \\ \vdots \\ y_k^{(r)}(t-\beta_s) \end{bmatrix},\ \bar{\mathbf{X}}(t)=\begin{bmatrix} \mathbf{X}(t) & 0 & \dots & 0 \\ 0 & \mathbf{X}(t) & \dots & 0 \\ \vdots & \vdots & \ddots & \vdots \\ 0 & 0 & \dots & \mathbf{X}(t) \end{bmatrix},$$

$$\bar{\mathbf{B}}(-\beta_s)=\begin{bmatrix} \mathbf{B}(-\beta_s) & 0 & \dots & 0 \\ 0 & \mathbf{B}(-\beta_s) & \dots & 0 \\ \vdots & \vdots & \ddots & \vdots \\ 0 & 0 & \dots & \mathbf{B}(-\beta_s) \end{bmatrix},\ \bar{\mathbf{B}}^r=\begin{bmatrix} \mathbf{B}^r & 0 & \dots & 0 \\ 0 & \mathbf{B}^r & \dots & 0 \\ \vdots & \vdots & \ddots & \vdots \\ 0 & 0 & \dots & \mathbf{B}^r \end{bmatrix},\ \bar{\mathbf{H}}=\begin{bmatrix} \mathbf{H} & 0 & \dots & 0 \\ 0 & \mathbf{H} & \dots & 0 \\ \vdots & \vdots & \ddots & \vdots \\ 0 & 0 & \dots & \mathbf{H} \end{bmatrix},$$

and $\mathbf{A}=[\mathbf{A}_1\ \mathbf{A}_2\ \dots\ \mathbf{A}_k]^T,\ i=1,2,\dots,k.$

4 Method of Solution

Now, we define the collocation points as

$$t_l=\frac{1}{N}l,\quad l=0,1,\dots,N. \tag{12}$$

First, the relations (10) and (11) are replaced into the Eq. (1) and we obtain the system

$$\sum_{r=0}^{m}\sum_{s=0}^{S}\mathbf{P}_{r,s}(t)\mathbf{y}^{(r)}(t-\beta_s)+\mathbf{Q}_r(t)\mathbf{y}^{(r)}(t)=\mathbf{g}(t), \tag{13}$$

where

$$\mathbf{P}_{r,s}(t)=\begin{bmatrix} P_{11}^{r,s}(t) & P_{12}^{r,s}(t) & \dots & P_{1k}^{r,s}(t)\\ P_{21}^{r,s}(t) & P_{22}^{r,s}(t) & \dots & P_{2k}^{r,s}(t)\\ \vdots & \vdots & \ddots & \vdots\\ P_{k1}^{r,s}(t) & P_{k2}^{r,s}(t) & \dots & P_{kk}^{r,s}(t)\end{bmatrix},\ \mathbf{Q}_r(t)=\begin{bmatrix} Q_{11}^{r}(t) & Q_{12}^{r}(t) & \dots & Q_{1k}^{r}(t)\\ Q_{21}^{r}(t) & Q_{22}^{r}(t) & \dots & Q_{2k}^{r}(t)\\ \vdots & \vdots & \ddots & \vdots\\ Q_{k1}^{r}(t) & Q_{k2}^{r}(t) & \dots & Q_{kk}^{r}(t)\end{bmatrix},$$

$$\mathbf{g}(t)=\begin{bmatrix} g_1(t)\\ g_2(t)\\ \vdots\\ g_k(t)\end{bmatrix}.$$

Then by using the collocation points (12) in the Eq. (13), we have

$$\sum_{r=0}^{m}\sum_{s=0}^{S}\mathbf{P}_{r,s}(t_l)\mathbf{y}^{(r)}(t_l-\beta_s)+\mathbf{Q}_r(t_l)\mathbf{y}^{(r)}(t_l)=\mathbf{g}(t_l), \tag{14}$$

or briefly

$$\sum_{r=0}^{m}\sum_{s=0}^{S}\mathbf{P}_{r,s}\bar{\mathbf{Y}}^{(r)}+\mathbf{Q}_r\mathbf{Y}^{(r)}=\mathbf{G}, \tag{15}$$

$$\mathbf{P}_{r,s}=\begin{bmatrix} \mathbf{P}_{r,s}(t_0) & 0 & \dots & 0\\ 0 & \mathbf{P}_{r,s}(t_1) & \dots & 0\\ \vdots & \vdots & \ddots & \vdots\\ 0 & 0 & \dots & \mathbf{P}_{r,s}(t_N)\end{bmatrix},\ \mathbf{Q}_r=\begin{bmatrix} \mathbf{Q}_r(t_0) & 0 & \dots & 0\\ 0 & \mathbf{Q}_r(t_1) & \dots & 0\\ \vdots & \vdots & \ddots & \vdots\\ 0 & 0 & \dots & \mathbf{Q}_r(t_N)\end{bmatrix},$$

$$\bar{\mathbf{Y}}^{(r)}=\begin{bmatrix} \mathbf{y}^{(r)}(t_0-\beta_s)\\ \mathbf{y}^{(r)}(t_1-\beta_s)\\ \vdots\\ \mathbf{y}^{(r)}(t_N-\beta_s)\end{bmatrix},\ \bar{\mathbf{Y}}^{(r)}=\begin{bmatrix} \mathbf{y}^{(r)}(t_0)\\ \mathbf{y}^{(r)}(t_1)\\ \vdots\\ \mathbf{y}^{(r)}(t_N)\end{bmatrix},\ \mathbf{G}=\begin{bmatrix} \mathbf{g}(t_0)\\ \mathbf{g}(t_1)\\ \vdots\\ \mathbf{g}(t_N)\end{bmatrix},\ \mathbf{X}=\begin{bmatrix} \bar{\mathbf{X}}(t_0)\\ \bar{\mathbf{X}}(t_1)\\ \vdots\\ \bar{\mathbf{X}}(t_N)\end{bmatrix}.$$

Then we obtain fundamental matrix equation by using (10), (11), and (15)

$$\left\{\sum_{r=0}^{m}\sum_{s=0}^{S}\mathbf{P}_{r,s}\mathbf{X}\bar{\mathbf{B}}_{(-\beta_s)}\bar{\mathbf{B}}^{r}\bar{\mathbf{H}}+\mathbf{Q}_r\mathbf{X}\bar{\mathbf{B}}^{r}\bar{\mathbf{H}}\right\}\mathbf{A}=\mathbf{G}, \tag{16}$$

We can write Eq. (16), briefly, in the following form:

$$\mathbf{W}\mathbf{A}=\mathbf{G}\Rightarrow[\mathbf{W};\mathbf{G}] \tag{17}$$

which corresponds to a system of the linear algebraic equations with the unknown Laguerre coefficients $a_{i,n}$, $i = 1, 2, \ldots, k$, $n = 0, 1, \ldots, N$.

Similarly, we consider the same procedure for the initial conditions and we obtain the matrix form:

$$\sum_{j=0}^{m-1} \left\{ \mathbf{a}_j \bar{\mathbf{X}}(0) \bar{\mathbf{B}}^j \bar{\mathbf{H}} \right\} \mathbf{A} = \boldsymbol{\lambda}, \tag{18}$$

where

$$\mathbf{a}_j = \begin{bmatrix} \mathbf{a}_j^1 & 0 & \ldots & 0 \\ 0 & \mathbf{a}_j^2 & \ldots & 0 \\ \vdots & \vdots & \ddots & \vdots \\ 0 & 0 & \ldots & \mathbf{a}_j^k \end{bmatrix}, \text{ and } \quad \begin{aligned} \boldsymbol{\lambda} &= \left[\lambda_1 \, \lambda_2 \ldots \lambda_k \right]^T, \\ \mathbf{a}_j^i &= \left[a_{0j}^i \, a_{1j}^i \ldots a_{m-1j}^i \right]^T. \end{aligned}$$

Summarily, we have the conditions, Eq. (18), in the form:

$$\mathbf{UA} = \boldsymbol{\lambda} \Rightarrow [\mathbf{U}; \boldsymbol{\lambda}]; \quad \mathbf{U} = \sum_{j=0}^{m-1} \left\{ \mathbf{a}_j \bar{\mathbf{X}}(0) \bar{\mathbf{B}}^j \bar{\mathbf{H}} \right\}. \tag{19}$$

By replacing the matrices (19) into last rows of the part $\mathbf{W}$ in Eq. (17), we have the new augmented matrix as $\left[\tilde{\mathbf{W}}; \tilde{\mathbf{G}} \right]$ [16]. By solving the system, therefore, from Eq. (3), approximate solution of the problem (1)–(2) is obtained in the following form:

$$y_i(t) \cong y_{i,N}(t) = \sum_{n=0}^{N} a_{i,n} L_n(t), \quad i = 1, 2, \ldots, k, \; 0 \le t \le 1.$$

5 Error Analysis

In this section, we check the accuracy of the present method. The approximate solutions $y_{i,N}(t)$ of Eq. (1), and their first derivatives are considered and substituted into Eq. (1). Then we obtain approximate results for $t = t_r \in [0, 1]$, $r = 0, 1, \ldots$.

$$E_{i,N}(t_r) = \left| \sum_{r=0}^{m} \sum_{i=1}^{k} \sum_{s=0}^{S} P_{ji}^{r,s}(t_r) y_i^{(r)}(t_r - \beta_s) + Q_{ji}^{r} y_i^{(r)}(t_r) - g_j(t_r) \right| \cong 0,$$

where $E_N(t_r) \le 10^{-k_{\alpha,\beta}} = 10^{-k}$($k$ is a positive integer) is prescribed, then the truncation limit N is increased until difference $E_N(t_r)$ becomes smaller than the prescribed 10^{-k} at each points.

6 Algorithm

Step 0. Input initial data: P, and Q.
Step 1. Set $m \leq N$ for $m \in \mathrm{N}$.
Step 2. Construct the matrices such as $\mathbf{P}_{r,s}(t)$, $\mathbf{Q}_r(t)$, $\bar{\mathbf{X}}(t)$, $\bar{\mathbf{B}}(-\beta_s)$, $\bar{\mathbf{B}}^r$, $\bar{\mathbf{H}}$, $\mathbf{g}(t)$.
Step 3. Replace in the fundamental equation.
Step 4. Apply the collocation points, $t_l = \frac{1}{N}l, \quad l = 0, 1, \ldots, N$ in S3.
Step 5. Compute $[\mathbf{W}; \mathbf{G}]$.
Step 6. Calculate initial condition matrices $[\mathbf{U}; \boldsymbol{\lambda}]$.
Step 7. Replace findings in S5 and get the new augmented matrix $\left[\tilde{\mathbf{W}}; \tilde{\mathbf{G}}\right]$.
Step 8. Solve the system in S7 where output: $y_{i,N}(t)$.
Step 9. Check the accuracy for the error function $E_N(t_r)$.
Step 10. If $E_N(t_r) \cong 0$. Else, then back S1.

7 Numerical Experiments

In this section, to show the accuracy and efficiency of the presented method, for the problem given at (1)–(2), is solved with it. Numerical calculations were performed using Maple software.

Example 1. Firstly, we deal with the system of first-order delay differential equations which is defined in Eq. (1) [7]

$$\begin{aligned} y_1'(t-1) + y_2'(t-1) &= 2t, \\ y_1'(t-1) - y_3'(t-1) &= 2t - 1, \\ y_1'(t-1) + y_3(t-1) &= t - 1, \quad 0 \leq t \leq 1, \end{aligned} \tag{20}$$

and initial conditions are given as

$$y_1(0) = 0, \; y_2(0) = 0, \text{ and } y_3(0) = 0. \tag{21}$$

where

$$\mathbf{P}_{0,1}(t) = \begin{bmatrix} 0\,0\,0 \\ 0\,0\,0 \\ 0\,0\,1 \end{bmatrix}, \; \mathbf{P}_{0,1}(t) = \begin{bmatrix} 1 & 1 & 0 \\ 1 & 0 & -1 \\ 1 & 0 & 1 \end{bmatrix}, \; \mathbf{g}(t) = \begin{bmatrix} 2t \\ 2t-1 \\ t-1 \end{bmatrix}$$

and the collocation points are $t_0 = 0$, $t_1 = 1/3$, $t_2 = 2/3$, and $t_3 = 1$. Then the fundamental matrix equation of the problem is

$$\underbrace{\left\{\mathbf{P}_{0,1}\mathbf{X}\bar{\mathbf{B}}_{(-1)}\bar{\mathbf{H}} + \mathbf{P}_{1,0}\mathbf{X}\bar{\mathbf{B}}_{(-1)}\bar{\mathbf{B}}\bar{\mathbf{H}} + \mathbf{P}_{2,0}\mathbf{X}\bar{\mathbf{B}}_{(-1)}\bar{\mathbf{B}}^2\bar{\mathbf{H}}\right\}}_{\mathbf{W}} \mathbf{A} = \mathbf{G},$$

where we get the augmented matrix $[\mathbf{W}; \mathbf{G}]$. Also, we consider the initial conditions given in Eq. (21) by the matrix form of Eq. (19). Then the new augmented matrix

$\left[\tilde{\mathbf{W}}; \tilde{\mathbf{G}}\right]$ is described by substituting the conditions. Finally, we have the solution of the system which gives us the exact solution of the system for $N = 3$ as:

$$y_1(t) = t^2, \; y_2(t) = 2t, \text{ and } y_3(t) = -t.$$

Example 2. We consider the following linear system of second-order retarded and advanced differential equations [10]

$$y_1''\left(t - \frac{1}{2}\right) + 2ty_1'\left(t + \frac{1}{3}\right) - ty_2\left(t - \frac{1}{2}\right) + t^2 y_3(t - 1) = g_1(t),$$

$$y_2''\left(t - \frac{1}{4}\right) - ty_1'\left(t + \frac{1}{5}\right) - ty_2\left(t - \frac{1}{6}\right) + 5y_3\left(t - \frac{1}{2}\right) = g_2(t),$$

$$y_3''\left(t + \frac{1}{3}\right) - ty_1'\left(t - \frac{1}{6}\right) + ty_3'\left(t - \frac{1}{3}\right) + 3y_1\left(t + \frac{1}{4}\right) + 2ty_2\left(t + \frac{1}{3}\right) = g_3(t), \quad 0 \le t \le 1,$$

with the initial conditions

$$y_1(0) = 0, \; y_1'(0) = 1, \; y_2(0) = 1, \; y_2'(0) = 0, \; y_3(0) = 1, \text{ and } y_3'(0) = 1.$$

where

$$\begin{cases} g_1(t) = -\sin\left(t - \frac{1}{2}\right) + 2t\cos\left(t - \frac{1}{3}\right) - t\cos\left(t - \frac{1}{2}\right) + t^2 e^{t-1}, \\ g_2(t) = -\cos\left(t - \frac{1}{4}\right) - t\cos\left(t + \frac{1}{5}\right) - t\cos\left(t - \frac{1}{6}\right) + 5e^{t-1/2}, \\ g_3(t) = e^{t+1/3} - t\cos\left(t - \frac{1}{6}\right) + te^{t-1/3} + 3\sin\left(t + \frac{1}{4}\right) + 2t\cos\left(t + \frac{1}{3}\right). \end{cases}$$

The exact solution of the system is $y_1(t) = \sin t$, $y_2(t) = \cos t$, and $y_3(t) = e^t$ (Figs. 2, 3, 4) and (Tables 1, 2, 3).

Table 1. Absolute errors for $N = 8$ and 10 comparison for $y_1(t)$, in Example 2.

t	$E_N = E_8$	$E_N = E_{10}$
0.0	0.00000	0.00000
0.1	0.1569E−7	0.1003E−8
0.2	0.2560E−7	0.2021E−8
0.3	0.3248E−7	0.1032E−8
0.4	0.1039E−7	0.9820E−8
0.5	0.3591E−7	0.3097E−8
0.6	0.3719E−7	0.2065E−8
0.7	0.2052E−6	0.1082E−7
0.8	0.5066E−6	0.5194E−7
0.9	0.1042E−5	0.1112E−6
1.0	0.3702E−5	0.3003E−6

Table 2. Absolute errors for $N = 8$ and 10 comparison for $y_2(t)$, in Example 2.

t	$E_N = E_8$	$E_N = E_{10}$
0.0	0.00000	0.00000
0.1	0.2981E−7	0.1973E−8
0.2	0.6330E−7	0.2371E−8
0.3	0.8120E−7	0.9873E−8
0.4	0.9001E−7	0.3205E−8
0.5	0.5361E−7	0.7903E−8
0.6	0.9137E−6	0.5608E−8
0.7	0.5209E−6	0.8379E−7
0.8	0.6650E−5	0.9970E−7
0.9	0.2401E−5	0.7260E−6
1.0	0.2703E−5	0.7780E−6

Table 3. Absolute errors for $N = 8$ and 10 comparison for $y_3(t)$, in Example 2.

t	$E_N = E_8$	$E_N = E_{10}$
0.0	0.00000	0.00000
0.1	0.2981E−7	0.1973E−8
0.2	0.6330E−7	0.2371E−8
0.3	0.8120E−7	0.9873E−8
0.4	0.9001E−7	0.3205E−8
0.5	0.5361E−7	0.7903E−8
0.6	0.9137E−6	0.5608E−8
0.7	0.5209E−6	0.8379E−7
0.8	0.6650E−5	0.9970E−7
0.9	0.2401E−5	0.7260E−6
1.0	0.2703E−5	0.7780E−6

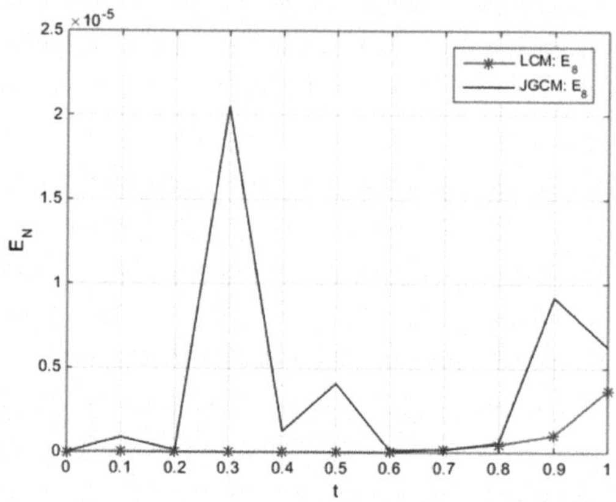

Fig. 2. Comparison of the error function for Laguerre collocation method (LCM), Jacobi Gauss collocation method (JGCM) for $y_1(t)$, $N = 8$ in Example 2.

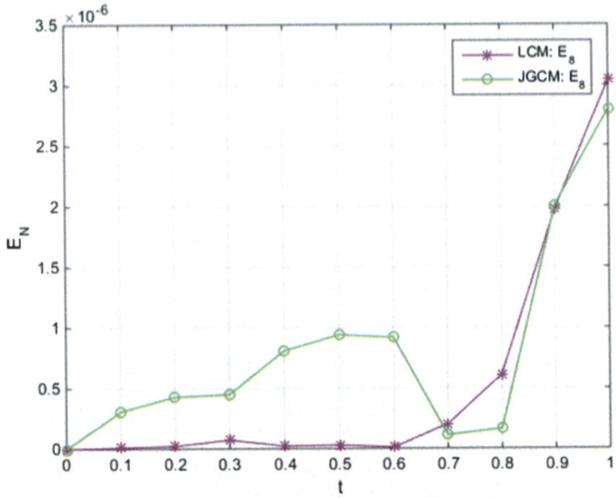

Fig. 3. Comparison of the error function for Laguerre collocation method (LCM), Jacobi Gauss collocation method (JGCM) for $y_2(t)$, $N = 8$ in Example 2.

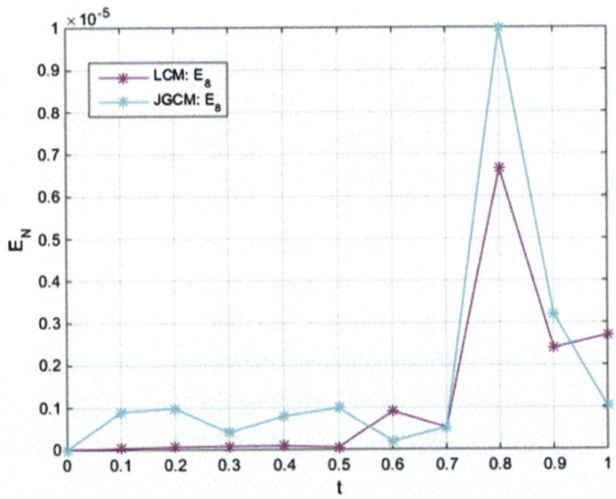

Fig. 4. Comparison of the error function for Laguerre collocation method (LCM), Jacobi Gauss collocation method (JGCM) for $y_3(t)$, $N = 8$ in Example 2.

Example 3. Let us consider a homogenous system in three dependent variables

$$y_1'(t) = \tfrac{2}{t}y_1(t) + \left(1 - \tfrac{2}{t}\right)y_2(t) - y_3(t),$$
$$y_2'(t) = \tfrac{1}{t}y_1(t) + \left(1 - \tfrac{1}{t}\right)y_2(t) - y_3(t),$$
$$y_3'(t) = \tfrac{1}{t}y_1(t) - \tfrac{1}{t}y_2(t), \quad 0 \le t \le 1,$$

with the initial conditions [8, 9]

$$y_1(1) = 0, \ y_2(1) = 1, \text{ and } y_3(1) = 1.$$

Applying the introduced technique for $N = 4$, then we have the following solutions

$$y_1(t) = 2 - 2t,\ y_2(t) = 2 - t,\ \text{and}\ y_3(t) = 2 - t,$$

which give us the exact solutions of the problem.

8 Conclusion

In this study, we introduce a matrix method depending on Laguerre polynomials in order to solve systems of high-order pantograph type delay differential equations with variable coefficients numerically. Furthermore, the error analysis is given to show the accuracy of the method. The present method and its error analysis are applied on illustrative examples which have been shown by figures and tables. The method has significant importance such as; the present method has a short and concise computing procedure by writing the algorithm in Maple18. The present method has sufficient results when N is chosen large enough. The method also can be extended to different types of mathematical models with respect to some modifications [17].

References

1. Murray, J.D.: Mathematical Biology. Interdisciplinary Applied Mathematics (2003)
2. Li, J., Kuang, Y., Mason, C.C.: Modeling the glucose-insulin regulatory system and ultradian insulin secretory oscillations with two explicit time delays. J. Theor. Biol. **242**(3), 722–735 (2006)
3. Sturis, J., Polonsky, K.S., Mosekilde, E., Van Cauter, E.: Computer model for mechanisms underlying ultradian oscillations of insulin and glucose. Am. J. Physiol. Endocrinol. Metab. **260**(5), E801–E809 (1991)
4. Tolić, I.M., Mosekilde, E., Sturis, J.: Modeling the insulin-glucose feedback system: the significance of pulsatile insulin secretion. J. Theor. Biol. **207**(3), 361–375 (2000)
5. Abdel-Halim Hassan, I.H.: Application to differential transformation method for solving systems of differential equations. Appl. Math. Model. **32**(12), 2552–2559 (2008)
6. Yüzbaşı, Ş.: On the solutions of a system of linear retarded and advanced differential equations by the Bessel collocation approximation. Comput. Math Appl. **63**(10), 1442–1455 (2012)
7. Gökmen, E., Sezer, M.: Taylor collocation method for systems of high-order linear differential-difference equations with variable coefficients. Ain Shams Eng. J. **4**, 117–125 (2013)
8. Akyüz, A., Sezer, M.: Chebyshev polynomial solutions of systems of high-order linear differential equations with variable coefficients. Appl. Math. Comput. **144**(2–3), 237–247 (2003)
9. Oğuz, C., Sezer, M., Oğuz, A.D.: Chelyshkov collocation approach to solve the systems of linear functional differential equations. NTMSCI **3**(4), 83–97 (2015)
10. Bhrawy, A.H., Doha, E.H., Baleanu, D., Hafez, R.M.: A highly accurate Jacobi collocation algorithm for systems of high-order linear differential-difference equations with mixed initial conditions. Math. Methods Appl. Sci. **38**(14), 3022–3032 (2015)
11. Ali, I., Brunner, H., Tang, T.: A spectral method for pantograph-type delay differential equations and its convergence analysis. J. Comput. Math. **27**, 254–265 (2009)
12. Biazar, J., Babolian, E., Islam, R.: Solution of the system of ordinary differential equations by Adomian decomposition method. Appl. Math. Comput. **147**(3), 713–719 (2004)

13. Bennett, D.L., Gourley, S.A.: Asymptotic properties of a delay differential equation model for the interaction of glucose with plasma and interstitial insülin. Appl. Math. Comput. **151**(1), 189–207 (2004)
14. Liu, L., Kalmár-Nagy, T.: High-dimensional harmonic balance analysis for second-order delay-differential equations. J. Vib. Control **16**(7–8), 1189–1208 (2010)
15. Gu, K., Niculescu, S.-I.: Survey on recent results in the stability and control of time-delay systems. J. Dyn. Syst. Meas. Contr. **125**(2), 158–165 (2003)
16. Gülsu, M., Gürbüz, B., Öztürk, Y., Sezer, M.: Laguerre polynomial approach for solving linear delay difference equations. Appl. Math. Comput. **217**(15), 6765–6776 (2011)
17. Biazar, J.: Solution of the epidemic model by Adomian decomposition method. Appl. Math. Comput. **173**, 1101–1106 (2006)

Solving the Fuzzy Fractional Differential Wave Equation by Mean Fuzzy Fourier Transform

S. Melliani(✉), M. Elomari, and L. S. Chadli

Sultan Moulay Slimane University, 523 Beni Mellal, Morocco
{s.melliani,m.elomari}@usms.ma, sa.chadli@yahoo.fr

Abstract. In this work, we will study the integral solution of wave fractional differential equations with fuzzy initial data by using Fourier transform, the exact solution is given in the case of $\gamma = 2$. Some examples are presented to illustrate the results.

1 Introduction

The present paper investigates the analytic solution of the following problem

$$\begin{cases} {}_{gH}D_t^{\gamma} u(t,x) -_g c^2 \frac{\partial^2}{\partial x^2} u(t,x) = 0, -\infty < x < \infty,\ t \geq 0,\ 1 < \gamma < 2 \\ u(0,x) = a(x) \\ \frac{\partial}{\partial t} u(0,x) = b(x) \end{cases}$$

where a and b are two absolutely valued-functions in E^1. $-g$ is the generalized Hukuhara difference. ${}_{gH}D$ is the generalized fuzzy fractional caputo's derivative.

In 1965 Zadeh [13] introduced the basic ideas of the fuzzy set theory, as an extension of the classical notion of set. The authors in [6] give a generalization of the Hukuhara difference which guaranteed the existence of this is for two segments in $\mathbb{R}$. As consequence in the same work Bede and Stefanini presented the generalized derivative of a set valued-functions. Agarwal et al. [1] are the pioneers working in fuzzy fractional (DEs). They formulated the Riemann-Liouville differentiability notion as the base to define the concept of fuzzy fractional DEs. After that, they proved the existence of solutions of fuzzy fractional integral equations (IEs) under compactness type conditions using the Hausdorff measure of noncompactness in the paper [2]. Allahviranloo et al. in [3] presented two new results on the existence of two kinds of $gH-$weak solutions of these problems and indicated the boundedness and continuous dependence of solutions on the initial data of the problems. In [5] the authors prove the existence and uniqueness theorems for non-linear fuzzy fractional Fredholm integro-differential equations under fractional generalized Hukuhara derivatives in the Caputo sense. From the idea of [5] we will try to prove the existence and uniqueness of fuzzy fractional wave equation.

This paper is organized as follows. In Sect. 2 we recall some concepts concerning the fuzzy metric space. The generalized derivative take place in the Sect. 3. In Sect. 4 we give the concept of fuzzy Fourier transform and we presented some properties. We presented the solution of the fuzzy wave equation in Sect. 5. Finally in Sect. 6 two examples are given to illustrate the usefulness of our main results.

H. Dutta et al. (Eds.): CMES 2019, AISC 1111, pp. 133–147, 2020.
https://doi.org/10.1007/978-3-030-39112-6_9

2 Preliminaries

In this section, we present some definitions and introduce the necessary notation, which will be used throughout the paper.

We denote E^1 the class of function defined as follows:

$$E^1 = \Big\{u : \mathbb{R} \to [0,1], \quad u \quad \text{satisfies } (1-4) \text{ below}\Big\}$$

1. u is normal, i.e. there is a $x_0 \in \mathbb{R}$ such that $u(x_0) = 1$;
2. u is a fuzzy convex set;
3. u is upper semi-continuous;
4. u closure of $\{x \in \mathbb{R}^n, \quad u(x) > 0\}$ is compact

For all $\alpha \in (0,1]$ the α-cut of an element of E^1 is defined by

$$u^\alpha = \Big\{x \in \mathbb{R},\ u(x) \geq \alpha\Big\}$$

By the previous properties we can write

$$u^\alpha = [\underline{u}(\alpha), \overline{u}(\alpha)]$$

By the extension principal of Zadeh we have

$$\begin{aligned}(u+v)^\alpha &= u^\alpha + v^\alpha;\\ (\lambda u)^\alpha &= \lambda u^\alpha\end{aligned}$$

For all $u, v \in E^1$ and $\lambda \in \mathbb{R}$

The distance between two element of E^1 is given by (see [4])

$$d(u,v) = \sup_{\alpha \in (0,1]} \max\Big\{|\overline{u}(\alpha) - \overline{v}(\alpha)|, |\underline{u}(\alpha) - \underline{v}(\alpha)|\Big\}$$

The metric space (E^1, d) is complete, separable and locally compact and the following properties for metric d are valid:

1. $d(u+v, u+w) = d(u,v)$;
2. $d(\lambda u, \lambda v) = |\lambda| d(u,v)$;
3. $d(u+v, w+z) \leq d(u,w) + d(v,z)$;

Remark 2.1. *The space (E^1, d) is a linear normed space with $\|u\| = d(u,0)$.*

Definition 2.2. *[10] A complex fuzzy number is a mapping $z : \mathbb{C} \to [0,1]$ with the following properties:*

1. *z is continuous;*
2. *z^α, $\alpha \in (0,1]$ is open, bounded, connected and simply connected;*
3. *z^1 is non-empty, compact, arcwise connected and simply connected.*

We denote the set of all fuzzy complex number by $\mathbb{C}^1$.

Definition 2.3. *[6] The generalized Hukuhara difference of two fuzzy numbers $u, v \in E^1$ is defined as follows*

$$u -_g v = w \Leftrightarrow \begin{cases} u = v + w \\ or \quad v = u + (-1)w \end{cases}$$

In terms of α-levels we have

$$\left(u -_g v\right)^\alpha = \left[\min\{\underline{u}(\alpha) - \underline{v}(\alpha), \overline{u}(\alpha) - \overline{v}(\alpha)\}, \max\{\underline{u}(\alpha) - \underline{v}(\alpha), \overline{u}(\alpha) - \overline{v}(\alpha)\}\right]$$

and the conditions for the existence of $w = u -_g v \in E^1$ are

$$\text{case (i)} \quad \begin{cases} \underline{w}(\alpha) = \underline{u}(\alpha) - \underline{v}(\alpha) \text{ and } \overline{w}(\alpha) = \overline{u}(\alpha) - \overline{v}(\alpha) \\ \text{with} \quad \underline{w}(\alpha) \text{ increasing, } \overline{w}(\alpha) \text{ decreasing, } \underline{w}(\alpha) \leq \overline{w}(\alpha) \end{cases} \quad (1)$$

$$\text{case (ii)} \quad \begin{cases} \underline{w}(\alpha) = \overline{u}(\alpha) - \overline{v}(\alpha) \text{ and } \overline{w}(\alpha) = \underline{u}(\alpha) - \underline{v}(\alpha) \\ \text{with} \quad \underline{w}(\alpha) \text{ increasing, } \overline{w}(\alpha) \text{ decreasing, } \underline{w}(\alpha) \leq \overline{w}(\alpha) \end{cases} \quad (2)$$

for all $\alpha \in [0, 1]$.

Throughout the rest of this paper, we assume that $u -_g v \in E^1$

Proposition 2.4. *[11]*

$$\|u -_g v\| = d(u, v)$$

Since $\|.\|$ is a norm on E^n and by the Proposition (2.4) we have

Proposition 2.5.

$$\|\lambda u -_g \mu u\| = |\lambda - \mu| \|u\|$$

Let $f : [a, b] \subset \mathbb{R} \to E^1$ a fuzzy-valued function. The α-level of f is given by

$$f(x, \alpha) = \left[\underline{f}(x, \alpha), \overline{f}(x, \alpha)\right], \ \forall x \in [a, b], \ \forall \alpha \in [0, 1].$$

Definition 2.6. *[6] Let $x_0 \in (a, b)$ and h be such that $x_0 + h \in (a, b)$, then the generalized Hukuhara derivative of a fuzzy value function $f : (a, b) \to E^1$ at x_0 is defined as*

$$\lim_{h \to 0} \left\| \frac{f(x_0 + h) -_g f(x_0)}{h} -_g f'_{gH}(x_0) \right\| = 0 \quad (3)$$

If $f_{gH}(x_0) \in E^1$ satisfying 3 exists, we say that f is generalized Hukuhara differentiable (gH-differentiable for short) at x_0.

Definition 2.7. *[6] Let $f : [a,b] \to E^1$ and $x_0 \in (a,b)$, with $\underline{f}(x,\alpha)$ and $\overline{f}(x,\alpha)$ both differentiable at x_0.*
We say that

1. *f is $[(i)-gH]$-differentiable at x_0 if*

$$f'_{i,gH}(x_0) = \left[\underline{f}'(x,\alpha), \overline{f}'(x,\alpha)\right] \tag{4}$$

2. *f is $[(ii)-gH]$-differentiable at x_0 if*

$$f'_{ii,gH}(x_0) = \left[\overline{f}'(x,\alpha), \underline{f}'(x,\alpha)\right] \tag{5}$$

Theorem 2.8. *Let $f : J \subset \mathbb{R} \to E^1$ and $g : J \to \mathbb{R}$ and $x \in J$. Suppose that $g(x)$ is differentiable function at x and the fuzzy-valued function $f(x)$ is gH-differentiable at x. So*

$$(fg)'_{gH} = (f'g)_{gH} + (fg')_{gH}$$

Proof. Using (2.5), for h enough small we get

$$\left\|\frac{f(x+h)g(x+h) -_g f(x)g(x)}{h} -_g ((f'(x)g(x))_{gH} + (f(x)g'(x))_{gH})\right\|$$

$$= \left\|\frac{f(x+h)g(x+h) -_g f(x)g(x+h) + f(x)g(x+h) -_g f(x)g(x)}{h} -_g ((f'(x)g(x))_{gH} + (f(x)g'(x))_{gH})\right\|$$

$$= \left\|\frac{(f(x+h) -_g f(x))g(x+h) + f(x)(g(x+h) -_g g(x))}{h} -_g ((f'(x)g(x))_{gH} + (f(x)g'(x))_{gH})\right\|$$

$$\leq \left\|\frac{(f(x+h) -_g f(x))g(x+h)}{h} -_g ((f'(x)g(x))_{gH})\right\| + \left\|\frac{(f(x)(g(x+h) -_g g(x))}{h} -_g ((f(x)g'(x))_{gH})\right\|$$

$$\leq \left\|\frac{(f(x+h) -_g f(x))}{h} g(x+h) -_g ((f'(x)g(x))_{gH})\right\| + \left\|f(x)\frac{((g(x+h) -_g g(x))}{h} -_g ((f(x)g'(x))_{gH})\right\|$$

which complete the proof by passing to limit.

Definition 2.9. *[6] We say that a point $x_0 \in (a,b)$, is a switching point for the differentiability of f, if in any neighborhood V of x_0 there exist points $x_1 < x_0 < x_2$ such that*

1. *type (1). at x_1 (4) holds while (5) does not hold and at x_2 (5) holds and (4) does not hold, or*
2. *type (2). at x_1 (5) holds while (4) does not hold and at x_2 (4) holds and (5) does not hold.*

Definition 2.10. *[3] Let $f : (a,b) \to E^1$. We say that $f(x)$ is gH-differentiable of the 2nd-order at x_0 whenever the function $f(x)$ is gH-differentiable of the order $i, i = 0,1$, at x_0, $((f(x_0))(i)_{gH} \in E^1)$, moreover there isn't any switching point on (a,b). Then there exists $(f)''_{gH}(x_0) \in E^1$ such that*

$$\lim_{h\to 0}\left\|\frac{f'(x_0+h) -_g f'(x_0)}{h} -_g f''_{gH}(x_0)\right\| = 0$$

Definition 2.11. *[3] Let $f : [a,b] \to E^1$ and $f'_g H(x)$ be gH-differentiable at $x_0 \in (a,b)$, moreover there isn't any switching point on (a,b) and $\underline{f}(x,\alpha)$ and $\overline{f}(x,\alpha)$ both differentiable at x_0. We say that*

- *f' is $[(i) - gH]$-differentiable at x_0 if*

$$f''_{i,gH}(x_0) = \left[\underline{f}''(x,\alpha), \overline{f}''(x,\alpha)\right]$$

- *f' is $[(ii) - gH]$-differentiable at x_0 if*

$$f''_{ii,gH}(x_0) = \left[\overline{f}''(x,\alpha), \underline{f}''(x,\alpha)\right]$$

Definition 2.12. *[8] Let $f : [a,b] \to E^1$. We say that $f(x)$ is fuzzy Riemann integrable to $I \in E^1$ if for any $\varepsilon > 0$, there exists $\delta > 0$ such that for any division $P = \{[u,v];\xi\}$ with the norms $\Delta(P) < \delta$, we have*

$$d\left(\sum_{p}^{*}(v-u)f(\xi), I\right) < \varepsilon$$

where $\sum_p^*$ denotes the fuzzy summation. We choose to write $I = \int_a^b f(x)dx$.

Theorem 2.13. *[6] If f is gH-differentiable with no switching point in the interval $[a,b]$ then we have*

$$\int_a^b f(t)dt = f(b) -_g f(a)$$

Theorem 2.14. *[12] Let $f(x)$ be a fuzzy-valued function on $(-\infty,\infty)$ and it is represented by $f(x,\alpha) = \left[\underline{f}(x,\alpha), \overline{f}(x,\alpha)\right]$ for any fixed $\alpha \in [0,1]$. Assume that $|\underline{f}(x,\alpha)|$ and $|\overline{f}(x,\alpha)|$ are Riemann integrable on $(-\infty,\infty)$ for all $\alpha \in [0,1]$. Then $f(x)$ is improper fuzzy Riemann-integrable on $(-\infty,\infty)$ and the improper fuzzy Riemann integral is a fuzzy number. Furthermore, we have*

$$\int_{-\infty}^{\infty} f(x)dx = \left[\int_{-\infty}^{\infty} \underline{f}(x,\alpha)dx, \int_{-\infty}^{\infty} \overline{f}(x,\alpha)dx\right]$$

From this theorem we can discuss the Fuzzy Riemann's improper integral

Lemma 2.15. *Let $f : \mathbb{R} \times \mathbb{R}^+ \to E^1$, given by $f(x,t;\alpha) = [\underline{f}(x,t;\alpha), \overline{f}(x,t;\alpha)]$, and let $a \in \mathbb{R}^+$*
If $\int_a^\infty \underline{f}(x,t;\alpha)dt$ and $\int_a^\infty \overline{f}(x,t;\alpha)dt$ are converges then

$$\int_a^\infty f(x,t;\alpha)dt \in E^1$$

Proof. Just use the conditions (1).

Theorem 2.16. *Let $f : \mathbb{R} \times \mathbb{R}^+ \to E^1$ be fuzzy-valued function such that $f(x,t;\alpha) = [\underline{f}(x,t;\alpha), \overline{f}(x,t;\alpha)]$. Suppose that for each $x \in [a,\infty)$, the fuzzy integral $\int_c^\infty f(x,t)dt$ is convergent and moreover $\int_a^\infty f(x,t)dx$ as a function of t is convergent on $[c,\infty)$. Then*

$$\int_c^\infty \int_a^\infty f(x,t)dxdt = \int_a^\infty \int_a^\infty f(x,t)dtdx$$

Proof. Applying the theorem of Fubini-Tonelli [7] to these two functions $\underline{f}(x,t;\alpha)$ and $\overline{f}(x,t;\alpha)$, and use the conditions (1)

Theorem 2.17. *Suppose both, $f(x,t)$ and $\partial_{x_{gH}} f(x,t)$, are fuzzy continuous in $[a,b] \times [c,\infty)$. Suppose also that the integral converges for $x \in \mathbb{R}$, and the integral $\int_c^\infty f(x,t)dt$ converges uniformly on $[a,b]$. Then F is gH-differentiable on $[a,b]$ and*

$$F'_{gH}(x) = \int_c^\infty \partial_{x_{gH}} f(x,t)dt$$

Proof. The continuity of $\partial_{x_{gH}} f(x,t)$ on $[a,b]$ by the convergence domain theorem of to $\underline{f}(x,t;\alpha)$ and $\overline{f}(x,t;\alpha)$ and use the condition (1).

According to the Theorem (2.8) we get

Theorem 2.18. *Let $f : [a,b] \to E^1$ and $g : [a,b] \to \mathbb{R}$ are two differentiable functions (f is gH-differentiable), then*

$$\int_a^b f'_{gH}(x)g(x)dx = f(b)g(b) -_g f(a)g(a) -_g \int_a^b f(x)g'(x)dx$$

Remark 2.19. *If $f,g \in A^{E^1}$ with $\lim\limits_{|x|\to\infty} f(x) = 0$, $\lim_{|x|\to\infty} g(x) = 0$ then*

$$\int_{-\infty}^\infty f'_{gH}(x)g(x)dx = \int_{-\infty}^\infty f(x)g'(x)dx$$

3 Fuzzy Generalized Hukuhara Partial Differentiation

In this section $f : \mathbb{D} \subset \mathbb{R} \times \mathbb{R}^+ \to E^1$ is called the two variable fuzzy-valued function. The parametric representation of the fuzzy-valued function fis expressed by $f(x,t,\alpha) = \left[\underline{f}(x,t,\alpha), \overline{f}(x,t,\alpha)\right]$

Definition 3.1. *[3] Let $f : \mathbb{D} \subset \mathbb{R} \times \mathbb{R}^+ \to E^1$ and $(x_0,t_0) \in \mathbb{D}$. Then first generalized Hukuhara partial derivative ($[gH-p]$-derivative for short) of f with respect to variables x,t are the functions $\partial_{x_{gH}} f(x_0,t_0)$ and $\partial_{t_{gH}} f(x_0,t_0)$ given by*

$$\lim_{h\to 0} \left\| \frac{f(x_0+h,t_0) -_g f(x_0,t_0)}{h} -_g \partial_{x_{gH}} f(x_0,t_0) \right\| = 0$$

and

$$\lim_{h\to 0} \left\| \frac{f(x_0,t_0+h) -_g f(x_0,t_0)}{h}, \partial_{x_{gH}} f(x_0,t_0) \right\| = 0$$

provided that $\partial_{x_{gH}} f(x_0,t_0), \partial_{t_{gH}} f(x_0,t_0) \in E^1$.

Definition 3.2. *[3] Let $f(x,t) : \mathbb{D} \to E^1$, $(x_0,t_0) \in \mathbb{D}$ and $\underline{f}(x,t;\alpha)$ and $\overline{f}(x,t;\alpha)$ both partial differentiable w.r.t. t at (x_0,t_0). We say that*

- *$f(x,t)$ is $[(i)-p]$-differentiable w.r.t. t at (x_0,t_0) if*

$$\partial_{t_{i,gH}} f(x_0,t_0) = \left[\partial_t \underline{f}(x_0,t_0;\alpha), \partial_t \overline{f}(x_0,t_0;\alpha)\right] \tag{1}$$

$$\partial_{t_{ii,gH}} f(x_0,t_0) = \left[\partial_t \overline{f}(x_0,t_0;\alpha), \partial_t \underline{f}(x_0,t_0;\alpha)\right] \tag{2}$$

We inspired of the Definition (2.10) we presented the following definition

Definition 3.3. *$f : \mathbb{R} \times \mathbb{R}^+ R \to E^1$. We say that the function $t = h(x)$, is switching boundary for the differentiability of $f(x,t)$ with respect to t, if for all x belongs to domain of $h(x)$ and for all $t \in \mathbb{R}^+$, there exist points $t_0 < t_1 < t_2$ such that*

1. *at (x,t_1) (1) holds while (2) does not hold and at (x,t_2) (2) holds and (1) does not hold, or*
2. *at (x,t_1) (2) holds while (1) does not hold and at (x,t_2) (1) holds and (2) does not hold.*

Theorem 3.4. *Consider $f : \mathbb{R} \times \mathbb{R}^+ \to E^1$ and $u : \mathbb{R} \to E^1$ are fuzzy-valued functions such that $u(x;\alpha) = [\underline{u}(x;\alpha), \overline{u}(x;\alpha)]$. Suppose that $h : \mathbb{R} \to \mathbb{R}$ and $p : \mathbb{R} \times \mathbb{R}^+ \to \mathbb{R}^+$ is a differentiable function w.r.t. t and*

$$\partial_t p(x,t) = \begin{cases} \partial_t p(x,t) \geq 0, & h_1(t) < x < h_2(t); \\ \partial_t p(x,t) < 0, & h_2(t) < x < h_3(t) \end{cases}$$

and $f(x,t) = p(x,t)u(x)$. Then $\partial_{t_{gH}} f(x,t)$ exists and

$$\partial_{t_{gH}} p(x,t) = \begin{cases} \partial_{t_{i,gH}} p(x,t) \geq 0, & h_1(t) < x < h_2(t); \\ \partial_{t_{ii,gH}} p(x,t) < 0, & h_2(t) < x < h_3(t) \end{cases}$$

In fact, the function $h_2(t)$ is switching boundary type 1 for differentiability of $f(x,t)$ with respect to t.

Proof. Since p is valued in $\mathbb{R}^+$ then we can set $f(x,t;\alpha) = p(x,t)[\underline{u}(x;\alpha), \overline{u}(x;\alpha)]$, which implies that

$$\partial_{t_{gH}} = \partial_t p(x,t)[\underline{u}(x;\alpha), \overline{u}(x;\alpha)]$$

.

If $h_1(t) < x < h_2(t)$ then

$$\partial_{t_{gH}} = [\partial_t p(x,t)\underline{u}(x;\alpha), \partial_t p(x,t)\overline{u}(x;\alpha)]$$

then $f(x,t)$ is $[(i)$-differentiable$]$ by report at t. In the same if $h_2(t) < x < h_3(t)$ we get

$$\partial_{t_{gH}} = [\partial_t p(x,t)\overline{u}(x;\alpha), \partial_t p(x,t)\underline{u}(x;\alpha)]$$

thus $f(x,t)$ is $[(ii)$-differentiable$]$ by report at t.

4 Generalized Fuzzy Fractional Derivative

We present generalized fuzzy fractional derivative and their properties.

Definition 4.1. *[5] Let $f \in L^{E^1}([a,b])$. The fuzzy Riemann-Liouville integral of fuzzy-valued function f is defined as following:*

$$I^q f(t) = \frac{1}{\Gamma(1-q)} \int_a^t (t-s)^{q-1} f(s) ds, \quad a < s < t, \quad 0 < q < 1$$

Definition 4.2. *[5] Let $f(x,t;\alpha) = [\underline{f}(x,t;\alpha), \overline{f}(x,t;\alpha)]$ be a valued-fuzzy function. The fuzzy Riemann-Liouville integral of $\overline{f}$ is defined as following:*

$$_{gH}D_t^q f(t,x;\alpha) = \frac{1}{\Gamma(1-q)} \int_a^t (t-s)^q f'_{gH}(s) ds, \quad a < s < t, \quad 0 < q < 1$$

Also we say that f is $[(i) - gH]$-differentiable at t_0 if

$$_{gH}D_t^q f(x,t;\alpha) = [D^q \underline{f}(x,t;\alpha), D^q \overline{f}(x,t;\alpha)]$$

and f is $[(ii) - gH]$-differentiable at t_0 if

$$_{gH}D_t^q f(x,t;\alpha) = [D^q \overline{f}(x,t;\alpha), \underline{f}(x,t;\alpha)]$$

Lemma 4.3. *Let $f \in A^{E^1}$ and $r \in (0,1)$, then*

1. *If f is $[(i) - gH]$-differentiable at t_0 then $D^r f$ is $[(i) - gH]$-differentiable at t_0.*
2. *If f is $[(ii) - gH]$-differentiable at t_0 then $D^r f$ is $[(ii) - gH]$-differentiable at t_0*

Proof. Note that

$$_{gH}D^q f(t) = \frac{1}{\Gamma(1-q)} \int_0^t (t-s)^{-q} f'_{gH}(s) ds$$

Since $\frac{1}{\Gamma(1-q)}(t-s)^{-q}$ is a nonegative quantity whenever $0 < t < s$.

Theorem 4.4. *Let $f \in A^{E^1}$ and $q \in (1,2)$, then*

$$_{gH}D^q f(t) = {}_{gH}D^{q-1} f'_{gH}(t)$$

Proof. We set $f(t) = [\underline{f}(t;\alpha), \overline{f}(t;\alpha)]$ and use Lemma (4.3)
If f is $[(i)$-differentiable$]$ then

$$f(t)' = [\underline{f}'(t;\alpha), \overline{f}'(t;\alpha)]$$

and

$$D^{q-1} f(t)' = [D^{q-1} \underline{f}'(t;\alpha), D^{q-1} \overline{f}'(t;\alpha)]$$

If f is $[(i)$-differentiable$]$ then

$$f(t)' = [\overline{f}'(t;\alpha), \underline{f}'(t;\alpha)]$$

and

$$D^{q-1} f(t)' = [D^{q-1} \overline{f}'(t;\alpha), D^{q-1} \underline{f}'(t;\alpha)]$$

Proposition 4.5. *Let* $f : L^{E^1}$.
If $D^{\gamma-1}f(t) = g(t)$*, then* $f(t) = f(0) + tf'_{gH}(0) + I^{\gamma-1}g(t)$

Proof. We set $f(t) = [\underline{f}(t;\alpha), \overline{f}(t;\alpha)]$ and $g(t) = [\underline{g}(t;\alpha), \overline{g}(t;\alpha)]$.

1. If f is [(i)-differentiable] by Theorem (4.4)

$$\begin{aligned} D^{\gamma-1}f(t) &= [D^{\gamma-1}\underline{f}(t;\alpha), D^{\gamma-1}\overline{f}(t;\alpha)] \\ &= [\underline{g}(t;\alpha), \overline{g}(t;\alpha)] \end{aligned}$$

Which implies that

$$\begin{cases} D^{\gamma-1}\underline{f}(t;\alpha) = \underline{g}(t;\alpha) \\ D^{\gamma-1}\overline{f}(t;\alpha) = \overline{g}(t;\alpha) \end{cases}$$

By [9] we get

$$\begin{cases} \underline{f}(t;\alpha) = \underline{f}(0;\alpha) + t\underline{f}'(0;\alpha) + I^{\gamma-1}\underline{g}(t;\alpha) \\ \overline{f}(t;\alpha) = \overline{f}(0;\alpha) + t\overline{f}'(0;\alpha) + I^{\gamma-1}\overline{g}(t;\alpha) \end{cases}$$

in the same if f is [(ii)-differentiable] then

$$\begin{cases} \underline{f}(t;\alpha) = \underline{f}(0;\alpha) + t\overline{f}'(0;\alpha) + I^{\gamma-1}\underline{g}(t;\alpha) \\ \overline{f}(t;\alpha) = \overline{f}(0;\alpha) + t\underline{f}'(0;\alpha) + I^{\gamma-1}\overline{g}(t;\alpha) \end{cases}$$

Thus

$$f(t) = f(0) + tf'_{gH}(0) + I^{\gamma-1}g(t)$$

5 Fuzzy Fourier Transform

In this section we discuss the Fourier transform in the fuzzy case

Lemma 5.1. *If* $f \in A^{E^1}$ *then the map*

$$\begin{aligned} F : \qquad & \mathbb{R} \longmapsto \mathbb{C}^1 \\ & \omega \to \textstyle\int_{-\infty}^{\infty} f(x)e^{-i\omega w}dx \end{aligned}$$

is well defined

Proof. We have

$$\left\| f(x)e^{-i\omega w} \right\| = \left\| f(x) \right\|$$

Since $f \in A^{E^1}$ then $f(x)e^{-i\omega w} \in A^{\mathbb{C}^1}$, which complete the proof.

Remark 5.2. *In the same the map and under same assumption*

$$F: \quad \mathbb{R} \longmapsto \mathbb{C}^1$$
$$\omega \to \int_{-\infty}^{\infty} f(x)e^{i\omega w}dx$$

is well defined

By the previous lemma and remark we can give a definition of the fuzzy Fourier transform

Definition 5.3. *Let $f : \mathbb{R} \to E^1$ a fuzzy-valued function. The fuzzy Fourier transform of f, denote $\mathscr{F}(f) : \mathbb{R} \to \mathbb{C}^1$, is given by*

$$\mathscr{F}(f(x)) = \frac{1}{\sqrt{2\pi}} \int_{-\infty}^{\infty} f(x)e^{-i\omega w}dx = F(\omega)$$

Also the fuzzy inverse Fourier transform of $F(\omega)$ is given by

$$\mathscr{F}^{-1}(F(\omega)) = \frac{1}{\sqrt{2\pi}} \int_{-\infty}^{\infty} f(x)e^{i\omega w}dx = f(x)$$

By the conditions (1) we have

Remark 5.4. *Let $f \in A^{\mathbb{C}^1}$.*
If $f(x,t;\alpha) = [\underline{f}(x,t;\alpha), \overline{f}(x,t;\alpha)]$, then we can denote

$$\mathscr{F}(f(x,t;\alpha)) = [\mathscr{F}(\underline{f}(x,t;\alpha)), \mathscr{F}(\overline{f}(x,t;\alpha))]$$

with

$$[z_1, z_2] = [Re(z_1), Re(z_2)] \times [Im(z_1), Im(z_2)]$$

and

$$\mathscr{F}^{-1}(f(x,t;\alpha)) = [\mathscr{F}^{-1}(\underline{f}(x,t;\alpha)), \mathscr{F}^{-1}(\overline{f}(x,t;\alpha))]$$

Using the conditions (1) and the linearity of Fourier transform on a "crisp" function we get for all $a, b > 0$

$$a\mathscr{F}(f(x,t;\alpha)) + b\mathscr{F}(g(x,t;\alpha)) = \mathscr{F}(af(x,t;\alpha) + bg(x,t;\alpha))$$

Theorem 5.5. *Let $f \in A^{E^1}$ such that $\lim_{|x| \to \infty} f(x) = 0$. suppose that $f'_{gH} \in A^{E^1}$. Then*

$$\mathscr{F}(f'_{gH}(x)) = i\omega\mathscr{F}(f(x))$$

Proof. Using Theorem (2.18) we get

$$\mathscr{F}(f'_{gH}(x)) = \frac{1}{\sqrt{2\pi}} \left[[f(x)e^{i\omega x}]_{-\infty}^{\infty} -_g (-i\omega) \int_{-\infty}^{\infty} f(x)e^{i\omega x}dx \right]$$

Using the limit $\lim_{|x| \to \infty} f(x) = 0$ we get the result.

Corollary 5.6. *If $f_{gH}^{(k)} \in A^{E^1}$ and $\lim_{|x|\to\infty} f^{(k)(x)} = 0$ for $k = 0, 1, 2$, then*

$$\mathscr{F}\left(f''_{gH}(x)\right) = -\omega^2 \mathscr{F}\left(f(x)\right)$$

By the Theorems (2.17) and (4.4) we have

Theorem 5.7.

$$\mathscr{F}\left({}_{gH}D_t^{\gamma} f(x,t)\right) = {}_{gH}D_t^{\gamma} \mathscr{F}\left(f(x,t)\right)$$

6 The Solution of the Fuzzy Fractional Wave Equation

In this section consider the following problem

$$\begin{cases} {}_{gH}D_t^{\gamma} u(t,x) -_g c^2 \frac{\partial^2}{\partial x^2} u(t,x) = 0 \quad 0 < x,t < 1,\ 0 < \gamma < 1 \\ u(0,x) = a(x), \\ \frac{\partial}{\partial t} u(0,x) = b(x) \end{cases} \tag{1}$$

where a and b are belongs to A^{E^1},

Proposition 6.1. *The problem (1) has a unique solution.*

Proof. Let $u(x,t)$ is fuzzy absolutely integrable, we define the fuzzy Fourier transform of $u(x,t)$ and its inverse by

$$\mathscr{F}\left(u(x,t)\right) = \frac{1}{\sqrt{2\pi}} \int_{-\infty}^{\infty} u(x,t) e^{-i\omega t} dx = U(\omega,t)$$

$$\mathscr{F}^{-1}\left(U(\omega,t)\right) = \frac{1}{\sqrt{2\pi}} \int_{-\infty}^{\infty} U(\omega,t) e^{i\omega t} d\omega = u(x,t)$$

If $D_{t_{gH}}^{\gamma} u(x,t)$, $\partial_{x_{gH}} u(x,t)$ and $\partial_{xx_{gH}} u(x,t)$ are fuzzy absolutely integrable in $(-\infty,\infty)$ by using

$$\mathscr{F}\left({}_{gH}D_t^{\gamma} u(t,x)\right) -_g \mathscr{F}\left(c^2 \frac{\partial^2}{\partial x^2} u(t,x)\right) = 0$$

It follows from the Corollary (5.6) that

$$\mathscr{F}\left(c^2 \frac{\partial^2}{\partial x^2} u(t,x)\right) = -c^2 \omega^2 U(\omega,t)$$

$$\mathscr{F}\left({}_{gH}D_t^{\gamma} u(t,x)\right) = D_t^{\gamma} U(\omega,t)$$

We get

$${}_{gH}D_t^{\gamma} U(\omega,t) = -c^2 U(\omega,t)$$

It follows that

$$_{gH}D_t^{\gamma-1}U'_{gH}(\omega,t) = -c^2U(\omega,t)$$

Thus we have the following problem

$$_{gH}D_t^{\gamma-1}U'_{gH}(\omega,t) = -c^2U(\omega,t) \quad (2)$$
$$U(\omega,0) = \mathscr{F}(a(x)) \quad (3)$$
$$\frac{\partial}{\partial t}U(\omega,0) = \mathscr{F}(b(x)) \quad (4)$$

by Lemma 3.2 [5] this problem has a unique solution given by

$$U(\omega,t) = U(\omega,0) + t\frac{\partial}{\partial t}U(\omega,0) -_g \frac{c^2}{\Gamma(\gamma-1)}\int_0^t\int_0^s (s-\tau)^{\gamma-2}U(\omega,\tau)d\tau ds$$

if u' is [(i)-differentiable], and

$$U(\omega,t) = U(\omega,0) + t\frac{\partial}{\partial t}U(\omega,0) + \frac{c^2}{\Gamma(\gamma-1)}\int_0^t\int_0^s (s-\tau)^{\gamma-2}U(\omega,\tau)d\tau ds$$

if u' is [(ii)-differentiable].

Which implies the existence and uniqueness of the solution of the problem (2) and by the inverse of Fourier transform we get the existence and uniqueness of the solution of (1).

7 Case $\gamma = 2$

In this section we set

$$u(x,t;\alpha) = [\underline{u}(x,t;\alpha), \overline{u}(x,t;\alpha)]$$
$$a(x;\alpha) = [\underline{a}(x;\alpha), \overline{u}(x;\alpha)]$$
$$b(x;\alpha) = [\underline{b}(x;\alpha), \overline{b}(x;\alpha)]$$

If u' is [(i)-differentiable] then

$$\frac{\partial^2}{\partial t^2}\underline{u}(x,t;\alpha) = c^2\frac{\partial^2}{\partial x^2}\underline{u}(x,t;\alpha)$$
$$\frac{\partial^2}{\partial t^2}\overline{u}(x,t;\alpha) = c^2\frac{\partial^2}{\partial x^2}\overline{u}(x,t;\alpha)$$

which implies

$$\underline{u}(x,t;\alpha) = \underline{F}(x-ct;\alpha) + \underline{G}(x+ct;\alpha)$$
$$\overline{u}(x,t;\alpha) = \overline{F}(x-ct;\alpha) + \overline{G}(x+ct;\alpha)$$

where

$$\underline{a}(x;\alpha) = \underline{F}(x-ct;\alpha) + \underline{G}(x+ct;\alpha) \quad (1)$$

$$\overline{a}(x,t;\alpha) = \overline{F}(x-ct;\alpha) + \overline{G}(x+ct;\alpha) \quad (2)$$

and

$$\underline{b}(x;\alpha) = \underline{F'}(x-ct;\alpha) + \underline{G'}(x+ct;\alpha) \quad (3)$$

$$\overline{b}(x,t;\alpha) = \overline{F'}(x-ct;\alpha) + \overline{G'}(x+ct;\alpha) \quad (4)$$

By the conditions (1) the solution is given by

$$u(x,t) = F(x-ct) + G(x+ct)$$

where F and G are given by the above formula (7.1) – (7.4).

8 Examples

In this section we will give some examples to illustrate the previous results.

Example 8.1.

$$\begin{cases} {}_{gH}D_t^{\frac{3}{2}} u(t,x) -_g c^2 \frac{\partial^2}{\partial x^2} u(t,x) = 0 & 0 < x,t < 1,\ 0 < \gamma < 1 \\ u(0,x;\alpha) = [(1+\alpha)e^{-x^2}, (3-\alpha)e^{-x^2}], \\ \frac{\partial}{\partial t} u(0,x) = 0 \end{cases} \quad (1)$$

the solution is given by $u(x,t) = \mathscr{F}^{-1}(U(\omega,t))$ *with*

$$U(\omega,t) = \left[\frac{\alpha+1}{\sqrt{2}} e^{-\omega^2}, \frac{-\alpha+3}{\sqrt{2}} e^{-\omega^2}\right] + \frac{c^2}{\Gamma(\frac{1}{2})} \int_0^t \int_0^s (s-\tau)^{-\frac{1}{4}} U(\omega,\tau) d\tau ds$$

Example 8.2.

$$\begin{cases} {}_{gH}D_t^2 u(t,x) -_g c^2 \frac{\partial^2}{\partial x^2} u(t,x) = 0 & 0 < x,t < 1,\ 0 < \gamma < 1 \\ u(x,0;\alpha) = [\alpha e^{-x^2}, (2-\alpha)e^{-x^2}], \\ \frac{\partial}{\partial t} u(x,0) = 0 \end{cases} \quad (2)$$

the solution is given by

$$u(x,t) = [\alpha, 1 - \frac{\alpha}{2}] e^{-x} \cosh(ct)$$

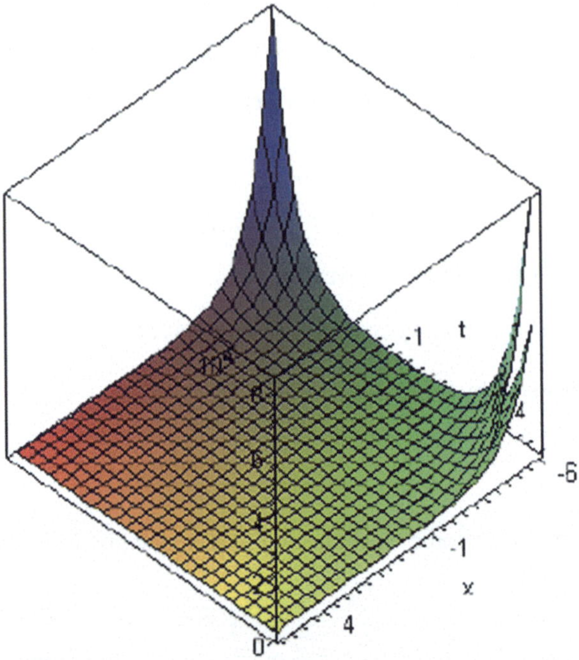

Fig. 1. Lower and upper branch of $u(x,t)$with $\alpha = 1$

9 Conclusions

This study makes it possible to explain the wave phenomena with uncertainty in experimental data.

References

1. Agarwal, R.P., Arshad, S., O'Regan, D., Lupulescu, V.: Fuzzy fractional integral equations under compactness type condition. Fract. Calc. Appl. Anal. **15**, 572–590 (2012)
2. Agarwal, R.P., Lakshmikantham, V., Nieto, J.J.: On the concept of solution for fractional differential equations with uncertainty. Nonlinear Anal. **72**, 2859–2862 (2010)
3. Allahviranloo, T., Gouyandeh, Z., Armand, A., Hasanoglu, A.: On fuzzy solutions for heat equation based on generalized hukuhara differentiability. Fuzzy Sets Syst. **265**, 1–23 (2015)
4. Anastassiou, G.A.: Fuzzy Mathematics: Approximation Theory. Studies in Fuzziness and Soft Computing. Springer, Heidelberg (2010)
5. Armand, A., Gouyandeh, Z.: Fuzzy fractional integro-differential equations under generalized caputo differentiability. Ann. Fuzzy Math. Inform. **10**(5), 789–798 (2015)
6. Bede, B., Stefanini, L.: Generalized differentiability of fuzzy-valued functions. Fuzzy Sets Syst. **230**, 119–141 (2013)
7. Brezis, H.: Functional Analysis, Sobolev Spaces and Partial Differential Equations. Springer, New York (2010)

8. Congxin, W., Ming, M.: Embedding problem of fuzzy number space: part III. Fuzzy Sets Syst. **46**(2), 281–286 (1992)
9. Kilbas, A.A., Srivastava, H.M., Trujillo, J.J.: Theory and Applications of Fractional Differential Equations. Elsevier, New York (2006)
10. Buckley, J.: Fuzzy complex numbers. Fuzzy Sets Syst. **33**(3), 333–345 (1989)
11. Stefanini, L., Bede, B.: Generalized hukuhara differentiability of interval-valued functions and interval differential equations. Nonlinear Anal. **71**(34), 1311–1328 (2009)
12. Wu, H.C.: The improper fuzzy Riemann integral and its numerical integration. Inf. Sci. **111**(14), 109–137 (1998)
13. Zadeh, L.A.: Fuzzy sets. Inf. Control **8**, 338–353 (1965)

An Efficient High Order Algorithm for Solving Regularized Long Wave Equation

Dursun Irk(✉) and Melis Zorşahin Görgülü

Department of Mathematics and Computer Science, University of Eskişehir Osmangazi, Eskişehir, Turkey
{dirk,mzorsahin}@ogu.edu.tr

Abstract. The Galerkin finite element method, based on cubic trigonometric B-spline for the space discretization and fourth order Runge Kutta method for time discretization is proposed for numerical solution of the Regularized Long Wave (RLW) equation. The numerical example related to single solitary wave is considered as the test problem. To see the accuracy for the proposed method, the maximum error norm L_∞ is computed and conservation property of the RLW equation will be validated by calculating the three conservation quantities, corresponding to mass, momentum and energy.

Keywords: Galerkin finite element method · Cubic trigonometric B-spline · Regularized long wave equation · Solitary waves

1 Introduction

Nonlinear partial differential equations are used in the modeling of many events in nature. There is no general method for the exact solution of nonlinear partial differential equations. Therefore, approximate solutions of such equations are frequently studied. The RLW equation, which can be used as an alternative to the KdV equation, has an important role in soliton theory. RLW equation has only a limited number of analytical solutions for boundary and initial conditions. The nonlinear partial differential equation which may be written in the following form as

$$u_t + u_x + \varepsilon u u_x - \mu u_{xxt} = 0, \tag{1}$$

is known as RLW equation formulated by Peregrine for studying soliton phenomenon [1, 2].

Due to the fact that the Solitary wave's boundary conditions are zero at the boundaries namely $u \to 0$ as $x \to \pm\infty$, the position range will be selected to be $u(a, t) = u(b, t) \approx 0$. Therefore, the boundary and initial conditions will be considered as

$$\begin{aligned} u(a, t) = u(b, t) = 0, \\ u_x(a, t) = u_x(b, t) = 0, \end{aligned} \quad t \in (0, T] \tag{2}$$

and

$$u(x, 0) = f(x). \tag{3}$$

H. Dutta et al. (Eds.): CMES 2019, AISC 1111, pp. 148–155, 2020.
https://doi.org/10.1007/978-3-030-39112-6_10

To solve numerically the RLW equation, various numerical methods including different degrees of spline functions are proposed [3–13]. Generally in these methods, Crank Nicolson method having second order accurate in time has been used for the time discretization for the RLW equation. By this study, it is aimed to get a numerical solution for the interested equation with a high accurately numerical method based on the use of the fourth order Runge Kutta method for time discretization. Trigonometric B-spline functions are rarely used in literature compared to B-spline functions. In this study, Galerkin finite element method based on cubic trigonometric B-spline functions will be used when time discretization is performed.

2 Application of the Method

When making calculations, the space-time plane $[a, b] \times [0, T]$ will be discretized by grids with Δt and h. Thus,

$$u(x_p, t_n) = u_p^n, \ \ p = 0, 1, \ldots, N; \ \ n = 0, 1, 2, \ldots$$

will be used for the exact solution and U_p^n will be used for the approximate solution of the exact solution at the points (x_p, t_n) where $x_p = a + ph$ and $t_n = n\Delta t$. The space interval $[a, b]$ will be divided into equal length N sub-interval as

$$a = x_0 < x_1 < \ldots < x_{N-1} < x_N = b.$$

The definition of cubic trigonometric B-spline functions in these knots is as follows:

$$T_p(x) = \frac{1}{\theta}\begin{cases} g^3(x_{p-2}) & , x \in [x_{p-2}, x_{p-1}) \\ -g^2(x_{p-2})\, g(x_p) \\ -g(x_{p-2})\, g(x_{p+1})\, g(x_{p-1}) - g(x_{p+2})\, g^2(x_{p-1}) & , x \in [x_{p-1}, x_p) \\ g(x_{p-2})\, g^2(x_{p+1}) \\ +g(x_{p+2})\, g(x_{p-1})\, g(x_{p+1}) + g^2(x_{p+2})\, g(x_p) & , x \in [x_p, x_{p+1}) \\ -g^3(x_{p+2}) & , x \in [x_{p+1}, x_{p+2}) \\ 0 & \text{otherwise} \end{cases} \tag{4}$$

where

$$\theta = \sin\left(\frac{h}{2}\right) \sin(h) \sin\left(\frac{3h}{2}\right),$$

$$g(x_p) = \sin\left(\frac{x - x_p}{2}\right).$$

An approach $U(x, t)$ to the unknown function $u(x, t)$ as a linear combination of cubic trigonometric functions can be performed as follows:

$$u(x, t) \approx U(x, t) = \sum_{p=-1}^{N+1} T_p(x)\delta_p(t). \tag{5}$$

Since the cubic trigonometric B-splines have local supports, there is no need to evaluate basis functions where they are zero. Therefore, by editing the trial solution (5), the following form can be written over the sub-elements $[x_p, x_{p+1}]$:

$$U(x, t) = \sum_{j=p-1}^{p+2} T_j(x)\delta_j(t) \tag{6}$$

Using cubic trigonometric B-spline (4) and the solution (6), unknown function $U_p = U(x_p, t)$ and the space derivatives at knots are written as

$$\begin{aligned} U_p &= \alpha_1\delta_{p-1} + \alpha_2\delta_p + \alpha_1\delta_{p+1}, \\ U'_p &= \alpha_3(-\delta_{p-1} + \delta_{p+1}), \\ U''_p &= \alpha_4\delta_{p-1} + \alpha_5\delta_p + \alpha_4\delta_{p+1}, \end{aligned} \tag{7}$$

where

$$\begin{aligned} \alpha_1 &= \sin^2\left(\tfrac{h}{2}\right)\csc(h)\csc\left(\tfrac{3h}{2}\right), \quad \alpha_2 = \left(\frac{2}{1+2\cos(h)}\right), \\ \alpha_3 &= \left(\tfrac{3}{4}\csc\left(\tfrac{3h}{2}\right)\right), \\ \alpha_4 &= \left(\frac{3\left(3\cos^2\left(\frac{h}{2}\right)-1\right)}{4\left(\sin(h)\sin\left(\frac{3h}{2}\right)\right)}\right), \qquad \alpha_5 = -\left(\frac{3\cot^2\left(\frac{h}{2}\right)}{(2+4\cos(h))}\right). \end{aligned}$$

By arranging, the Eq. (1) can be rewritten in the following form as

$$(u - \mu u_{xx})_t = -(u_x + \varepsilon u u_x). \tag{8}$$

If both sides of the equation are multiplied by the weight function $W(x)$ and integrating over the space interval $[a, b]$, we get the following equation:

$$\int_a^b W(x)(u_t - \mu u_{xxt})\,dx = -\int_a^b W(x)(u_x + \varepsilon u u_x)\,dx. \tag{9}$$

Now, weight function and unknown functions in Eq. (9) are taken as cubic trigonometric B-spline shape function and an approach for unknown function $u(x, t)$ given in (6) respectively. Thus, the fully discretized approximation form can be obtained over the sub-element $[x_p, x_{p+1}]$ as

$$\sum_{j=p-1}^{p+2}\left[\int_{x_p}^{x_{p+1}}\left(T_iT_j - \mu T_iT_j''\right)dx\right](\delta_t)_j - \sum_{j=p-1}^{p+2}\left[\int_{x_p}^{x_{p+1}}\left(-T_iT_j' - \varepsilon T_i\sum_{r=p-1}^{p+2}(T_r\delta_r)\,T_j'\right)dx\right]\delta_j. \tag{10}$$

The approximation (10) can be written as

$$[\boldsymbol{A}^e - \mu\boldsymbol{B}^e)]\,\delta_t^e - [-\boldsymbol{C}^e - \varepsilon\boldsymbol{D}^e(\delta^e)]\,\delta^e, \quad p = 0, 1, \ldots, N-1 \tag{11}$$

where

$$A_{ij}^e = \int_{x_p}^{x_{p+1}} T_iT_j\,dx, \quad B_{ij}^e = \int_{x_p}^{x_{p+1}} T_iT_j''\,dx, \quad C_{ij}^e = \int_{x_p}^{x_{p+1}} T_iT_j'\,dx$$

$$D_{ij}^e = \int_{x_p}^{x_{p+1}} T_i T_r \delta_r T_j' dx, \ \left(\delta^e\right) = \left(\delta_{p-1}, \ldots, \delta_{p+2}\right)^T.$$

Combining contributions from all elements lead to the matrix equation

$$[\mathbf{A} - \mu\mathbf{B}]\,\delta_t = [-\mathbf{C} - \varepsilon\mathbf{D}]\,\delta \tag{12}$$

or

$$\delta_t = \mathbf{E}\delta \tag{13}$$

where

$$\mathbf{E} = [\mathbf{A} - \mu\mathbf{B}]^{-1}[-\mathbf{C} - \varepsilon\mathbf{D}],$$
$$\delta = (\delta_{-1}, \delta_0, \ldots, \delta_N, \delta_{N+1})^T.$$

After initial vector

$$\mathbf{d}^0 = (\delta_{-1}^0, \ldots, \delta_{N-1}^0, \delta_{N+1}^0)$$

is found using the conditions (2) and (3), unknown vectors

$$\mathbf{d}^{n+1} = (\delta_{-1}^{n+1}, \ldots, \delta_N^{n+1}, \delta_{N+1}^{n+1})$$

is found by using fourth order Runge Kutta method for a system of ODEs (13).

3 Propagation of Solitary Wave

The conservation quantities of the RLW equation

$$\begin{aligned}
I_1 &= \int_{-\infty}^{\infty} u dx \approx \int_a^b U dx, \\
I_2 &= \int_{-\infty}^{\infty} (u^2 + \mu(u_x)^2) dx \approx \int_a^b (U^2 + \mu(U_x)^2) dx, \\
I_3 &= \int_{-\infty}^{\infty} \left(u^3 + 3u^2\right) dx \approx \int_a^b (U^3 + 3U^2) dx.
\end{aligned} \tag{14}$$

corresponding to the mass, energy and momentum [14] will be calculated by approximately the trapezoid rule. Also error norm

$$L_\infty = \max\left|u_p - U_p\right| \tag{15}$$

and the order of convergence

$$\text{order} = \frac{\log\left|\left(u - U_{\Delta t_p}\right)/\left(u - U_{\Delta t_{p+1}}\right)\right|}{\log\left|\Delta t_p/\Delta t_{p+1}\right|}, \tag{16}$$

will be calculated.

The solitary wave solution of the RLW equation is

$$u(x,\ t) = 3c\mathrm{sech}^2(k[x - \tilde{x}_0 - vt]). \tag{17}$$

Taking $t = 0$ in (17), the initial condition of the test problem is found as

$$u(x,\ 0) = 3c\mathrm{sech}^2(k[x - \tilde{x}_0]), \tag{18}$$

where the velocity and amplitude of the solitary wave are $v = 1+\varepsilon c$ and $3c$, respectively. $\tilde{x}_0$ is the point at which the peak of the initial wave and $k = \sqrt{\frac{\varepsilon c}{4\mu v}}$.

Three invariants for the RLW equation using the initial condition

$$u(x,\ 0) = 3c\mathrm{sech}^2(k[x - \tilde{x}_0]),$$

can be determined analytically as

$$\begin{aligned} I_1 &= \int_{-\infty}^{\infty} u dx = \tfrac{6c}{k}, \\ I_2 &= \int_{-\infty}^{\infty} (u^2 + \mu(u_x)^2)dx = \tfrac{12c^2}{k} + \tfrac{48kc^2\mu}{5}, \\ I_3 &= \int_{-\infty}^{\infty} \left(u^3 + 3u^2\right) dx = \tfrac{36c^2}{k}\left(1 + \tfrac{4c}{5}\right). \end{aligned}$$

The error norm L_∞, conservation quantities I_1, I_2, I_3 and order of convergence for the obtained algorithm over the space interval $-80 \le x \le 120$ by using the parameters $c = 0.1$, $h = 0.02$ and various time steps are documented in Table 1. As clearly seen from the Table 1 that the conservation constants are almost the same as the exact results, the error norm decreases as the time step decreases, and the convergence rate is almost 4, which is the accuracy of the Runge-Kutta method.

Table 1. Invariants, error norm and order of convergences with the amplitude = 0.3

Δt	L_∞	I_1	I_2	I_3	Order
2	2.21×10^{-3}	3.97994975	0.808728306	2.57327030	3.88
1	1.50×10^{-4}	3.97994975	0.810400790	2.57880375	3.97
0.5	9.61×10^{-6}	3.97994975	0.810460516	2.57900091	3.99
0.2	2.48×10^{-7}	3.97994975	0.810462474	2.57900737	3.99
0.1	1.56×10^{-8}	3.97994975	0.810462494	2.57900744	3.97
0.05	9.92×10^{-10}	3.97994975	0.810462494	2.57900744	
Exact	0	3.97994975	0.810462494	2.57900744	

The propagation of the solitary wave simulation is performed throughout the interval $-80 \le x \le 120$ up to time $t = 20$ using the parameters $\varepsilon = \mu = 1$, $\tilde{x}_0 = 0$, $c = 0.1$ and $h = 0.02$, $\Delta t = 0.05$. Using these parameters, the state of the solitary wave at the time of the start and at certain times are drawn in Fig. 1 up to time $t = 20$. It is seen from the figure that the solitary wave in the beginning is moving while maintaining its shape and the solitary wave travels 22 units between $t = 0$ and $t = 20$ since the velocity of the solitary wave is $v = 1 + \varepsilon c = 1.1$.

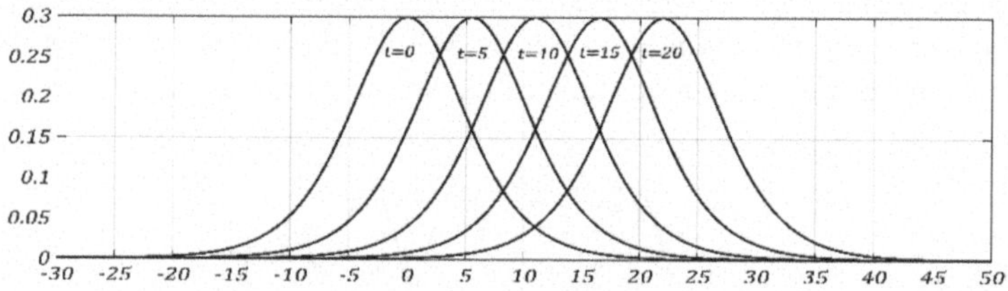

Fig. 1. Solitary waves

For the proposed method, the graph of the absolute error is plotted at time $t = 20$ in Fig. 2 by using $h = 0.02$ and $\Delta t = 0.05$. As can be seen from Fig. 2, the maximum error comes from the middle of the space interval.

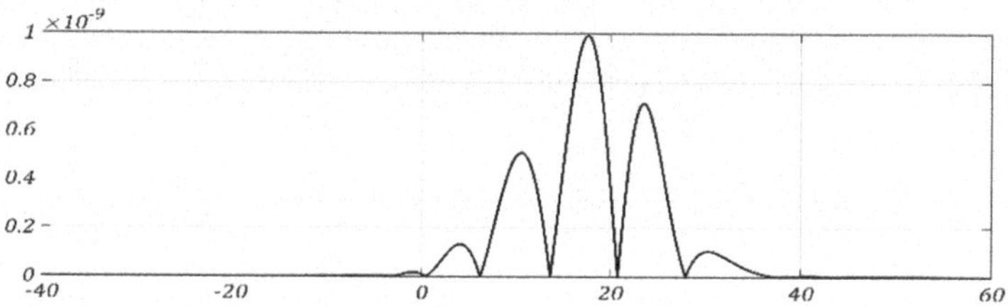

Fig. 2. Absolute error

Secondly, the error norm L_∞, conservation quantities I_3, I_2, I_3 and order of convergence for the obtained algorithm over the space interval $-80 \le x \le 120$ by using the parameters $c = 1/3$, $h = 0.02$ and various time steps are presented in Table 2.In this case, the test problem is repeated for a solitary wave with greater amplitude $3c = 1$. According to the Table 2, the error norm decreases as the time step decreases, and as in the previous table, the convergence rate is almost 4.

Table 2. Invariants, error norm and order of convergences with the amplitude = 1

Δt	L_∞	I_1	I_2	I_3	Order
2	7.92×10^{-2}	8.00000000	5.142783585	18.4296950	3.62
1	6.44×10^{-3}	8.00000000	5.572265396	20.1556551	3.86
0.5	4.44×10^{-4}	8.00000000	5.599022280	20.2627554	3.98
0.2	1.16×10^{-5}	8.00000000	5.599989708	20.2666255	4.00
0.1	7.29×10^{-7}	8.00000000	5.599999677	20.2666654	3.99
0.05	4.57×10^{-8}	8.00000000	5.599999990	20.2666666	
Exact	0	8.00000000	5.6	20.2666667	

The propagation of the solitary wave simulation is performed throughout the interval $-80 \leq x \leq 120$ up to time $t = 20$ using the parameters $\varepsilon = \mu = 1, \tilde{x}_0 = 0$, $c = 1/3$ and $h = 0.02$, $\Delta t = 0.05$. Using these parameters, the state of the solitary wave at the time of the start and at certain times are drawn in Fig. 3 up to time $t = 20$. It is seen from the figure that the solitary wave in the beginning is moving while maintaining its shape and the solitary wave travels $80/3 \simeq 26.6$ units between $t = 0$ and $t = 20$ since the velocity of the solitary wave is $v = 1 + \varepsilon c = 4/3$. It can be seen here that the solitary wave moves faster because the amplitude of the solitary wave is larger.

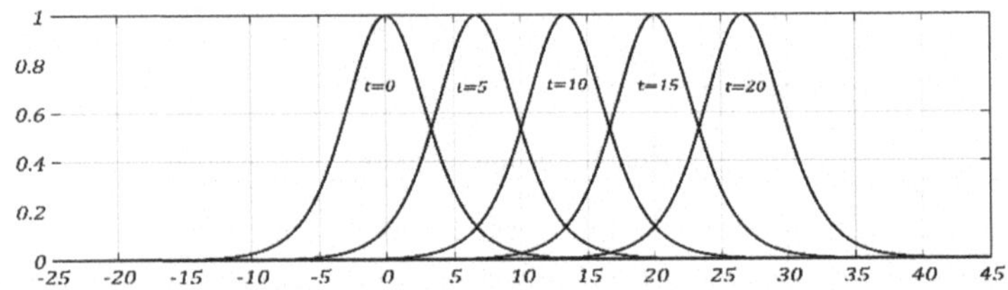

Fig. 3. Solitary waves

The graph of the absolute error for the proposed method is plotted at time $t = 20$ in Fig. 4 by using $h = 0.02$ and $\Delta t = 0.05$. When the figure is examined, it can be seen that the maximum error occurs again at the midpoints of the space interval.

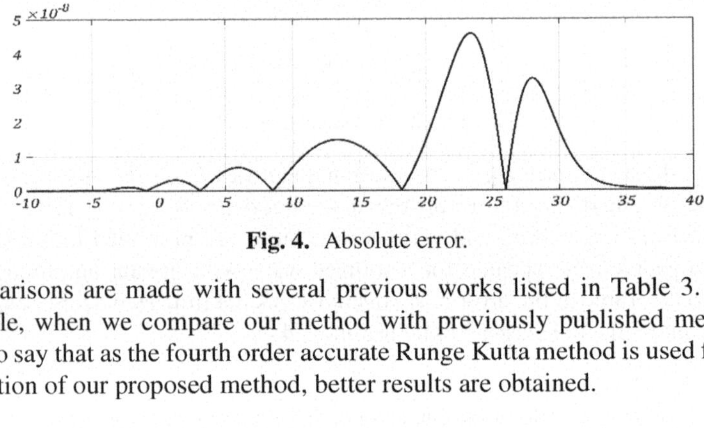

Fig. 4. Absolute error.

Comparisons are made with several previous works listed in Table 3. According to the table, when we compare our method with previously published methods, it is possible to say that as the fourth order accurate Runge Kutta method is used for the time discretization of our proposed method, better results are obtained.

Table 3. c =0.1, $h = 0.125$, $\Delta t = 0.1$.

	$L_\infty \times 10^5$	I_1	I_2	I_3
Present	0.00437	3.9799497	0.8104625	2.5790074
[12] (CN)	8.78896	3.9799498	0.8104273	2.5790075
[12] (AM)	0.20615	3.9799498	0.8104625	2.5790074
[13] (CN)	8.78967	3.9799497	0.8104624	2.5790074
Exact		3.9799497	0.8104625	2.5790074

4 Conclusion

In this study, the Galerkin method based on cubic trigonometric B-splines for space discretization and fourth order Runge Kutta method for time discretization are presented for the numerical solution of the RLW equation. The proposed method has been tested on the propagation of a single solitary wave. To see the accuracy of the methods, the error norms and the conservation quantities are documented according to the obtained results. According to these results, it can be said that the proposed numerical solution algorithm is effective in maintaining both the error norm and conservation constants, and has high accuracy because the error is much smaller when compared to other studies.

References

1. Peregrine, D.H.: Calculations of the development of an undular bore. J. Fluid Mech. **25**, 321–330 (1966)
2. Benjamin, T.B., Bona, J.L., Mahony, J.J.: Model equations for long waves in non-linear dispersive systems. Philos. Trans. R. Soc. London A. **272**, 47–78 (1972)
3. Gardner, L.R.T., Gardner, G.A., Dağ, İ.: A B-spline finite element method for the regularized long wave equation. Commun. Numer. Meth. Eng. **11**, 59–68 (1995)
4. Gardner, L.R.T., Dağ, İ.: The boundary-forced regularized long wave equation. II. Nuova Cimento **110B**, 1487–1496 (1995)
5. Zaki, S.I.: Solitary waves of the splitted RLW equation. Comput. Phys. Comm. **138**, 80–91 (2001)
6. Avilez-Valente, P., Seabra-Santos, F.J.: A Petrov-Galerkin finite element scheme for the regularized long wave equation. Comput. Mech. **34**, 256–270 (2004)
7. Saka, B., Dağ, İ., Doğan, A.: Galerkin method for the numerical solution of the RLW equation using quadratic B-splines. Int. J. Comput. Math. **81**, 727–739 (2004)
8. Esen, A., Kutluay, S.: Application of a lumped Galerkin method to the regularized long wave equation. Appl. Math. Comput. **174**, 833–845 (2006)
9. Dağ, İ., Saka, B., Irk, D.: Galerkin methods for the numerical solution of the RLW equation using quintic B-splines. J. Comput. Appl. Math. **190**(1–2), 532–547 (2006)
10. Saka, B., Dağ, İ.: A numerical solution of the RLW equation by Galerkin method using quartic B-splines. Commun. Numer. Methods Eng. **24**, 1339–1361 (2008)
11. Doğan, A.: Numerical solution of RLW equation using linear finite elements within Galerkin's method. Appl. Math. Model. **26–7**, 771–783 (2002)
12. Irk, D., Keskin, P.: Quadratic trigonometric B-spline Galerkin methods for the regularized long wave equation. J. Appl. Anal. Comput. **7**, 617–631 (2017)
13. Irk, D., Keskin, P.: Cubic trigonometric B-spline Galerkin methods for the regularized long wave equation. J. Phys: Conf. Ser. **766**(1), 012032 (2016)
14. Olver, P.J.: Euler operators and conservation laws of the BBM equation. Math. Proc. Camb. Phil. Soc. **85**, 143–159 (1979)

On the Solitary Wave Solutions to the (2+1)-Dimensional Davey-Stewartson Equations

Hajar F. Ismael[1,2(✉)] and Hasan Bulut[2]

[1] Department of Mathematics, University of Zakho, Zakho, Iraq
hajar.ismael@uoz.edu.krd
[2] Department of Mathematics, University of Firat, Elazig, Turkey
hbulut@firat.edu.tr

Abstract. In this article, by using the Bernoulli sub-equation, we build the analytical traveling wave solution of the (2+1)-dimensional Davey-Stewartson equation system. First of all, the imaginary (2+1)-dimensional Davey-Stewatson system is transformed into a system of nonlinear differential equations, After getting the resultant equation, the homogeneous method of balance between the highest power and the highest derivative of the ordinary differential equation is authorized and finally the outcomes equations are solved in order to achieve some new analytical solutions. Wolfram Mathematica Package is used for different cases as well as for different values of constants to investigate the solutions of the resulting system of a nonlinear differential equation. The results of this study are shown in 2D and 3D dimensions graphically.

Keywords: Bernoulli sub-equation · Davey-Stewatson equations

1 Introduction

Progressing of soliton formation and its application in differential systems has been remarkable in recent years. Disputing modes of solitary energy propagating on behalf of a chain of other biological molecules has pulled forward interesting. New attainment of topological, nontopological solitons as well as transformation phenomena in polyacetylene chains with the action of an electrical field [1]. The physical phenomena of nonlinear partial differential equations (NLPDEs) are involved in many fields of physics, for example, plasma physics, optical fibers, nonlinear optics, fluid mechanics, chemistry, biology, geochemistry as well as engineering sciences [2].

Researchers have been reported an assorted numerical and analytical techniques to seek solutions of NLPDEs for example a homotopy analysis method [3,4], a finite forward difference method [5,6], homotopy perturbation method [7,8], spectral methods [9], Adomian decomposition method [10,11], Adams-Bashforth scheme [12], Adams-Bashforth-Moulton scheme [13], shooting scheme

H. Dutta et al. (Eds.): CMES 2019, AISC 1111, pp. 156–165, 2020.
https://doi.org/10.1007/978-3-030-39112-6_11

[14–17], the sine-Gordon expansion method [18,19], the inverse scattering method [20], functional variable method [21], the Bernoulli sub-ODE function method [22,23], the modified auxiliary expansion method [24], the modified $\exp(-\varphi(\xi))$-expansion function method [25–27], the $\tan(\phi(\xi)/2)$-expansion method [28, 29], G'/G-expansion method [30,31], the decomposition-Sumudu-like-integral-transform method [32], the extended sinh-Gordon expansion method [33,34] and the generalized exponential rational function method [35,36].

Scholars have been used different methods to find some kind of solution like exact, analytical, numerical and semi-analytic solutions of Davey-Stewartson equations for instance, the G'/G method [37], the improved $\tan(\phi(\xi)/2)$-expansion method with generalized G'/G-expansion method [32], the rational expansion method [38], time splitting spectral method [39], the Gram-type determinant solution and Casorati-type determinant solution [40]. Also, different analytical approaches such as, the method of multiple scales combined with a quasi discreteness approximation [41], sine-Gordon expansion method [42], the new generalized G'/G-expansion method [43], the extended Weierstrass transformation method [44], the sine-cosine, tanh-coth and exp-function methods [45] and the extended mapping method technique [46] have been developed to investigate analytical solutions for the different types of NLPDEs.

In this study, some novel soliton solution of Davey and Stewartson equations by using the Bernoulli sub-equation is investigated. The variable approach of the traveling wave changes the NLPDEs into nonlinear ordinary differential equations and it is solved for different physical nonzero parameters. Outcomes cases are present in 2D and 3D-dimensions.

2 Structures of Bernoulli Sub-equation Function Method

The mainly modified steps of this technique are [47,48]:

Let we have a nonlinear partial differential equation:

$$P\left(u_x, u_t, u_{xt}, u_{xx}, ...\right) = 0, \tag{1}$$

and defining the traveling wave transformation

$$u(x,t) = q(\eta), \eta = x + \gamma t, \tag{2}$$

where $\gamma \neq 0$. Applying Eq. (4) on Eq. (3) as a result, we get a nonlinear ordinary differential equation:

$$N(q, q^{'}, q^{''}, ...) = 0. \tag{3}$$

Using a trial equation of solution as follows:

$$q\left(\eta\right) = \sum_{i=0}^{n} a_i F^i = a_0 + a_1 F + a_2 F^2 + ... + a_n F^n, \tag{4}$$

and

$$F^{'} = bF + dF^M, b \neq 0, d \neq 0, M \in R - \{0, 1, 2\}. \tag{5}$$

here $F(\eta)$ is Bernoulli differential polynomial. Inserting Eq. (6) into Eq. (5) as well as using Eq. (7) produces:

$$\Omega(F(\eta)) = b_k F(\eta)^s + \cdots + b_1 F(\eta) + b_0 = 0, \tag{6}$$

via the balance principle, the connection of n and M will be evaluate.

By taking all the coefficients of $\Omega(F(\eta))$ to be zero, we get an algebraic equations system:

$$b_i = 0, \ \ i = 0, \cdots, k, \tag{7}$$

solving Eq. (9), we will find the values of $a_0, a_1, ..., a_n$.

Step 4. Solving Bernoulli Eq. (7), two cases are observed depending on the values of b and d:

$$F(\eta) = \left[\frac{-d}{b} + \frac{E}{e^{b(M-1)\eta}}\right]^{\frac{1}{1-M}}, b \neq d, \tag{8}$$

$$F(\eta) = \left[\frac{(E-1) + (E+1)\tanh\left(\frac{b(1-M)\eta}{2}\right)}{1 - \tanh\left(\frac{b(1-M)\eta}{2}\right)}\right]^{\frac{1}{1-M}}, b = d, E \in R. \tag{9}$$

Where E is the non-zero constant of integration, with the help of Mathematical packages, we gain the solutions to Eq. (5), using a complete polynomial discrimination system. Also, all the solutions gained in this method are plotted and the suitable parameter values on (1+1)-dimensional surfaces of solutions are taken into account.

3 The (2+1)-Dimensional Davey-Stewartson Equations

In this article, the Davey-Stewartson equations in dimensional [49,50] are considered

$$i\phi_t + \frac{1}{2}\sigma^2\left(\phi_{xx} + \sigma^2\phi_{yy}\right) + \lambda|\phi|^2\phi - \phi\,\psi_x = 0, \tag{10}$$

$$\psi_{xx} - \sigma^2\psi_{yy} - 2\lambda\Big(\,|\phi|^2\Big)_x = 0, \tag{11}$$

here $\phi(x, y, t)$ and $\psi(x, y, t)$ represents the dependent variables while, x and y are the independent variables axes as well as is represent a time-independent variable. Also, σ and λ represent constant coefficients. First of all we convert the (2+1)-dimensional imaginary Davey-Stewartson equations into a system of nonlinear ODE to study and analyze its exact solutions.

Using the following transformation:

$$\phi(x, y, t) = e^{i\theta}u(\xi), \ \psi(x, y, t) = v(\xi), \ \xi = \mu(x + y - \eta t), \ \theta = \kappa x + \lambda y + \beta t. \tag{12}$$

where $\mu, \eta, \kappa, \lambda$ and β are real constants. Applying Eq. (12), the (2+1)-dimensional Davey-Stewartson equations are changed to

$$\mu^2\left(1 - \sigma^2 - 2\sigma^4\right)u'' - \left(\beta + \kappa^2\sigma^2 + \lambda^2\right)u - uv + \kappa u^3 = 0, \tag{13}$$

$$\mu\left(\eta-2\kappa\sigma^2-2\lambda\right)i\,u'=0, \tag{14}$$

$$-\mu^2\left(1-\sigma^2\right)v''+4\kappa\mu^2\left(uu''+u'^2\right)=0. \tag{15}$$

Integrating Eq. (15) twice with respect to ξ and taking the constant of integration to be zero, one gets

$$v=\frac{2\kappa}{1-\sigma^2}u^2. \tag{16}$$

Finding the close solution, we find from Eq. (14) that

$$\eta=2\kappa\sigma^2+2\lambda. \tag{17}$$

Now substituting Eq. (16) into Eq. (13), we get

$$\mu^2\left(1-\sigma^2\right)\left(1-\sigma^2-2\sigma^4\right)u''-\left(1-\sigma^2\right)\left(\beta+\kappa^2\sigma^2+\lambda^2\right)u-\kappa\left(1+\sigma^2\right)u^3=0. \tag{18}$$

Now to evaluate the balances between and, the relationship between and can written

$$M=n+1. \tag{19}$$

Case 1. Using $n=2$, $M=3$ and then substituting them into Eq. (4) with using Eq. (5), the following equations are obtained:

$$u=a_0+a_1F+a_2F^2, \tag{20}$$

$$u'=a_1bF+a_1dF^3+2a_2bF^2+2a_2dF^4, \tag{21}$$

$$u''=a_1b^2F+4a_2b^2F^2+4a_1bdF^3+12a_2bdF^4+3a_1d^2F^5+8a_2d^2F^6, \tag{22}$$

where $a_2\neq 0$, $b\neq 0$, $d\neq 0$. Substituting Eqs. (20–22) into Eq. (18), a system of algebraic equations are found. Inserting Eqs. (8) or (9) into a system of algebraic equations, we can investigate the following solutions:

Case 1a. For $a_0=\frac{b\mu\sqrt{2-6\sigma^2+4\sigma^4}}{\sqrt{\kappa}}$, $a_1=0, a_2=\frac{2d\mu\sqrt{2-6\sigma^2+4\sigma^4}}{\sqrt{\kappa}}$, $\beta=-\lambda^2-\kappa^2\sigma^2+2b^2\mu^2\left(-1+\sigma^2+2\sigma^4\right)$, we get (Fig. 1)

$$\phi(x,y,t)=\frac{b\mu\sqrt{2-6\sigma^2+4\sigma^4}\mathrm{e}^{\mathrm{i}(\beta t+\kappa x+\lambda y)}\left(d\mathrm{e}^{2b\mu\left(x+y-2\left(\lambda+\kappa\sigma^2\right)t\right)}+bE\right)}{\sqrt{\kappa}\left(-d\mathrm{e}^{2b\mu(x+y-2(\lambda+\kappa\sigma^2)t)}+bE\right)}, \tag{23}$$

$$\psi(x,y,t)=-\frac{4b^2\mu^2\left(-1+2\sigma^2\right)\left(d\mathrm{e}^{2b\mu\left(x+y-2\left(\lambda+\kappa\sigma^2\right)t\right)}+bE\right)^2}{\left(d\mathrm{e}^{2b\mu(x+y-2(\lambda+\kappa\sigma^2)t)}-bE\right)^2}. \tag{24}$$

Case 1b. $\lambda=\sqrt{-\alpha+\kappa^2}$, $\sigma=i$, we get (Fig. 2)

$$\phi(x,y,t)=\mathrm{e}^{\mathrm{i}(\beta t+\kappa x+\lambda y)}\left(a_0+\frac{a_2}{-\frac{d}{b}+E\mathrm{e}^{-2b\xi}}+\frac{a_1}{\sqrt{-\frac{d}{b}+E\mathrm{e}^{-2b\xi}}}\right), \tag{25}$$

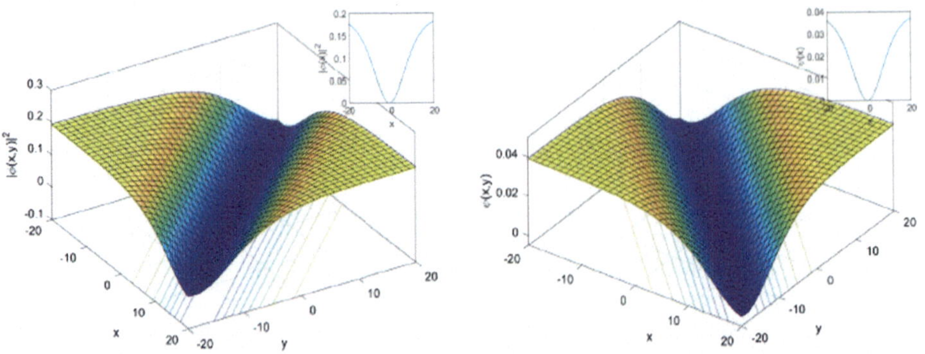

Fig. 1. 3D surfaces with its 2D figures of Eqs. (21) and (22) with values $b = 1, d = -1, E = 1, \sigma = 0.1, \kappa = 0.1, t = 0.5, \mu = 0.1, \lambda = 1$ and $y = 2$ for 2D surface.

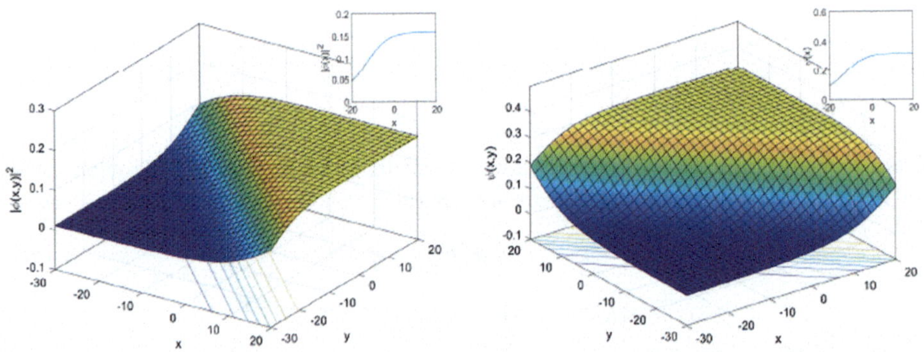

Fig. 2. 3D surfaces with its 2D figures of Eqs. (23) and (24) with values $b = 1, d = -1, E = 0.1, \kappa = 2, a_0 = 0.1, a_2 = 0.1, a_1 = 0.2, t = 0.5, \mu = 0.1, \alpha = -0.5$ and $y = 2$ for 2D surface.

$$\psi(x, y, t) = \kappa\left(a_0 + \frac{a_2}{-\frac{d}{b} + E\mathrm{e}^{-2b\xi}} + \frac{a_1}{\sqrt{-\frac{d}{b} + E\mathrm{e}^{-2b\xi}}}\right)^2. \tag{26}$$

Case 2. If taking $n = 3$ and $M = 4$ in Eq. (4) with using Eq. (5), the following equations are found:

$$u = a_0 + a_1F + a_2F^2 + a_3F^3, \tag{27}$$

$$u' = a_1bF + 2a_2bF^2 + 3a_3bF^3 + a_1dF^4 + 2a_2dF^5 + 3a_3dF^6, \tag{28}$$

$$\begin{aligned} u'' = {} & a_1b^2F + 4a_2b^2F^2 + 9a_3b^2F^3 + 5a_1bdF^4 + 14a_2bdF^5 \\ & + 27a_3bdF^6 + 4a_1d^2F^7 + 10a_2d^2F^8 + 18a_3d^2F^9, \end{aligned} \tag{29}$$

where $a_3 \neq 0, b \neq 0, d \neq 0$. putting Eqs. (27–29) into Eq. (18), a system of algebraic equations is evaluated. Solving this system the following cases and solutions have resulted:

Case 2a. When $a_0 = -\frac{\mu\sqrt{\beta+\lambda^2+\kappa^2\sigma^2}\sqrt{1-3\sigma^2+2\sigma^4}}{\sqrt{\kappa}\sqrt{\mu^2(-1+\sigma^2+2\sigma^4)}}$, $a_1 = 0$, $a_2 = 0$, $a_3 = -\frac{3d\mu\sqrt{2-6\sigma^2+4\sigma^4}}{\sqrt{\kappa}}$, $b = \frac{\sqrt{2}\sqrt{\beta+\lambda^2+\kappa^2\sigma^2}}{3\sqrt{\mu^2(-1+\sigma^2+2\sigma^4)}}$, we obtain (Fig. 3)

$$\phi(x,y,t) = \mathrm{e}^{\mathrm{i}(\beta t+\kappa x+y\lambda)}\frac{\mu\sqrt{B}}{\sqrt{\kappa}}\left(-\frac{3\sqrt{2}d}{-\frac{d}{b}+E\mathrm{e}^{-3b\xi}}-\frac{\sqrt{A}}{\sqrt{\mu^2 B}}\right), \tag{30}$$

$$\psi(x,y,t) = -\frac{2\left(bE\sqrt{A}-d\mathrm{e}^{3b\xi}\left(\sqrt{A}-3\sqrt{2}b\sqrt{\mu^2\left(-1+\sigma^2+2\sigma^4\right)}\right)\right)^2}{\left(d\mathrm{e}^{3b\xi}-bE\right)^2\left(1+\sigma^2\right)}, \tag{31}$$

where $A = \beta+\lambda^2+\kappa^2\sigma^2$ and $B = 1-3\sigma^2+2\sigma^4$.

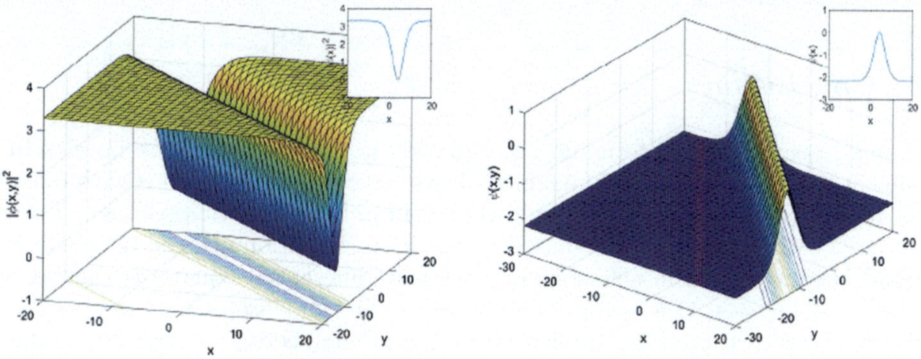

Fig. 3. 3D surfaces with its 2D figures of Eqs. (30) and (31) with values $d=-1, E=1, \sigma=2, \kappa=1, t=1/2, \mu=0.1, \alpha=0.5, \lambda=1$ and $y=2$ for 2D surface.

Case 2b. When $a_0 = -\frac{\mu\sqrt{\beta+\lambda^2+\kappa^2\sigma^2}\sqrt{1-3\sigma^2+2\sigma^4}}{\sqrt{\kappa}\sqrt{\mu^2(-1+\sigma^2+2\sigma^4)}}$, $a_1 = 0$, $a_2 = 0$, $a_3 = -\frac{3d\mu\sqrt{2-6\sigma^2+4\sigma^4}}{\sqrt{\kappa}}$, $b = \frac{\sqrt{2}\sqrt{\beta+\lambda^2+\kappa^2\sigma^2}}{3\sqrt{\mu^2(-1+\sigma^2+2\sigma^4)}}$, we obtain (Fig. 4)

$$\phi(x,y,t) = \mathrm{e}^{\mathrm{i}(\beta t+\kappa x+\lambda y)}\left(\begin{array}{l} a_0+\frac{a_3}{-\frac{d}{b}+E\mathrm{e}^{-3b\xi}}+\frac{a_2}{\left(-\frac{d}{b}+E\mathrm{e}^{-3b\xi}\right)^{2/3}}+ \\ \frac{a_1}{\left(-\frac{d}{b}+E\mathrm{e}^{-3b\xi}\right)^{1/3}} \end{array}\right), \tag{32}$$

$$\psi(x,y,t) = \kappa\left(a_0+\frac{a_3}{-\frac{d}{b}+E\mathrm{e}^{-3b\xi}}+\frac{a_2}{\left(-\frac{d}{b}+E\mathrm{e}^{-3b\xi}\right)^{2/3}}+\frac{a_1}{\left(-\frac{d}{b}+E\mathrm{e}^{-3b\xi}\right)^{1/3}}\right)^2. \tag{33}$$

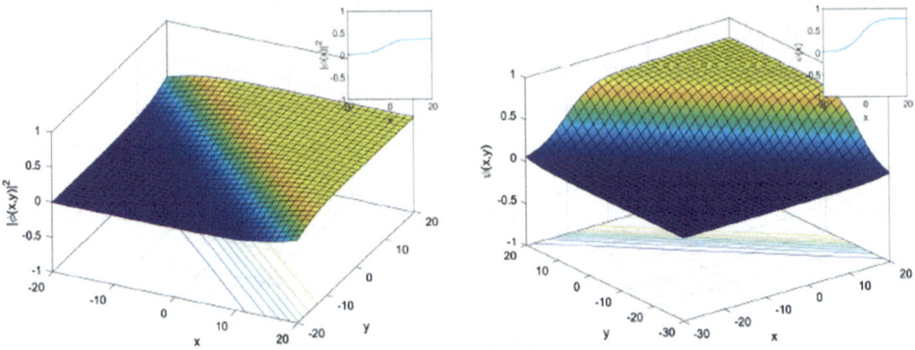

Fig. 4. 3D surfaces with its 2D figures of Eqs. (32) and (33) with values $b = 1, d = -1, a_0 = 0, a_1 = 0.4, a_2 = 0.1, a_3 = 0.1, E = 1, t = 0.5, \mu = 0.1, \alpha = 0.5, \lambda = 2$ and $y = 2$ for 2D surface.

4 Conclusion

In this researcher, the Bernoulli sub-equation is used to find some novel solutions of (2 + 1)-dimensional imaginary Davey-Stewartson equations with different physical parameters by utilizing the Wolfram Mathematica package. These methods with using computer-based symbolic computation utilized to construct broad classes of soliton solutions of nonlinear differential equations that arise in applied physics. Our resultant may appreciate and useful in some sciences like mathematical physics, applied physics, and engineering in terms of nonlinear science. Moreover, the method proposed in this paper, should be reliable, effective, provide more solutions as well. These methods may be applied to other nonlinear partial differential equations.

References

1. Ilhan, O.A., Esen, A., Bulut, H., Baskonus, H.M.: Singular solitons in the pseudo-parabolic model arising in nonlinear surface waves. Results Phys. (2019). https://doi.org/10.1016/j.rinp.2019.01.059
2. Aktürk, T., Gürefe, Y., Bulut, H.: New function method to the (n+1)-dimensional nonlinear problems. Int. J. Optim. Control Theor. Appl. (2017). https://doi.org/10.11121/ijocta.01.2017.00489
3. Kocak, Z. F., Bulut, H., Yel, G.: The solution of fractional wave equation by using modified trial equation method and homotopy analysis method. In AIP Conference Proceedings (2014)
4. Nofal, T.A.: An approximation of the analytical solution of the Jeffery-Hamel flow by homotopy analysis method. Appl. Math. Sci. **5**(53), 2603–2615 (2011)
5. Sulaiman, T.A., Bulut, H., Yokus, A., Baskonus, H.M.: On the exact and numerical solutions to the coupled Boussinesq equation arising in ocean engineering. Indian J. Phys. (2019). https://doi.org/10.1007/s12648-018-1322-1

6. Yousif, M.A., Mahmood, B.A., Ali, K.K., Ismael, H.F.: Numerical simulation using the homotopy perturbation method for a thin liquid film over an unsteady stretching sheet. Int. J. Pure Appl. Math. **107**(2) (2016). https://doi.org/10.12732/ijpam.v107i2.1
7. Yokus, A., Baskonus, H.M., Sulaiman, T.A., Bulut, H.: Numerical simulation and solutions of the two-component second order KdV evolutionarysystem. Numer. Methods Partial Differ. Equ. (2018). https://doi.org/10.1002/num.22192
8. Atangana, A., Ahmed, A., Oukouomi Noutchie, S.C.: On the Hamilton-Jacobi-Bellman equation by the homotopy perturbation method. Abstr. Appl. Anal. **2014**, 8 (2014)
9. Bueno-Orovio, A., Pérez-García, V.M., Fenton, F.H.: Spectral methods for partial differential equations in irregular domains: the spectral smoothed boundary method. SIAM J. Sci. Comput. **28**(3), 886–900 (2006)
10. Bulut, H., Ergüt, M., Asil, V., Bokor, R.H.: Numerical solution of a viscous incompressible flow problem through an orifice by Adomian decomposition method. Appl. Math. Comput. **153**(3), 733–741 (2004)
11. Ismael, H.F., Ali, K.K.: MHD casson flow over an unsteady stretching sheet. Adv. Appl. Fluid Mech. (2017). https://doi.org/10.17654/FM020040533
12. Owolabi, K.M., Atangana, A.: On the formulation of Adams-Bashforth scheme with Atangana-Baleanu-Caputo fractional derivative to model chaotic problems. Chaos Interdiscip. J. Nonlinear Sci. **29**(2), 23111 (2019)
13. Baskonus, H.M., Bulut, H.: On the numerical solutions of some fractional ordinary differential equations by fractional Adams-Bashforth-Moulton method. Open Math. (2015). https://doi.org/10.1515/math-2015-0052
14. Ismael, H.F.: Carreau-Casson fluids flow and heat transfer over stretching plate with internal heat source/sink and radiation. Int. J. Adv. Appl. Sci. J. **6**(2), 81–86 (2017). https://doi.org/10.1371/journal.pone.0002559
15. Ali, K.K., Ismael, H.F., Mahmood, B.A., Yousif, M.A.: MHD Casson fluid with heat transfer in a liquid film over unsteady stretching plate. Int. J. Adv. Appl. Sci. **4**(1), 55–58 (2017)
16. Ismael, H.F., Arifin, N.M.: Flow and heat transfer in a Maxwell liquid sheet over a stretching surface with thermal radiation and viscous dissipation. JP J. Heat Mass Transf. **15**(4) (2018). https://doi.org/10.17654/HM015040847
17. Zeeshan, A., Ismael, H.F., Yousif, M.A., Mahmood, T., Rahman, S.U.: Simultaneous effects of slip and wall stretching/shrinking on radiative flow of magneto nanofluid through porous medium. J. Magn. **23**(4), 491–498 (2018). https://doi.org/10.4283/JMAG.2018.23.4.491
18. Baskonus, H.M., Bulut, H., Sulaiman, T.A.: New complex hyperbolic structures to the Lonngren-wave equation by using sine-Gordon expansion method. Appl. Math. Nonlinear Sci. **4**(1), 141–150 (2019)
19. Eskitaşçıoğlu, EI., Aktaş, M.B., Baskonus, H.M.: New complex and hyperbolic forms for Ablowitz-Kaup-Newell-Segur wave equation with fourth order. Appl. Math. Nonlinear Sci. **4**(1), 105–112 (2019)
20. Vakhnenko, V.O., Parkes, E.J., Morrison, A.J.: A Bäcklund transformation and the inverse scattering transform method for the generalised Vakhnenko equation. Chaos Solitons Fractals (2003). https://doi.org/10.1016/S0960-0779(02)00483-6
21. Hammouch, Z., Mekkaoui, T.: Traveling-wave solutions of the generalized Zakharov equation with time-space fractional derivatives. J. MESA **5**(4), 489–498 (2014)
22. Baskonus, H.M., Bulut, H.: An effective schema for solving some nonlinear partial differential equation arising in nonlinear physics. Open Phys. (2015). https://doi.org/10.1515/phys-2015-0035

23. Baskonus, H.M., Bulut, H.: Exponential prototype structures for (2+1)-dimensional Boiti-Leon-Pempinelli systems in mathematical physics. Waves Random Complex Media (2016). https://doi.org/10.1080/17455030.2015.1132860
24. Wei, G., Ismael, H.F., Bulut, H., Baskonus, H.M.: Instability modulation for the (2+1)-dimension paraxial wave equation and its new optical soliton solutions in Kerr media. Phys. Scr. (2019). http://iopscience.iop.org/10.1088/1402-4896/ab4a50
25. Ilhan, O.A., Bulut, H., Sulaiman, T.A., Baskonus, H.M.: Dynamic of solitary wave solutions in some nonlinear pseudoparabolic models and Dodd–Bullough–Mikhailov equation. Indian J. Phys. (2018). https://doi.org/10.1007/s12648-018-1187-3
26. Cattani, C., Sulaiman, T.A., Baskonus, H.M., Bulut, H.: Solitons in an inhomogeneous Murnaghan's rod. Eur. Phys. J. Plus (2018). https://doi.org/10.1140/epjp/i2018-12085-y
27. Houwe, A., Hammouch, Z., Bienvenue, D., Nestor, S., Betchewe, G.: Nonlinear Schrödingers equations with cubic nonlinearity: M-derivative soliton solutions by $\exp(-\Phi(\xi))$-expansion method (2019)
28. Manafian, J., Aghdaei, M.F.: Abundant soliton solutions for the coupled Schrödinger-Boussinesq system via an analytical method. Eur. Phys. J. Plus (2016). https://doi.org/10.1140/epjp/i2016-16097-3
29. Hammouch, Z., Mekkaoui, T., Agarwal, P.: Optical solitons for the Calogero-Bogoyavlenskii-Schiff equation in (2 + 1) dimensions with time-fractional conformable derivative. Eur. Phys. J. Plus (2018). https://doi.org/10.1140/epjp/i2018-12096-8
30. Khalique, C.M., Mhlanga, I.E.: Travelling waves and conservation laws of a (2+1)-dimensional coupling system with Korteweg-de Vries equation. Appl. Math. Nonlinear Sci. (2018). https://doi.org/10.21042/amns.2018.1.00018
31. Aghdaei, M.F., Manafian, J.: Optical soliton wave solutions to the resonant davey-stewartson system. Opt. Quantum Electron. (2016). https://doi.org/10.1007/s11082-016-0681-0
32. Yang, X., Yang, Y., Cattani, C., Zhu, C.M.: A new technique for solving the 1-D Burgers equation. Therm. Sci. (2017). https://doi.org/10.2298/TSCI17S1129Y
33. Bulut, H., Sulaiman, T.A., Baskonus, H.M.: Dark, bright optical and other solitons with conformable space-time fractional second-order spatiotemporal dispersion. Optik (Stuttg). (2018). https://doi.org/10.1016/j.ijleo.2018.02.086
34. Cattani, C., Sulaiman, T.A., Baskonus, H.M., Bulut, H.: On the soliton solutions to the Nizhnik-Novikov-Veselov and the Drinfel'd-Sokolov systems. Opt. Quantum Electron. (2018). https://doi.org/10.1007/s11082-018-1406-3
35. Osman, M.S., Ghanbari, B.: New optical solitary wave solutions of Fokas-Lenells equation in presence of perturbation terms by a novel approach. Optik (Stuttg). (2018). https://doi.org/10.1016/j.ijleo.2018.08.007
36. Ghanbari, B., Kuo, C.-K.: New exact wave solutions of the variable-coefficient (1 + 1)-dimensional Benjamin-Bona-Mahony and (2 + 1)-dimensional asymmetric Nizhnik-Novikov-Veselov equations via the generalized exponential rational function method. Eur. Phys. J. Plus **134**(7), 334 (2019)
37. Ebadi, G., Biswas, A.: The G'/G method and 1-soliton solution of the Davey-Stewartson equation. Math. Comput. Model. **53**(5–6), 694–698 (2011)
38. Zedan, H.A., Al Saedi, A.: Periodic and solitary wave solutions of the Davey-Stewartson equation. Appl. Math. Inf. Sci. **4**(2), 253–260 (2010)
39. Besse, C., Mauser, N.J., Stimming, H.P.: Numerical study of the Davey-Stewartson system. ESAIM Math. Model. Numer. Anal. **38**(6), 1035–1054 (2004)

40. Ye, X.: On the fully discrete Davey-Stewartson system with self-consistent sources. Pacific J. Appl. Math. **7**(3), 163 (2015)
41. Li, Z.-F., Ruan, H.-Y.: (2+1)-dimensional Davey-Stewartson II equation for a two-dimensional nonlinear monatomic lattice. Zeitschrift für Naturforsch. A **61**(1–2), 45–52 (2006)
42. Baskonus, H.M.: New acoustic wave behaviors to the Davey-Stewartson equation with power-law nonlinearity arising in fluid dynamics. Nonlinear Dyn. (2016). https://doi.org/10.1007/s11071-016-2880-4
43. Abdelaziz, M.A.M., Moussa, A.E., Alrahal, D.M.: Exact solutions for the nonlinear (2+1)-dimensional Davey-Stewartson equation using the generalized (G'/G)-expansion method. J. Math. Res. **6**(2) (2014)
44. Gurefe, Y., Misirli, E., Pandir, Y., Sonmezoglu, A., Ekici, M.: New exact solutions of the Davey-Stewartson equation with power-law nonlinearity. Bull. Malaysian Math. Sci. Soc. **38**(3), 1223–1234 (2015)
45. Cevikel, A.C., Bekir, A.: New solitons and periodic solutions for (2+1)-dimensional Davey-Stewartson equations. Chin. J. Phys. **51**(1), 1–13 (2013)
46. El-Kalaawy, O.H., Ibrahim, R.S.: Solitary wave solution of the two-dimensional regularized long-wave and Davey-Stewartson equations in fluids and plasmas. Appl. Math. **3**(08), 833 (2012)
47. Baskonus, H.M., Bulut, H.: On the complex structures of Kundu-Eckhaus equation via improved Bernoulli sub-equation function method. Waves Random Complex Media (2015). https://doi.org/10.1080/17455030.2015.1080392
48. Baskonus, H.M., Bulut, H.: An effective schema for solving some nonlinear partial differential equation arising in nonlinear physics. Open Phys. (2015).https://doi.org/10.1515/phys-2015-0035
49. Anker, D., Freeman, N.C.: On the soliton solutions of the Davey-Stewartson equation for long waves. Proc. R. Soc. London Ser. A (1978). https://doi.org/10.1098/rspa.1978.0083
50. Mirzazadeh, M.: Soliton solutions of Davey-Stewartson equation by trial equation method and ansatz approach. Nonlinear Dyn. **82**(4), 1775–1780 (2015)

Radiative MHD Flow of Third-Grade Fluid Towards a Stretched Cylinder

Anum Shafiq[1,2], Z. Hammouch[3(✉)], and Hakan F. Oztop[4]

[1] Department of Mathematics, Preston University Kohat, Islamabad Campus, Kohat, Pakistan
anumshafiq@ymail.com

[2] School of Mathematics and Statistics, Nanjing University of Information Science and Technology, Nanjing 210044, China

[3] FST Errachidia, Moulay Ismail University of Meknes, BP 509 Boutalamine, 52000 Errachidia, Morocco
z.hammouch@fste.umi.ac.ma

[4] Faculty of Technology, Firat University, Elazig, Turkey
hakanfoztop@firat.edu.tr

Abstract. This work is concerned with magnetohydrodynamic (MHD) stagnation point flow of third-grade fluid due to a stretching cylinder. Thermal radiation effects are considered in the analysis of heat transfer phenomenon. Joule heating and viscous dissipation effects are also retained. The resulting nonlinear system is computed for the series solutions. Influence of various physical parameters on the velocity and temperature profiles are scrutinized graphically. Comparison between Newtonian and third-grade fluids is made. Velocity and temperature profiles in the presence/absence of stagnation point are discussed graphically. Numerical values of skin friction and Nusselt number are also computed and interpreted. The comparison is conducted between Newtonian and third-grade fluids velocities in the presence of magnetohydrodynamic over a flat plat for two cases (i) without stagnation point (ii) with stagnation point. It is observed that in the presence of MHD, the velocity profile is higher for third-grade fluid for both the cases. On the other hand it is also examined that the velocity profile is higher for both Newtonian and third-grade fluids with stagnation point.

Keywords: Viscous dissipation · Stagnation point · Thermal radiation · Magnetohydrodynamics (MHD) · Third-grade fluid · Stretching cylinder

1 Introduction

The boundary layer theory has been effectively applied to non-Newtonian fluid models and has gained a lot of attention during the last few decades. Such flows play very important role in reducing the drag forces and increase the heat transfer rate. The viscoelastic features of non-Newtonian fluids in general are

H. Dutta et al. (Eds.): CMES 2019, AISC 1111, pp. 166–185, 2020.
https://doi.org/10.1007/978-3-030-39112-6_12

classified by three categories namely the differential, rate and integral types. The simplest subclass of differential type materials is second-grade. It should be noted that second-grade fluid captures the normal stress effect whereas the shear thinning and shear thickening properties even in steady flow situation can be only analyzed by third-grade fluid. Aforesaid analysis with magnetohydrodynamic phenomenon has ample applications in industries and technology. It has many practical applications such as to design the MHD generators, accelerators, cooling systems, geothermal energy extractions etc. Some relevant literature on this topic can be surveyed through recent studies [1–16] and many few references therein.

The thermal radiation effect has a pivotal role for controlling heat transfer processes in polymer processing industry. The final product's quality depends on controlling heat factors to a certain extent. The design of several advanced energy convection systems which function at high temperature are due to the effects of thermal radiation in flow and heat transfer processes. Thermal radiation occurs in these systems, due to the emission by the hot walls and the working fluid. Its effects become more significant with the increase in the difference between the surface and the ambient temperature. Therefore, this proves the fact that thermal radiation is indeed one of the vital factors in controlling the heat transfer process. The knowledge of radiation heat transfer in the system may even lead to a desired product with sought features. Hayat et al. [17] reported the effect of thermal radiation on the boundary layer flow with heat transfer in third grade fluid along a permeable stretching surface. Mushtaq et al. [18] investigated the effect of thermal radiation on the two-dimensional incompressible flow of upper-convected Maxwell (UCM) fluid and solved numerically by the shooting method using a fourth-order Runge-Kutta integration technique. Bhattacharyya et al. [19] analyzed the flow of micropolar fluid and heat transfer past a porous shrinking surface with thermal radiation and get the dual solution for the several values of different parameters. Mukhopadhyay [20] examined the boundary layer flow and heat transfer along a porous exponential stretching surface in presence of a magnetic field and thermal radiation. Velocity slip and thermal slip are taken into account and solved by shooting technique. Rashidi et al. [21] reported the analytical solution of free convective heat and mass transfer in a steady non-uniform magnetohydrodynamic fluid flow over a stretching vertical surface embedded in a porous medium with thermal radiation effects. Mukhopadhyay [22] is performed similarity analysis to study the structure of the boundary layer stagnation point flow and heat transfer over a stretching plate subject to suction with variable viscosity and thermal radiation. Heat transfer of a steady, incompressible water based nanofluid flow over a stretching surface in the presence of transverse magnetic field with thermal radiation and buoyancy effect are numerically investigated by Rashidi et al. [23]. Chamkha et al. [24] analyzed the problem of steady mixed convection boundary layer flow over an isothermal vertical wedge embedded in a porous medium saturated with a nanofluid and thermal radiation effects are also considered. Sheikholeslami et al. [25] studied the effect of thermal radiation on magnetohydrodynamics nanofluid flow between two horizontal rotating plates. Seini and Makinde [26] have been investigated the magnetohydrodynamic boundary layer flow due to exponential stretching surface with radiation and chemical reaction.

Many researchers have been focused on stagnation point flow and heat transfer along a stretching surface due to its widespread applications. Its applications contain the flow over the tips of the submarines, rockets, oil ships and air crafts. Bhattacharyya [27] examined the problem of unsteady boundary layer flow along a shrinking/stretching surface in the region of stagnation point. Turkyilmazoglu and Pop [28] described the stagnation point flow of Jeffrey fluid over a stretching/shrinking plate with parallel external flow. Nandy and Mahapatra [29] analyzed influences of velocity slip and heat generation/absorption in MHD flow and heat transfer over convectively heated stretching/shrinking surface in the presence of nanoparticle fractions. The mixed convection stagnation point flow of a non-Newtonian fluid along stretching surface with convective boundary condition was reported by Hayat et al. [30]. Bhattacharyya [31] described the stagnation point flow of Casson fluid over a shrinking/stretching sheet with heat transfer. Rashidi and Freidoonimehr [32] studied the entropy generation in MHD stagnation point flow through a porous medium. Bhattacharyya et al. [33] investigated the reactive solute distribution for the laminar stagnation point boundary layer flow past a stretching sheet subject to suction or blowing. Hayat et al. [34] describe the boundary layer stagnation point flow of Jeffrey fluid near a stretching plate in the presence of Soret and Dufour effects and melting heat transfer.

It is clear from the previous study that proper attention has not been focused to the boundary layer flow of non-Newtonian fluids by a stretching cylinder. So our intention here is to analyze the heat transfer in MHD stagnation point flow of third-grade fluid along a stretching cylinder. The analysis has been carried out in the presence of thermal radiation and heat generation/absorption. The viscous dissipation and Joule heating effects are also taken into account. Hence the governing mathematical problems are solved for the convergent series solutions by using analytical technique named as homotopy analysis method (HAM) [35–41]. Nusselt number is computed. Graphical results and numerical values are interpreted.

2 Mathematical Formulation

Consider magnetohydrodynamic stagnation point flow of an electrically conducting third-grade fluid due to a stretching cylinder with thermal radiation. Heat transfer is analyzed in the presence of Joule heating and viscous dissipation effects. Cylindrical coordinates are chosen in such a way that $z-$axis is along the axis of stretching cylinder and $r-$axis normal to it (Fig. 1). Under the boundary layer approximations (i.e, $u = O(\delta)$, $r = O(\delta)$, $w = O(1)$ and $z = O(1)$) the laws of conservation of mass and momentum give

$$\frac{\partial u}{\partial r} + \frac{u}{r} + \frac{\partial w}{\partial z} = 0, \tag{1}$$

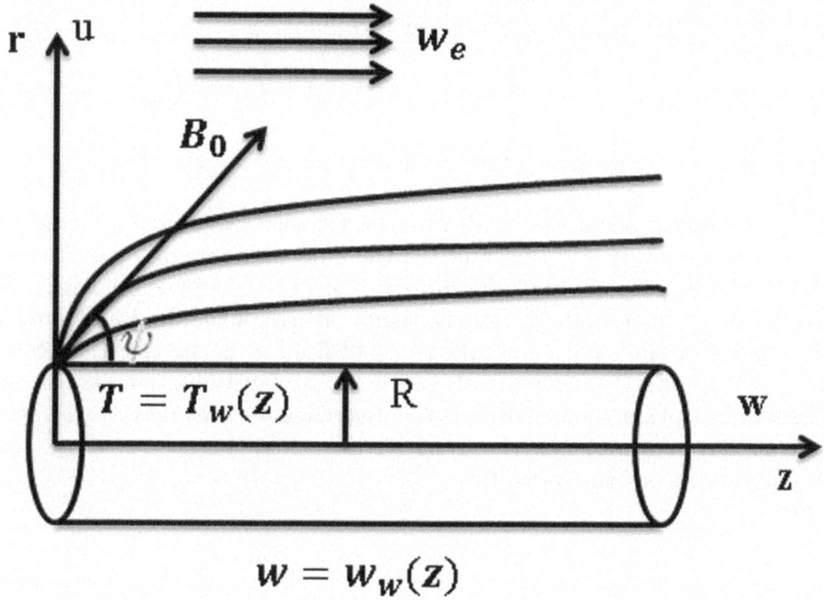

Fig. 1. Schematic diagram of the problem.

$$
\begin{aligned}
u\frac{\partial w}{\partial r} + w\frac{\partial w}{\partial z} &= W_e\frac{dW_e}{dz} + \nu\left(\frac{\partial^2 w}{\partial r^2} + \frac{1}{r}\frac{\partial w}{\partial r}\right) + \frac{\alpha_1^*}{\rho}\left[\frac{w}{r}\frac{\partial^2 w}{\partial r\partial z} + \frac{u}{r}\frac{\partial^2 w}{\partial r^2} + \frac{3}{r}\frac{\partial w}{\partial r}\frac{\partial w}{\partial z}\right.\\
&\quad + \frac{1}{r}\frac{\partial u}{\partial r}\frac{\partial w}{\partial r} + 4\frac{\partial w}{\partial r}\frac{\partial^2 w}{\partial r\partial z} + w\frac{\partial^3 w}{\partial r^2\partial z} + 2\frac{\partial u}{\partial r}\frac{\partial^2 w}{\partial r^2}\\
&\quad \left. + u\frac{\partial^3 w}{\partial r^3} + 3\frac{\partial^2 w}{\partial r^2}\frac{\partial w}{\partial z} + \frac{\partial^2 u}{\partial r^2}\frac{\partial w}{\partial r}\right] + \frac{\alpha_2^*}{\rho}\left[\frac{2}{r}\frac{\partial u}{\partial r}\frac{\partial w}{\partial r} + \frac{2}{r}\frac{\partial w}{\partial r}\frac{\partial w}{\partial z}\right.\\
&\quad \left. + 2\frac{\partial^2 u}{\partial r^2}\frac{\partial w}{\partial r} + 2\frac{\partial u}{\partial r}\frac{\partial^2 w}{\partial r^2} + 2\frac{\partial^2 w}{\partial r^2}\frac{\partial w}{\partial z} + 4\frac{\partial w}{\partial r}\frac{\partial^2 w}{\partial r\partial z}\right]\\
&\quad + \frac{\beta_3^*}{\rho}\left[\frac{2}{r}\left(\frac{\partial w}{\partial r}\right)^3 + 6\left(\frac{\partial w}{\partial r}\right)^2\frac{\partial^2 w}{\partial r^2}\right] + \frac{\sigma B_0^2}{\rho}\sin^2\psi\,(W_e - w). \qquad (2)
\end{aligned}
$$

$$
\begin{aligned}
u\frac{\partial T}{\partial r} + w\frac{\partial T}{\partial z} &= \frac{k}{\rho c_p}\left(\frac{\partial^2 T}{\partial r^2} + \frac{1}{r}\frac{\partial T}{\partial r}\right) + \frac{16\sigma^* T_\infty^3}{3k^*\rho c_p}\left(\frac{\partial^2 T}{\partial r^2} + \frac{1}{r}\frac{\partial T}{\partial r}\right) + \frac{\nu}{c_p}\left(\frac{\partial w}{\partial r}\right)^2\\
&\quad + \frac{\alpha_1^*}{\rho c_p}\left[3\frac{\partial u}{\partial r}\left(\frac{\partial w}{\partial r}\right)^2 + u\frac{\partial w}{\partial r}\frac{\partial^2 w}{\partial r^2} + w\frac{\partial w}{\partial r}\frac{\partial^2 w}{\partial r\partial z} + 3\left(\frac{\partial w}{\partial r}\right)^2\frac{\partial w}{\partial z}\right]\\
&\quad + 2\frac{\beta_3^*}{\rho c_p}\left(\frac{\partial w}{\partial r}\right)^4 + \frac{\sigma B_0^2}{\rho c_p}\sin^2\psi\,(w - W_e)^2, \qquad (3)
\end{aligned}
$$

The subjected conditions can be mentioned as follows:

$$w(r,z) = W_w(z) = \frac{W_0 z}{l},\ u(r,z) = 0,\ T(r,z) = T_\infty + b\left(\frac{z}{l}\right),\ \text{at } r = R,$$

$$w(r,z) \longrightarrow W_e(z) = \frac{W_\infty z}{l},\quad T(r,z) \longrightarrow T_\infty,\ \text{at } r \longrightarrow \infty. \tag{4}$$

In the above expressions u and w denote the velocity components in the r and z directions respectively, (α_1^*, α_2^* and β_3) the fluid parameters, ν the kinematic viscosity, ρ the density of fluid, W_0 and W_∞ are the reference velocities, l the characteristic length, T and T_∞ are the temperatures of the fluid and surrounding respectively, k the thermal conductivity of fluid, σ is the electrical conductivity, B_0 is the applied magnetic field, σ^* is the Stefan-Boltzmann constant, k^* is the mean absorption coefficient, b is the dimensional constants, c_p the specific heat at constant pressure, q_r is the radiative heat flux, W_w is stretching velocity and W_∞ is the free stream velocity.

We use

$$\eta = \sqrt{\frac{W_0}{\nu l}}\left(\frac{r^2 - R^2}{2R}\right),\quad w(r,z) = \frac{W_0\ z}{l} f'(\eta),\ u(r,z) = -\sqrt{\frac{\nu W_0}{l}}\frac{R}{r} f(\eta),\ \theta = \frac{T - T_\infty}{T_w - T_\infty}, \tag{5}$$

Incompressibility condition is identically satisfied and the Eqs. (2)–(4) can be written as

$$\begin{aligned}
&(1+2\gamma\eta) f''' + A^2 + 2\gamma f'' - f'^2 + f f'' + \alpha_1\left[(1+2\gamma\eta)\left\{2f' f''' - f f^{(iv)} + 3f''^2\right\}\right.\\
&\left. +\gamma\left(6f' f'' - 2f f'''\right)\right] + \alpha_2\left[2(1+2\gamma\eta) f''^2 + \gamma\left(2f' f'' + 2f f'''\right)\right]\\
&+\beta\ \mathrm{Re}\left[6(1+2\gamma\eta)^2 f''^2 f''' + 8\gamma(1+2\gamma\eta) f''^3\right] + M^2 \sin^2\psi\,(A - f') = 0,
\end{aligned} \tag{6}$$

$$\begin{aligned}
&\left(1+\frac{4}{3}R_d\right)(1+2\gamma\eta)\theta'' + 2\gamma\left(1+\frac{4}{3}R_d\right)\theta' + \Pr Ec\ (1+2\gamma\eta) f''^2\\
&+ \Pr\left(f'\ \theta - f\ \theta'\right) + \alpha_1 \Pr Ec\left[2\gamma f f''^2 + (1+2\gamma\eta) f' f''^2 - (1+2\gamma\eta) f f'' f'''\right]\\
&+ \beta \Pr Ec\ \mathrm{Re}\,(1+2\gamma\eta) f''^4 + M^2 \sin^2\psi \Pr Ec\,(f' - A)^2 = 0,
\end{aligned} \tag{7}$$

$$f(0) = 0,\ f'(0) = 1,\ f'(\infty) \to A,\ \theta(0) = 1,\ \theta(\infty) \to 0, \tag{8}$$

where γ is the curvature parameter, Re is the Reynolds number, M is the magnetic parameter, (α_1, α_2, β) are the fluid parameters, A is the ratio of velocities, R_d is the radiation parameter, Pr is the Prandtl number and Ec is the Eckert number. These parameters are defined as follows:

$$\begin{aligned}
&\gamma = \left(\frac{\nu l}{W_0 R^2}\right)^{1/2},\quad \mathrm{Re} = \frac{W z}{\nu},\quad M^2 = \frac{\sigma B_0^2 l}{\rho W_0},\quad \alpha_1 = \frac{\alpha_1^* W_0}{l\mu}, \alpha_2 = \frac{\alpha_2^* W_0}{l\mu},\\
&\beta = \frac{\beta_3 W_0^2}{l^2\mu},\quad A = \frac{W_\infty}{W_0},\quad R_d = \frac{4\sigma^* T_\infty^3}{k^* k},\quad \Pr = \frac{\mu c_p}{k},\quad Ec = \frac{W_0^2 (z/l)^2}{c_p (T_w - T_\infty)}.
\end{aligned} \tag{9}$$

The local skin friction coefficient is defined as

$$C_f = \frac{\tau_{rz}}{\rho U_w^2} = \frac{\tau_{rz}|_{r=R}}{\rho U_w^2}$$

$$\underset{z}{\overset{1/2}{\mathrm{Re}}}\, C_f = \left[f''(0) + 3\alpha_1 f''(0) + 2\beta \text{ Re } f''^3(0) \right]. \tag{10}$$

The Nusselt number is given by

$$Nu_z = \frac{zq_w}{k(T_w - T_\infty)} = -\frac{z\left(k + \frac{16\sigma^* T_\infty^3}{3k_1}\right)\frac{\partial T}{\partial y}\Big|_{r=R}}{k(T_w - T_\infty)}$$

$$\underset{z}{\overset{-1/2}{\mathrm{Re}}}\; Nu_z = -\left(1 + \frac{4}{3}R_d\right)\theta'(0), \tag{11}$$

in which $\mathrm{Re}_z = \frac{Wz}{\nu}$ is the local Reynolds number.

3 Homotopic Solutions

The velocity and temperature can be expressed in the set of base functions

$$\left\{ \eta^k \exp(-n\eta) \big| \, k \geq 0, \; n \geq 0 \right\}, \tag{12}$$

can be expressed as follows

$$f(\eta) = a_{0,0}^0 + \sum_{n=0}^{\infty}\sum_{k=0}^{\infty} a_{m,n}^k \eta^k \exp(-n\eta), \tag{13}$$

$$\theta(\eta) = \sum_{n=0}^{\infty}\sum_{k=0}^{\infty} b_{m,n}^k \eta^k \exp(-n\eta), \tag{14}$$

where $a_{m,n}^k$ and $b_{m,n}^k$ are the coefficients. The initial guesses and linear operators for the dimensionless momentum and energy equations are (f_0, θ_0) and $(\mathcal{L}_f, \mathcal{L}_\theta)$. The chosen initial guesses and linear operators are given by

$$f_0(\eta) = A\eta + (1 - A)(1 - \exp(-\eta)), \tag{15}$$

$$\theta_0(\eta) = \exp(-\eta), \tag{16}$$

$$\mathcal{L}_f(f) = \frac{d^3 f}{d\eta^3} - \frac{df}{d\eta}, \; \mathcal{L}_\theta(\theta) = \frac{d^2\theta}{d\eta^2} - \theta, \tag{17}$$

satisfy the following properties

$$\mathcal{L}_f \; [C_1 + C_2 \exp(\eta) + C_3 \exp(-\eta)] = 0, \tag{18}$$

$$\mathcal{L}_\theta [C_4 \exp(\eta) + C_5 \exp(-\eta)] = 0, \tag{19}$$

where $C_i (i = 1 - 5)$ depict the arbitrary constants.

3.1 Zeroth Order Problems

The zeroth order problems are

$$(1-p)\mathcal{L}_f[\hat{f}(\eta,p)-f_0(\eta)] = p\hbar_f\mathcal{N}_f\left[\hat{f}(\eta,p)\right], \tag{20}$$

$$\hat{f}(\eta;p)\Big|_{\eta=0}=0,\quad \frac{\partial \hat{f}'(\eta;p)}{\partial\eta}\Bigg|_{\eta=0}=1,\quad \frac{\partial \hat{f}'(\eta;p)}{\partial\eta}\Bigg|_{\eta=\infty}=A, \tag{21}$$

$$(1-p)\mathcal{L}_\theta[\hat{\theta}(\eta,p)-\theta_0(\eta)] = p\hbar_\theta\mathcal{N}_\theta\left[\hat{f}(\eta,p),\hat{\theta}(\eta,p)\right], \tag{22}$$

$$\hat{\theta}(\eta;p)\Big|_{\eta=0}=1,\quad \hat{\theta}(\eta;p)\Big|_{\eta=\infty}=0, \tag{23}$$

with non-linear operators $\mathcal{N}_f\left[\hat{f}(\eta,p)\right]$ and $\mathcal{N}_\theta\left[\hat{f}(\eta,p),\hat{\theta}(\eta,p)\right]$ defined by

$$\begin{aligned}\mathcal{N}_f\left[\hat{f}(\eta;p)\right] &= (1+2\gamma\eta)\frac{\partial^3\hat{f}(\eta,p)}{\partial\eta^3}+A^2+2\gamma\frac{\partial^2\hat{f}(\eta,p)}{\partial\eta^2}-\left(\frac{\partial\hat{f}(\eta,p)}{\partial\eta}\right)^2+\hat{f}(\eta,p)\frac{\partial^2\hat{f}(\eta,p)}{\partial\eta^2}\\
&+\alpha_1\left\{(1+2\gamma\eta)\left(2\frac{\partial\hat{f}(\eta,p)}{\partial\eta}\frac{\partial^3\hat{f}(\eta;q)}{\partial\eta^3}-\hat{f}(\eta,p)\frac{\partial^4\hat{f}(\eta,p)}{\partial\eta^4}+3\left(\frac{\partial^2\hat{f}(\eta,p)}{\partial\eta^2}\right)^2\right)\right.\\
&\left.+\gamma\left(6\frac{\partial\hat{f}(\eta,p)}{\partial\eta}\frac{\partial^2\hat{f}(\eta,p)}{\partial\eta^2}-2\hat{f}(\eta,p)\frac{\partial^3\hat{f}(\eta;q)}{\partial\eta^3}\right)\right\}\\
&+\alpha_2\left\{2(1+2\gamma\eta)\left(\frac{\partial^2\hat{f}(\eta,p)}{\partial\eta^2}\right)^2+\gamma\left(2\frac{\partial\hat{f}(\eta;p)}{\partial\eta}\frac{\partial^2\hat{f}(\eta,p)}{\partial\eta^2}+2\hat{f}(\eta,p)\frac{\partial^3\hat{f}(\eta;q)}{\partial\eta^3}\right)\right\}\\
&+\beta\,\mathrm{Re}\left[6(1+2\gamma\eta)^2\left(\frac{\partial^2\hat{f}(\eta,p)}{\partial\eta^2}\right)^2\frac{\partial^3\hat{f}(\eta;q)}{\partial\eta^3}+8\gamma(1+2\gamma\eta)\left(\frac{\partial^2\hat{f}(\eta,p)}{\partial\eta^2}\right)^3\right]\\
&+M^2\sin^2\psi\left\{A-\frac{\partial\hat{f}(\eta,p)}{\partial\eta}\right\},\end{aligned} \tag{24}$$

$$\begin{aligned}\mathcal{N}_\theta\left[\hat{f}(\eta;p),\hat{\theta}(\eta;p)\right] &= \left(1+\frac{4}{3}R_d\right)(1+2\gamma\eta)\frac{\partial^2\hat{\theta}(\eta;p)}{\partial\eta^2}+2\gamma\left(1+\frac{4}{3}R_d\right)\frac{\partial\hat{\theta}(\eta;p)}{\partial\eta}\\
&+\Pr Ec\,(1+2\gamma\eta)\left(\frac{\partial^2\hat{f}(\eta,p)}{\partial\eta^2}\right)^2+\Pr\left(\frac{\partial f(\eta;q)}{\partial\eta}\hat{\theta}(\eta;p)-\hat{f}(\eta,p)\frac{\partial\hat{\theta}(\eta;p)}{\partial\eta}\right)\\
&+\alpha_1\Pr Ec\left[2\gamma\hat{f}(\eta,p)\left(\frac{\partial^2\hat{f}(\eta,p)}{\partial\eta^2}\right)^2+(1+2\gamma\eta)\frac{\partial f(\eta;q)}{\partial\eta}\left(\frac{\partial^2\hat{f}(\eta,p)}{\partial\eta^2}\right)^2\right.\\
&\left.-(1+2\gamma\eta)\hat{f}(\eta,p)\frac{\partial^2\hat{f}(\eta,p)}{\partial\eta^2}\frac{\partial^3\hat{f}(\eta;q)}{\partial\eta^3}\right]+\beta\Pr Ec\,\mathrm{Re}\,(1+2\gamma\eta)\left(\frac{\partial^2\hat{f}(\eta;q)}{\partial\eta^2}\right)^4\\
&+M^2\sin^2\psi\Pr Ec\left[\frac{\partial\hat{f}(\eta;q)}{\partial\eta}-A\right]^2,\end{aligned} \tag{25}$$

in which $p\in[0,1]$ indicates the embedding parameter and $\hbar_f$ and $\hbar_\theta$ the nonzero auxiliary parameters.

3.2 mth-Order Deformation Problems

The mth-order deformation problems are

$$\mathcal{L}_f\left[f_m(\eta)-\chi_m f_{m-1}(\eta)\right]=\hbar_f\mathcal{R}_m^f(\eta), \tag{26}$$

$$\hat{f}_m(\eta;p)\Big|_{\eta=0}=0,\quad \frac{\partial\hat{f}_m'(\eta;p)}{\partial\eta}\Big|_{\eta=0}=0,\quad \frac{\partial\hat{f}_m'(\eta;p)}{\partial\eta}\Big|_{\eta=\infty}=0, \tag{27}$$

$$\mathcal{L}_\theta\left[\theta_m(\eta)-\chi_m\theta_{m-1}(\eta)\right]=\hbar_\theta\mathcal{R}_m^\theta(\eta), \tag{28}$$

$$\hat{\theta}_m(\eta;p)\Big|_{\eta=0}=0,\quad \hat{\theta}_m(\eta;p)\Big|_{\eta=\infty}=0, \tag{29}$$

$$\begin{aligned}\mathcal{R}_m^f(\eta)&=(1+2\gamma\eta)f_{m-1}'''(\eta)+A^2(1-\chi_m)+2\gamma f_{m-1}''+\sum_{k=0}^{m-1}\left(f_{m-1-k}f_k''-f_{m-1-k}'f_k'\right)\\&+\sum_{k=0}^{m-1}\alpha_1\Big[(1+2\gamma\eta)\left(2f_{m-1-k}'f_k'''-f_{m-1-k}f_k^{(iv)}+3f_{m-1-k}''f_k''\right)\\&+\gamma\left(6f_{m-1-k}'f_k''-2f_{m-1-k}f_k'''\right)\Big]+\alpha_2\sum_{k=0}^{m-1}\left[2(1+2\gamma\eta)f_{m-1-k}''f_k''\right.\\&\left.+\gamma\left(2f_{m-1-k}'f_k''+2f_{m-1-k}f_k'''\right)\right]+\beta\,\mathrm{Re}\sum_{k=0}^{m-1}\sum_{l=0}^{k}\Big[6(1+2\gamma\eta)^2f_{m-1-k}''f_{k-l}''f_l'''\\&+8\gamma(1+2\gamma\eta)f_{m-1-k}''f_{k-l}''f_l''\Big]+M^2\sin^2\psi\left[A(1-\chi_m)-f_{m-1}'\right],\end{aligned} \tag{30}$$

$$\begin{aligned}\mathcal{R}_m^\theta(\eta)&=\left(1+\frac{4}{3}R_d\right)(1+2\gamma\eta)\theta_{m-1}''+2\gamma\left(1+\frac{4}{3}R_d\right)\theta_{m-1}'+\Pr Ec(1+2\gamma\eta)\sum_{k=0}^{m-1}f_{m-1-k}''f_k''\\&+\Pr\sum_{k=0}^{m-1}\left(f_{m-1-k}'\theta_k-f_{m-1-k}\theta_k'\right)+\alpha_1\Pr Ec\Bigg[2\gamma\sum_{k=0}^{m-1}f_{m-1-k}\sum_{l=0}^{k}f_{k-l}''f_l''\\&+(1+2\gamma\eta)\sum_{k=0}^{m-1}f_{m-1-k}'\sum_{l=0}^{k}f_{k-l}''f_l''-(1+2\gamma\eta)\sum_{k=0}^{m-1}f_{m-1-k}\sum_{l=0}^{k}f_{k-l}''f_l'''\Bigg]\\&+\beta\Pr Ec\,\mathrm{Re}(1+2\gamma\eta)\sum_{k=0}^{m-1}f_{m-1-k}''\sum_{l=0}^{k}f_{k-l}''\sum_{s=0}^{l}f_{l-s}''f_s''\\&+M^2\sin^2\psi\Pr Ec\left[\sum_{k=0}^{m-1}f_{m-1-k}'f_k'-2Af_{m-1}'-A^2(1-\chi_m)\right]\end{aligned} \tag{31}$$

$$\chi_m=\begin{cases}0, & m\le 1\\ 1, & m>1\end{cases}. \tag{32}$$

Setting $p=0$ and $p=1$ then one has

$$\hat{f}(\eta;0)=f_0(\eta),\qquad \hat{f}(\eta;1)=f(\eta), \tag{33}$$

$$\hat{\theta}(\eta;0)=\theta_0(\eta),\qquad \hat{\theta}(\eta;1)=\theta(\eta). \tag{34}$$

When p varies from 0 to 1, $\hat{f}(\eta;p)$ and $\hat{\theta}(\eta;p)$ deforms from the initial solutions $f_0(\eta)$ and $\theta_0(\eta)$ to the final solutions $f(\eta)$ and $\theta(\eta)$, respectively. Taylor's series leads to the following relations

$$\hat{f}(\eta;p) = f_0(\eta) + \sum_{m=1}^{\infty} f_m(\eta)p^m, \qquad f_m(\eta) = \frac{1}{m!}\frac{\partial^m \hat{f}(\eta;p)}{\partial p^m}\bigg|_{p=0}, \tag{35}$$

$$\hat{\theta}(\eta;p) = \theta_0(\eta) + \sum_{m=1}^{\infty} \theta_m(\eta)p^m, \qquad \theta_m(\eta) = \frac{1}{m!}\frac{\partial^m \hat{\theta}(\eta;p)}{\partial p^m}\bigg|_{p=0}. \tag{36}$$

The auxiliary parameters are properly chosen such that the series solutions (35) and (36) converge at $p = 1$. Therefore

$$f(\eta) = f_0(\eta) + \sum_{m=1}^{\infty} f_m(\eta), \tag{37}$$

$$\theta(\eta) = \theta_0(\eta) + \sum_{m=1}^{\infty} \theta_m(\eta). \tag{38}$$

Denoting the special solutions by (f_m^*, θ_m^*) one can express the general solutions (f_m, θ_m) of Eqs. (26)–(29) as follows:

$$f_m(\eta) = f_m^*(\eta) + C_1 + C_2 \exp(\eta) + C_3 \exp(-\eta), \tag{39}$$

$$\theta_m(\eta) = \theta_m^*(\eta) + C_4 \exp(\eta) + C_5 \exp(-\eta), \tag{40}$$

in which the constants $C_i(i = 1 - 5)$ in veiw of the conditions (27) and (29) are

$$C_2 = 0 = C_4, \quad C_3 = \frac{\partial f_m^*(\eta)}{\partial \eta}\bigg|_{\eta = 0}, \quad C_1 = -C_3 - f_m^*(0), \quad C_5 = -\theta_m^*(0). \tag{41}$$

4 Convergence

To get the series solutions through homotopy analysis method it is important to check the convergence of the desired solutions. Such solutions involve the auxiliary parameters $\hbar_f$ and $\hbar_\theta$. These parameters are useful in adjusting and controlling the convergence region. Therefore $\hbar_f$ and $\hbar_\theta-$curves are plotted for 16th order of approximation in Fig. 2 for the suitable ranges of the auxiliary parameters. Here the suitable values for $\hbar_f$ and $\hbar_\theta$ are $-1.3 \leq \hbar_f < -0.4$ and $-0.9 \leq \hbar_\theta < -0.2$. Furthermore, convergence of series solution is checked and shown in Table 1. Note that the series solutions converge at 11th order of approximation up to 5 decimal places for the momentum equation and 12th order of approximation is enough for the temperature equation.

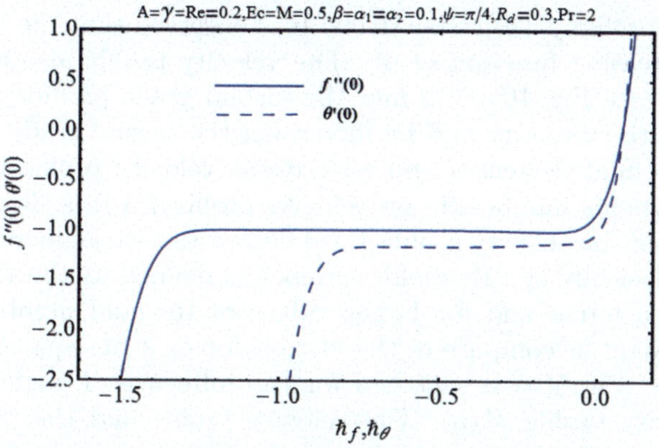

Fig. 2. $\hbar$–curves of the functions $f(\eta)$ and $\theta(\eta)$ at 16th order of approximation.

Table 1. Convergence of homotopy solutions when $\alpha_1 = \alpha_2 = \beta = 0.1$, $M = 0.5$, $\gamma = A = \mathrm{Re} = 0.2$, $\psi = \pi/4$, $R_d = 0.3$, $\mathrm{Pr} = 2$ and $Ec = 0.5$.

Order of approximation	$-f''(0)$	$-\theta'(0)$
1	0.98019	1.0966
2	1.06910	1.0973
5	1.13430	1.0458
11	1.13740	1.0316
12	1.13740	1.0315
14	1.13740	1.0315
50	1.13740	1.0315

5 Results and Discussion

This section illustrates the impact of physical parameters. The results are displayed graphically in the Figs. 3, 4, 5, 6, 7, 8, 9, 10 and 11. The conclusions for flow field and other physical quantities of interest are drawn. The numerical values of the skin friction coefficient and local Nusselt number are presented in the Tables 2 and 3 for various values of α_1, α_2, β, M, Re, R_d, Pr and Ec. Fig. 3(a) displays the effect of magnetic parameter M on velocity profile $f'(\eta)$ by keeping other physical parameter fixed. It is of interest to note that the velocity profile decreases with an increase in magnetic parameter M whereas the boundary layer thickness reduces. Clearly by increasing magnetic force, the Lorentz force increases which causes resistance in the fluid flow and consequently the velocity profile decreases. Fig. 3(b) shows the effect of third grade parameter β on the velocity profile $f'(\eta)$. Here it is examined that the velocity increases near the wall for larger values of β whereas it becomes vanishes away from the wall. Figs. 4(a) and (b) illustrate the behavior of second grade parameters α_1 and α_2

on the velocity profile $f'(\eta)$ respectively. It is observed that the velocity profile $f'(\eta)$ is an increasing function of α_1. The velocity profile also increases when α_2 is increased (see Fig. 4(b)). In fact the second grade parameter are directly proportional to the viscosity and by increasing the second grade parameter the viscosity of the fluid decreases and as a result velocity profile increased. The behavior of Reynolds number Re on velocity profile $f'(\eta)$ is shown in Fig. 5(a) It is observed that the velocity profile $f'(\eta)$ decreases with an increase in Reynold number Re. Physically the Reynolds number is defined as the ratio of inertial forces to viscous forces and for larger values of Reynold number the inertial forces are dominant as compare to the viscous forces. Consequently the velocity profile increases. Fig. 5(b) is sketched for the influence of angle of inclination ψ on the velocity profile $f'(\eta)$. The velocity profile and the thermal boundary layer decreases for larger values of ψ. In fact due to the larger values of angle of inclination the Lorentz forces are dominant and therefore the velocity profile decreases. The influence of curvature parameter γ is shown in Fig. 6(a) It is revealed that velocity and boundary layer thickness increase when curvature parameter γ increases. In fact with the increase of curvature parameter, the radius of curvature decreases which reduces the contact area of the cylinder with the fluid. Therefore resistance offered by the surface decreases and velocity of the fluid increases. The behavior of A on velocity profile $f'(\eta)$ is shown in Fig. 6(b) It is analyzed that velocity profile $f'(\eta)$ increases for both the cases $A > 1$ and $A < 1$. However the boundary layers in these two cases have opposite behavior. It is noticed that there is no boundary layer for $A = 1$.

Figure 7(a) is sketched for the behavior of angle of inclination ψ on temperature field $\theta(\eta)$. It is clear from the Fig. that temperature profile increases with an increase in angle of inclination ψ. Because Lorentz force increases with an increase in angle of inclination which is a resistive force. Hence more heat is produced due to the resistive forces. Therefore temperature profile $\theta(\eta)$ increases. Fig. 7(b) portrays the effects of curvature parameter γ on the temperature profile $\theta(\eta)$. It is depicted that temperature profile shows merging behavior near the surface of cylinder while it increases away from the cylinder when $0.5 < \eta < 6$ and become vanish when $\eta \geq 6$. The thermal boundary layer thickness increase with an increase in curvature parameter γ. The influence of ratio parameter A is analyzed in the Fig. 8(a) It is observed that temperature and thermal boundary layer thickness decrease for larger values of A. The effects of thermal radiation parameter R_d on temperature distribution $\theta(\eta)$ is shown in Fig. 8(b). Temperature and thermal boundary layer thickness increase when radiation parameter is increased. It is due the reason that with the increase of thermal radiation parameter the mean absorption coefficient decreases. This leads to enhancement of temperature profile. Fig. 9(a) is plotted to see the variation of Prandtl number Pr on the temperature field $\theta(\eta)$. It is revealed that both the temperature and thermal boundary layer thickness are increased for smaller values of Pr. Thermal diffusivity decreases with an increase in Prandtl number consequently temperature field decreases. Fluids with high Prandtl number have low thermal diffusivity and fluids subject to low Prandtl number have high Prandtl number.

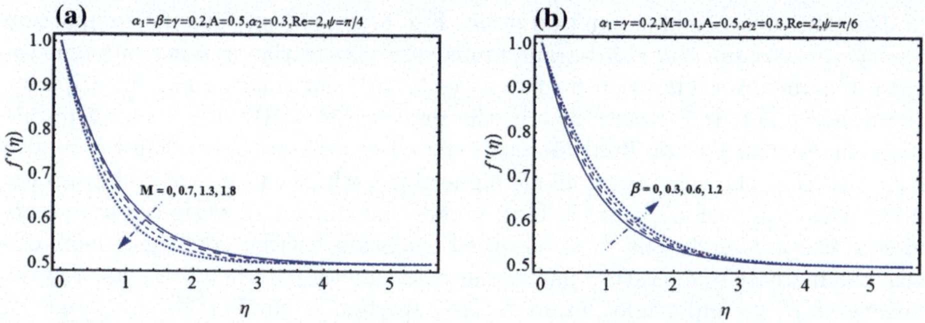

Fig. 3. (a) Influence of M on $f'(\eta)$ and (b) Influence of β on $f'(\eta)$.

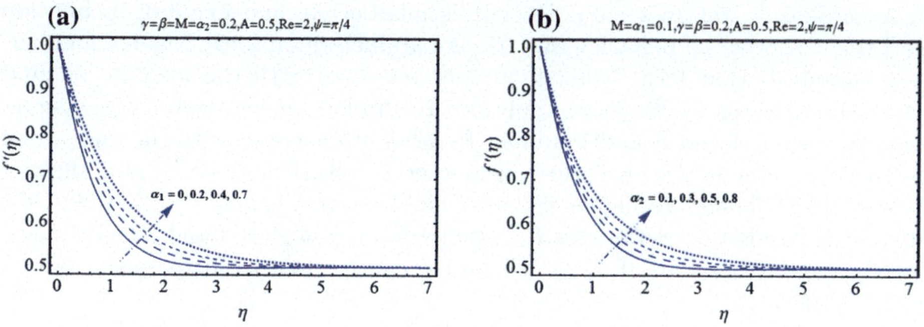

Fig. 4. (a) Influence of α_1 on $f'(\eta)$ and (b) Influence of α_2 on $f'(\eta)$.

We displayed the temperature field for various values of Eckert number Ec by keeping other parameters fixed in Fig. 9(b). The effect of Eckert number is to increase the temperature boundary layer thickness due to the frictional heating. Fig. 10(a) shows the comparison of Newtonian, second-grade and third-grade fluids velocities over a cylinder in the presence of magnetohydrodynamic. It is analyzed that the velocity for third-grade fluid is higher than the Newtonian and second-grade fluid. Further the momentum boundary layer thickness is higher for third-grade fluid. Fig. 10(b) is sketched to see the comparison of Newtonian, second-grade and third-grade fluids velocities over a cylinder in the absence of magnetohydrodynamic. It is analyzed that the velocity for third-grade fluid is higher than the Newtonian and second-grade fluid. Further the momentum boundary layer thickness is higher for third-grade fluid. Comparison between Newtonian and third-grade fluids velocities with magnetohydrodynamic over a cylinder is shown in Fig. 11(a) for two cases (i) without stagnation point (ii) with stagnation point. It is depicted that in the presence of magnetohydrodynamic, the velocity profile is higher for third-grade fluid for both the cases. Further it is also noted that the velocity profile is higher for both Newtonian and third-grade fluids in the presence of stagnation point. Fig. 11(b) is drawn for the comparison between Newtonian and third-grade fluids velocities in the presence of magneto-hydrodynamic over a flat plat for two cases (i) without stagnation point (ii) with stagnation point. It is noted that in the presence of MHD, the velocity profile is higher for third-grade fluid for both the cases. On the other hand it is also examined that the velocity profile is higher for both Newtonian and third-grade fluids with stagnation point. Table 2 shows the impact of various parameters on skin friction coefficient. It is observed that skin friction coefficient increases with the increase of curvature parameter γ, magnetic parameter M, third-grade parameter β, second-grade parameter α_1, Reynold's number Re and angle of inclination ψ while it decreases with the increase of second-grade parameter α_2 and ratio parameter A. Hence in order to reduce the value of skin friction coefficient which is very useful for industrial applications, one needs to reduce the radius of cylinder and decrease magnetic parameter M, third-grade parameter β, second-grade parameter α_1, Reynold's number Re and angle of inclination ψ. Table 3 shows the behavior of various parameters on local Nusselt number. It is examined that local Nusselt number increases with the increase of fluid parameter $(\alpha_1, \alpha_2, \beta)$, Reynold's number Re, radiation parameter R_d, stagnation parameter A and Prandtl number Pr while it decreases with the increase of magnetic parameter M, curvature parameter γ, Eckert number Ec and angle of inclination ψ. Therefore higher values of fluid parameter $(\alpha_1, \alpha_2, \beta)$, Reynold's number Re, radiation parameter R_d, stagnation parameter A and Prandtl number Pr and small values of M, γ, Ec and ψ can be used to increase the rate of heat transfer.

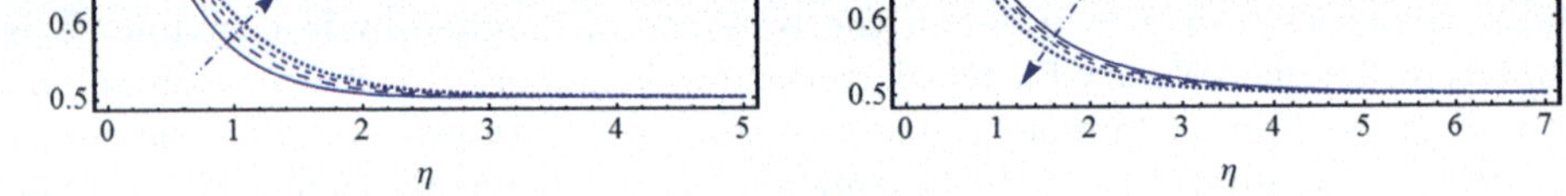

Fig. 5. (a) Influence of Re on $f'(\eta)$. (b) Influence of ψ on $f'(\eta)$.

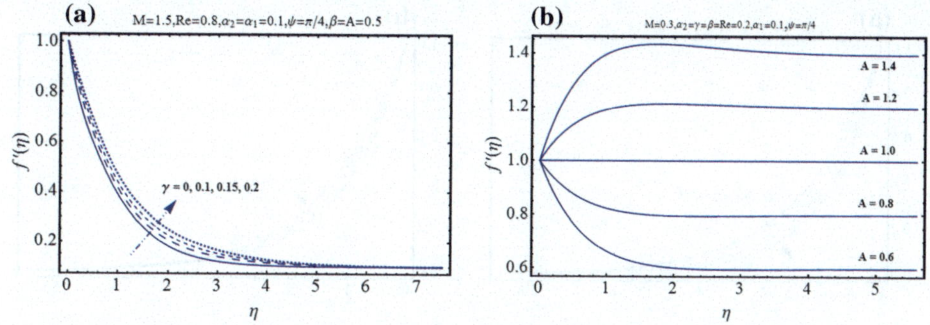

Fig. 6. (a) Influence of γ on $f'(\eta)$. (b) Influence of A on $f'(\eta)$.

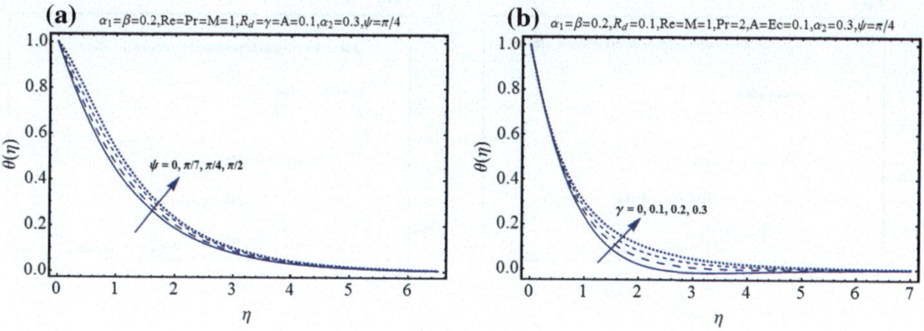

Fig. 7. (a) Influence of ψ on $\theta(\eta)$. (b)Influence of γ on $\theta(\eta)$.

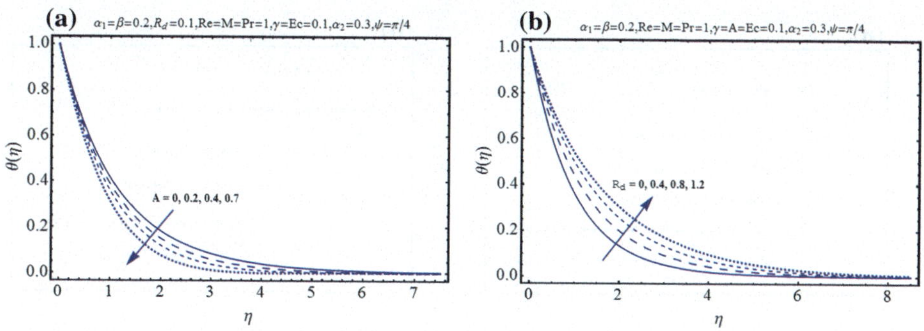

Fig. 8. (a) Influence of A on $\theta(\eta)$. (b) Influence of R_d on $\theta(\eta)$.

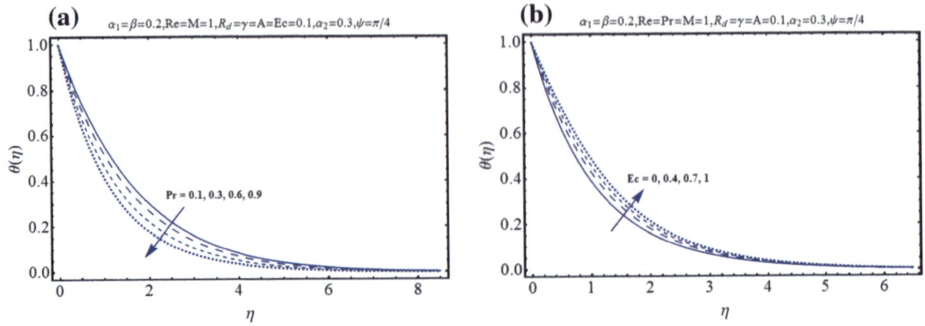

Fig. 9. (a) Influence of Pr on $\theta(\eta)$. (b) Influence of Ec on $\theta(\eta)$.

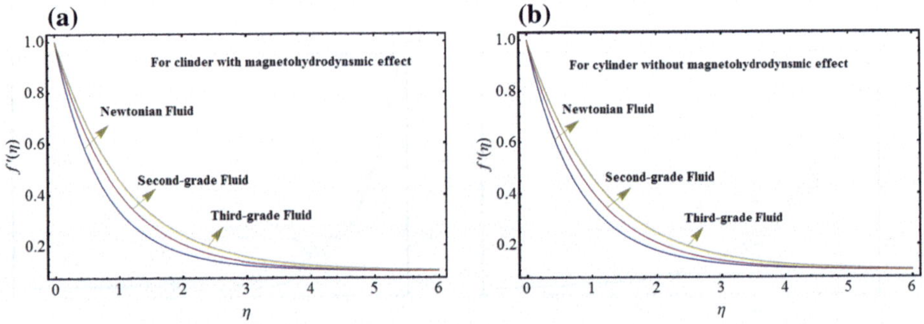

Fig. 10. (a) Comparison of velocity profiles for Newtonian, second-grade and third-grade fluids in the presence of MHD. (b)Comparison of velocity profiles for Newtonian, second-grade and third-grade fluids in the absence of MHD.

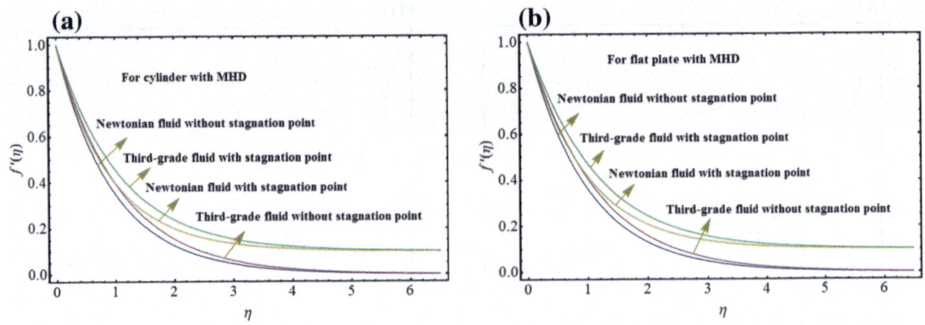

Fig. 11. (a) Comparison of velocity profiles for Newtonian and third-grade fluids for cylinder, (b) Comparison of velocity profiles for Newtonian and third-grade fluids for plate.

Table 2. Numerical values of skin friction coefficients $Re_x^{1/2}C_f$ for different values of physical parameters

α_1	α_2	β	M	Re	γ	ψ	A	$-Re_x^{1/2}C_f$
0.0	0.1	0.1	0.5	0.2	0.2	$\pi/4$	0.2	1.4029
0.1								1.5374
0.2								1.6550
0.1	0.0	0.1	0.5	0.2	0.2	$\pi/4$	0.2	1.6491
	0.1							1.5374
	0.2							1.4408
0.1	0.1	β0.0	0.5	0.2	0.2	$\pi/4$	0.2	1.5252
		0.1						1.5374
		0.2						1.5487
0.1	0.1	0.1	M0.3	0.2	0.2	$\pi/4$	0.2	1.5080
			0.5					1.5374
			0.7					1.5803
0.1	0.1	0.1	0.5	0.0Re	0.2	$\pi/4$	0.2	1.5252
				0.1				1.5314
				0.2				1.5374
0.1	0.1	0.1	0.5	0.2	0.0γ	$\pi/4$	0.2	1.0228
					0.1			1.2676
					0.2			1.5374
0.1	0.1	0.1	0.5	0.2	0.2	0	0.2	1.4910
						$\pi/4$		1.5374
						$\pi/2$		1.5820
0.1	0.1	0.1	0.5	0.2	0.2	$\pi/4$	0.0	1.6733
							0.1	1.6200
							0.2	1.5374

Table 3. Numerical values of Nusselt number $\mathrm{Re}_x^{-1/2} Nu_x$ for different values of physical parameters

α_1	α_2	β	M	Re	R_d	γ	ψ	A	Pr	Ec	$\mathrm{Re}_x^{-1/2} Nu_x$
0.0	0.1	0.1	0.5	0.2	0.3	0.2	$\pi/4$	0.2	2	0.5	1.3945
0.1											1.4440
0.2											1.4819
0.1	0.0	0.1	0.5	0.2	0.3	0.2	$\pi/4$	0.2	2	0.5	1.3860
	0.1										1.4440
	0.2										1.4957
0.1	0.1	0.0	0.5	0.2	0.3	0.2	$\pi/4$	0.2	2	0.5	1.4354
		0.1									1.4440
		0.2									1.4516
0.1	0.1	0.1	0.3	0.2	0.3	0.2	$\pi/4$	0.2	2	0.5	1.4732
			0.5								1.4440
			0.7								1.4018
0.1	0.1	0.1	0.5	0.0	0.3	0.2	$\pi/4$	0.2	2	0.5	1.4354
				0.1							1.4398
				0.2							1.4440
0.1	0.1	0.1	0.5	0.2	0.0	0.2	$\pi/4$	0.2	2	0.5	1.2022
					0.1						1.2876
					0.3						1.4440
0.1	0.1	0.1	0.5	0.2	0.3	0.0	$\pi/4$	0.2	2	0.5	1.5721
						0.1					1.5112
						0.2					1.4440
0.1	0.1	0.1	0.5	0.2	0.3	0.2	0	0.2	2	0.5	1.4899
							$\pi/4$				1.4440
							$\pi/2$				1.4002
0.1	0.1	0.1	0.5	0.2	0.3	0.2	$\pi/4$	0.0	2	0.5	1.2297
								0.1			1.3312
								0.2			1.4440
0.1	0.1	0.1	0.5	0.2	0.3	0.2	$\pi/4$	0.2	1.0	0.5	1.0444
									1.5		1.2635
									2		1.4440
0.1	0.1	0.1	0.5	0.2	0.3	0.2	$\pi/4$	0.2	2	0.5	1.4440
										0.7	1.2745
										0.9	1.1050

6 Concluding Remarks

The MHD stagnation point flow of third-grade fluid with heat transfer phenomenon in the presence of thermal radiation over a stretching cylinder are examined. Impact of involved parameters is seen. The following observations hold:

- The effect of third grade parameter β is to increase the boundary layer thickness.
- Velocity and temperature profiles shows merging behavior near the cylinder surface and these increase away from the cylinder with an increase in curvature parameter γ.
- Effect of fluid parameters and Reynolds number on boundary layer thickness is similar in a qualitative sense.
- Velocity profile decreases while temperature profile increases for larger values of angle of inclination ψ.
- With the increase in Pr the temperature profile and thermal boundary layer thickness decrease.
- Minimum values of skin friction coefficient are achieved for small values of Re, α_1, β, M, γ, ψ and large values of α_2 and A.
- Rate of heat transfer is higher for larger values of α_1, α_2, β, Re, R_d, A and Pr.

References

1. Shehzad, S.A., Hussain, T., Hayat, T., Ramzan, M., Alsaedi, A.: Boundary layer flow of third grade nanofluid with Newtonian heating and viscous dissipation. J. Central South Univ. **22**(1), 360–367 (2015)
2. Shafiq, A., Hammouch, Z., Sindhu, T.N.: Bioconvective MHD flow of tangent hyperbolic nanofluid with Newtonian heating. Int. J. Mech. Sci. **133**, 759–766 (2017)
3. Amkadni, M., Azzouzi, A., Hammouch, Z.: On the exact solutions of laminar MHD flow over a stretching flat plate. Commun. Nonlinear Sci. Numer. Simul. **13**(2), 359–368 (2008)
4. Zakia, H.: Multiple solutions of steady MHD flow of dilatant fluids. Eur. J. Pure Appl. Math. **1**(2), 11–20 (2008)
5. Hammouch, Z., Mekkaoui, T., Sadki, H.: Similarity solutions of a steady MHD flow over a semi-infinite surface. J.—MESA **8**(1), 109–117 (2017)
6. Haq, R.U., et al.: Heat exchange within the partially heated C-shape cavity filled with the water based SWCNTs. Int. J. Heat Mass Transf. **127**, 506–514 (2018)
7. Shafiq, A., Sindhu, T.N., Hammouch, Z.: Characteristics of homogeneous heterogeneous reaction on flow of walters' b liquid under the statistical paradigm. International workshop of Mathematical Modelling, Applied Analysis and Computation. Springer, Singapore (2018)
8. Bhattacharyya, K.: MHD stagnation point flow of Casson fluid and heat transfer over a stretching sheet with thermal radiation. J. Thermodyn. **2013**, 1–9 (2013)
9. Rostami, B., Rashidi, M.M., Rostami, P., Momoniat, E., Freidoonimehr, N.: Analytical Investigation of laminar viscoelastic fluid flow over a wedge in the presence of buoyancy force effects. Abstr. Appl. Analy. **2014**, 1–11 (2014)

10. Ellahi, R., Rahman, S.U., Gulzar, M.M., Nadeem, S., Vafai, K.: A mathematical study of non-Newtonian micropolar fluid in arterial blood flow through composite stenosis. Appl. Math. Inf. Sci. **8**(4), 1567–1573 (2014)
11. Fetecau, C., Imran, M.A., Sohail, A.: On some unsteady motions of second-grade fluids in a rectangular ducts edge. Ann. Acad. Rom. Sci. Ser. Math. Appl. **6**(1), 74–91 (2014)
12. Hayat, T., Shafiq, A., Alsaedi, A., Awais, M.: MHD axisymmetric flow of third-grade fluid between stretching sheets with heat transfer. Comput. Fluids **86**, 103–108 (2013)
13. Fetecau, C., Rana, M., Nigar, N., Fetecau, C.: First exact solutions for flows of rate type fluids in a circular duct that applies a constant couple to the fluid. Z. Naturforsch. **69a**, 232–238 (2014)
14. Mukhopadhyay, S.: Casson fluid flow and heat transfer over a nonlinearly stretching surface. Chin. Phys. B **22**(7), 074701 (2013)
15. Hayat, T., Shafiq, A., Nawaz, M., Alsaedi, A.: MHD axisymmetric flow of third-grade fluid between porous disks with heat transfer. Appl. Math. Mech. Engl. Ed. **33**(6), 749–764 (2012)
16. Shafiq, A., Nawaz, M., Hayat, T., Alsaedi, A.: Magnetohydrodynamic axisymmetric flow of a third-grade fluid between two porous disks. Brazilian J. Chem. Eng. **30**(3), 599–609 (2013)
17. Hayat, T., Shafiq, A., Alsaedi, A.: Effect of joule heating and thermal radiation in flow of third grade fluid over radiative surface. PLoS ONE **9**(1), e83153 (2014)
18. Mushtaq, A., Mustafa, M., Hayat, T., Alsaedi, A.: Effects of thermal radiation on the stagnation point flow of upper-convected Maxwell fluid over a stretching sheet. J. Aerosp. Eng. **27**(4), 04014015 (2014)
19. Bhattacharyya, K., Mukhopadhyay, S., Layek, G.C., Pop, I.: Effects of thermal radiation on micropolar fluid flow and heat transfer over a porous shrinking sheet. Int. J. Heat Mass Transf. **55**, 2945–2952 (2012)
20. Mukhopadhyay, S.: Slip effects on MHD boundary layer flow over an exponentially stretching sheet with suction/blowing and thermal radiation. Ain Shams Eng. J. **4**, 485–491 (2013)
21. Rashidi, M.M., Rostami, B., Freidoonimehr, N., Abbasbandy, S.: Free convective heat and mass transfer for MHD fluid flow over a permeable vertical stretching sheet in the presence of the radiation and buoyancy effects. Ain Shams Eng. J. **5**, 901–912 (2014)
22. Mukhopadhyay, S.: Effects of thermal radiation and variable fluid viscosity on stagnation point flow past a porous stretching sheet. Meccanica **48**, 1717–1730 (2013)
23. Rashidi, M.M., Ganesh, N.V., Hakeem, A.K.A., Ganga, B.: Buoyancy effect on MHD flow of nanofluid over a stretching sheet in the presence of thermal radiation. J. Mol. Liq. **198**, 234–238 (2014)
24. Chamkha, A.J., Abbasbandy, S., Rashad, A.M., Vajravelu, K.: Radiation effects on mixed convection over a wedge embedded in a porous medium filled with a nanofluid. Transp. Porous Media **91**, 261–279 (2012)
25. Sheikholeslami, M., Ganji, D.D., Javed, M.Y., Ellahi, R.: Effect of thermal radiation on magnetohydrodynamics nanofluid flow and heat transfer by means of two phase model. J. Magn. Magn. Mater. **374**, 36–43 (2015)
26. Seini, Y.I., Makinde, O.D.: MHD boundary layer flow due to exponential stretching surface with radiation and chemical reaction. Math. Probl. Eng. **2013**, 1–7 (2013)

27. Bhattacharyya, K.: Heat transfer analysis in unsteady boundary layer stagnation-point flow towards a shrinking/stretching sheet. Ain Shams Eng. J. **4**, 259–264 (2013)
28. Turkyilmazoglu, M., Pop, I.: Exact analytical solutions for the flow and heat transfer near the stagnation point on a stretching/shrinking sheet in a Jeffrey fluid. Int. J. Heat Mass Trans. **57**, 82–88 (2013)
29. Nandy, S.K., Mahapatra, T.R.: Effects of slip and heat generation/absorption on MHD stagnation flow of nanofluid past a stretching/shrinking surface with convective boundary conditions. Int. J. Heat Mass Trans. **64**, 1091–1100 (2013)
30. Hayat, T., Shehzad, S.A., Alsaedi, A., Alhothuali, M.S.: Mixed convection stagnation point flow of casson fluid with convective boundary conditions. Chin. Phys. Lett. **29**, 114704 (2012)
31. Bhattacharyya, K.: Boundary layer stagnation point flow of Casson fluid and heat transfer towards a shrinking/stretching sheet. Front. Heat Mass Trans. **4**, 023003 (2013)
32. Rashidi, M.M., Freidoonimehr, N.: Analysis of entropy generation in MHD stagnation point in porous media with heat transfer. Int. J. Comput. Methods Eng. Sci. Mech. **15**, 345–355 (2014)
33. Bhattacharyya, K., Mukhopadhyay, S., Layek, G.: Reactive solute transfer in magnetohydrodynamic boundary layer stagnation-point flow over a stretching sheet with suction/blowing. Chem. Eng. Commun. **199**, 368–383 (2012)
34. Hayat, T., Iqbal, Z., Mustafa, M., Alsaedi, A.: Stagnation-point flow of Jeffrey fluid with melting heat transfer and Soret and Dufour effects. Int. J. Numer. Methods Heat Fluid Flow **24**, 402–418 (2014)
35. Liao, S.J.: Homotopy Analysis Method in Nonlinear Differential Equations. Springer, Heidelberg (2012)
36. Hayat, T., Hussain, Z., Farooq, M., Alsaedi, A., Obaid, M.: Thermally stratified stagnation point flow of an Oldroyd-B fluid. Int. J. Nonlinear Sci. Numer. Simulat. **15**, 77–86 (2014)
37. Sheikholeslami, M., Ellahi, R., Ashorynejad, H.R., Domairry, G., Hayat, T.: Effect of heat transfer in flow of nanofluids over a permeable stretching wall in a porous medium. J. Comput. Theore. Nanoscience **11**, 486–496 (2014)
38. Rashidi, M.M., Rajvanshi, S.C., Kavyani, N., Keimanesh, M., Pop, I., Saini, B.S.: Investigation of heat transfer in a porous annulus with pulsating pressure gradient by homotopy analysis method. Arab. J. Sci. Engin. **39**(6), 5113–5128 (2014)
39. Turkyilmazoglu, M.: Solution of Thomas-Fermi equation with a convergent approach. Commun. Nonlinear Sci. Numer. Simulat. **17**, 4097–4103 (2012)
40. Abbasbandy, S., Hashemi, M.S., Hashim, I.: On convergence of homotopy analysis method and its application to fractional integro-differential equations. Quaest. Math. **36**(1), 93–105 (2013)
41. Hayat, T., Naseem, A., Farooq, M., Alsaedi, A.: Unsteady MHD three dimensional flow with viscous dissipation and Joule heating. Eur. Phys. J. Plus **128**, 158 (2013)

A Fractional Mixing Propagation Model of Computer Viruses and Countermeasures Involving Mittag-Leffler Type Kernel

Sümeyra Uçar[1], Necati Özdemir[1], and Zakia Hammouch[2](✉)

[1] Department of Mathematics, Faculty of Arts and Sciences, Balıkesir University, Balıkesir, Turkey
{sumeyraucar,nozdemir}@balikesir.edu.tr

[2] Department of Mathematics, Faculty of Sciences ans Techniques, Moulay Ismail University of Meknes, 52000 Errachidia, Morocco
hammouch.zakia@gmail.com

Abstract. Countermeasures are recognized as a remarkable effort to comprehend the computer virus problem and estimate its forthcoming actions. Countermeasure-Competing (CMC) strategy is a conception comprising viruses and countermeasures. The main point of this paper is to probe a mixing propagation model of computer viruses and countermeasures in the light of the newly fractional derivative introduced by Atangana and Baleanu. The existence and uniqueness of solutions for this fractionalized model is discussed by taking the fixed point theory into consideration.. The efficacious belongings of this fractional model are exhibited theoretically, confirmed by numerical graphics.

1 Introduction

With the immense usage of the Internet and increasing globalization throughout the world, detrimental software has turned out to be a major threat in terms of cyber security. Computer viruses are one of this harmful software that can multiply itself without any user interaction. An erupting virus in computer systems may carry out numerous unwanted activities by damaging programs, reformatting hard disks, limiting one's access to data, stealing one's personal information, etc. Therefore the question how to restrain the abundance of computer viruses on networks has become the centre of attention in both our work and lives.

In order to extensively investigate the propagation of computer viruses, mathematical models are required, and the analogy between the spreading behaviour of computer viruses and biological viruses has led researchers to develop a multitude of mathematical models in the world of computing [1–8]. Furthermore, it is considered that countermeasures, which can also be referred to as new virus definition files or software patches are extremely efficient ways to tackle the diffusion of computer viruses. For this reason, Chen and Carley [9] presented a novel approach named as CMC strategy (Countermeasure-Competing) where

H. Dutta et al. (Eds.): CMES 2019, AISC 1111, pp. 186–199, 2020.
https://doi.org/10.1007/978-3-030-39112-6_13

they argue that countermeasures and computer viruses propagate at distinct ratios, and computers equipped with countermeasures are permanently immunized. Moreover, they demonstrated that CMC is a more powerful strategy for dramatically reducing the extent of virus infection.

The concept of fractional calculus has recently come to be a subject of interest in different disciplines due to its ability to reflect system behaviour more realistically than integer order derivatives [10–18]. This is why there are several fractional derivatives which may be regarded as conventional, such as the Riemann-Liouville (RL) and Caputo operators, and these derivatives possess the memory properties for real-world phenomena. However, the complexity of their numerical process prompts the researchers to investigate new fractional derivatives because of their singular points. In this context, Atangana-Baleanu [19] defined a new non-singular derivative with Mittag-Leffler kernel named Atangana-Baleanu derivative (AB). It follows from the studies latterly carried out that this new operator has been employed in various complex problems, such as showing the applicability of AB derivative to Rubella disease in [20]. Avcı et al. [21] considers an advection-diffusion equation on a line segment using Mittag-Leffler kernel while Gomez-Aguilar [22] studies a nonlinear alcoholism model considering the influence of Twitter by use of Liouville-Caputo and AB fractional derivatives. For other notable studies, please see [23–30].

Based on the aforesaid CMC scheme, we focus on the mixing propagation model of computer viruses and countermeasures suggested by Zhu et al. [31]. In line with our current objective, we will denominate two types of computers by taking their connection to the Internet into consideration, and thereby refer to a computer as whether internal or external. Each and every existing internal computer falls into one of the three categories which we define below. The first category, which comprises the internal computers that are not infected but susceptible to infection as a result of countermeasure deficiency, is susceptible internal computers (S-computers). The second category, which encompasses the internal computers that are already infected, is called infective internal computers (I-computers). And the last category, which contains the internal computers that are not infected and also have temporary immunity as a result of the existing countermeasures, is named C-computers. We shall use $S(t)$ to specify the average number of S-computer at time t, and $I(t)$ for the same number of I-computer at time t, and $C(t)$ for the said number of C-computer at time t. In order to prevent any confusion, we will refer to them as S, I, and C, respectively. Linked with these notations the SICS (Susceptible-Infected-Countermeasure-Susceptible) model is handled by the below manner with classical derivative [31]:

$$\begin{aligned}\frac{dS(t)}{dt} &= \lambda - b_1SI - b_2SC + d_1I + d_2C - \phi S,\\ \frac{dI(t)}{dt} &= b_1SI - b_2IC - (d_1+\phi)I,\\ \frac{dC(t)}{dt} &= b_2(S+I)C - (d_2+\phi)C. \qquad (1.1)\end{aligned}$$

with the initial condition $(S(0), I(0), C(0)) \in \mathbb{R}^3_+$. In this model, there are some assumptions in order to obtain a view of the mixing propagation attitudes exhibited by computer viruses and countermeasures. These assumptions are: Each and every external computer is susceptible; the rate at which an external computer establishes internet connection is fixed on λ; the rate at which an internal computer loses internet connection is fixed on ϕ; the probability with which I -computers infect all S-computers is $b_1 I(t)$ where b_1 is a constant that is positive; the probability with which all S-computers or I-computers gain countermeasures is fixed at $b_2 C(t)$ when b_2 is a constant that is positive; the probability with which the infection is eliminated in all I-computers is fixed at d_1; since countermeasures are rendered invalid, the probability with which all C-computers become prone to loss of immunity remains fixed at d_2.

To deeply understand the combined impact of computer viruses and countermeasures, we study the above SICS model considering the AB derivative. Section 2, we summarize some basic definitions and theorems of the AB fractional derivative. In Sect. 3, we present the SICS model with AB fractional derivative and prove detailed the existence and uniqueness conditions of the solutions using the fixed point theory. Variable numerical results of our new model put in place so as to demonstrate the effect of this fractional derivative in Sect. 4. In Sect. 5, we finalize the study with the concluding remarks.

2 Preliminary Tools

In this section, we briefly give some basic definitions and properties of the AB fractional derivative and present an Implicit Euler scheme to solve the so-called problem.

Definition 1. Let $a < b$, $g \in H^1(a,b)$ be a function and $\eta \in [0,1]$. The Atangana-Baleanu derivative in Caputo sense of order η of g is defined as [19]

$$ {}^{ABC}_{a}D^{\eta}_{t}[g(t)] = \frac{F(\eta)}{1-\eta} \int_a^t g'(x) E_\eta \left[-\eta \frac{(t-x)^\eta}{1-\eta} \right] dx \tag{1.2} $$

where $F(\eta)$ is a normalization function with $F(0) = F(1) = 1$ and E_η is the Mittag-Leffler function.

Definition 2. Let $a < b$, $g \in H^1(a,b)$ be a function and $\eta \in [0,1]$. The Atangana-Baleanu derivative in Riemann-Liouville sense of order η of g is represented as [19]:

$$ {}^{ABR}_{a}D^{\eta}_{t}[g(t)] = \frac{F(\eta)}{1-\eta} \frac{d}{dt} \int_a^t g(x) E_\eta \left[-\eta \frac{(t-x)^\eta}{1-\eta} \right] dx. \tag{1.3} $$

Definition 3. The fractional integral connected to the fractional derivative is given as [19]:

$$ {}_{a}^{AB}I_{t}^{\eta}\left[g\left(t\right)\right]=\frac{1-\eta}{F\left(\eta\right)}g\left(t\right)+\frac{\eta}{F\left(\eta\right)\Gamma\left(\eta\right)}\int_{a}^{t}g\left(y\right)\left(t-y\right)^{\eta-1}dy. \tag{1.4} $$

Theorem 1. *Let g be a continuous function on $[a,b]$. The following inequality holds on $[a,b]$ [19]:*

$$ \left\|{}_{0}^{ABR}D_{t}^{\eta}\left[g\left(t\right)\right]\right\|<\frac{F\left(\eta\right)}{1-\eta}\left\|g\left(t\right)\right\|, \tag{1.5} $$

where $\left\|g\left(t\right)\right\|=\max_{a\leq t\leq b}\left|g\left(t\right)\right|$.

Theorem 2. *The Atangana-Baleanu derivative in Caputo and RL sense fulfill Lipschitz condition [19]:*

$$ \left\|{}_{0}^{ABC}D_{t}^{\eta}\left[g\left(t\right)\right]-{}_{0}^{ABC}D_{t}^{\eta}\left[h\left(t\right)\right]\right\|\leq H\left\|g\left(t\right)-h\left(t\right)\right\| \tag{1.6} $$

and

$$ \left\|{}_{0}^{ABR}D_{t}^{\eta}\left[g\left(t\right)\right]-{}_{0}^{ABR}D_{t}^{\eta}\left[h\left(t\right)\right]\right\|\leq H\left\|g\left(t\right)-h\left(t\right)\right\|. \tag{1.7} $$

Theorem 3. *[19] The fractional ordinary differential equation*

$$ {}_{a}^{ABC}D_{t}^{\eta}\left[g\left(t\right)\right]=u\left(t\right) $$

possess a unique solution given by

$$ g\left(t\right)=\frac{1-\eta}{F\left(\eta\right)}u\left(t\right)+\frac{\eta}{F\left(\eta\right)\Gamma\left(\eta\right)}\int_{a}^{t}u\left(y\right)\left(t-y\right)^{\eta-1}dy. $$

In what follows, we describe the Fractional Euler Method introduced by Baleanu et. al [32], which will be used for numerical simulations throughout the present work. To express this method, we consider the nonlinear differential equation with AB derivative in Caputo sense as follows:

$$ \begin{cases} {}_{0}^{ABC}D_{t}^{\eta}y\left(t\right)=f\left(t,y\left(t\right)\right),0<t\leq T<\infty. \\ y\left(0\right)=y_{0}, \end{cases} \tag{1.8} $$

where $0<\eta<1$. Let N is an arbitrary positive integer, we regard a uniform mesh on the interval $[0,T]$ and the nodes $0,1,...,N$ where the time step size $h=\frac{T}{N}$. y_i means the numerical approximation of $y\left(t_i\right)$.
According to [32], the following nonlinear Volterra integral equation is formulated

$$ y(t)=y_0+\frac{1-\eta}{F(\eta)}f(t,y(t))+\frac{\eta}{\Gamma(\eta)F(\eta)}\int_0^t f(\tau,y(\tau))d\tau, \tag{1.9} $$

A descritization of the integral equation (1.9) using the Euler Method, gives

$$y_{i+1} = y_0 + \frac{1-\eta}{F(\eta)} f(t_{i+1}, y_{i+1}) + \frac{\eta h_N^\eta}{F(\eta)\Gamma(\eta+1)} \sum_{j=0}^{i} b_{i+1,j}^{(\eta)} f(t_j, y_j), \tag{1.10}$$

where $i = 0, ..., N-1$ and $b_{i+1,j}^{(\eta)}$; $j = 0, ..., i$ are computed from

$$b_{i+1,j}^{(\eta)} = -(i-j)^\eta + (i-j+1)^\eta .$$

The stability and the error estimation were proved elegantly in [32].

3 SICS Model with AB Derivative

To extend and amend this model, we formulate the Eq. (1.1) by substituting the integer order time derivative by the fractional time derivative:

$$\begin{aligned}
{}_0^{ABC}D_t^\eta (S(t)) &= \lambda - b_1 SI - b_2 SC + d_1 I + d_2 C - \phi S, \\
{}_0^{ABC}D_t^\eta (I(t)) &= b_1 SI - b_2 IC - (d_1 + \phi) I, \\
{}_0^{ABC}D_t^\eta (C(t)) &= b_2 (S + I) C - (d_2 + \phi) C.
\end{aligned} \tag{1.11}$$

The related initial conditions are $S(0) \geq 0, I(0) \geq 0, C(0) \geq 0$.

3.1 Existence and Uniqueness Analysis

Providing the solution of nonlinear equations is one of the hard subjects in differential calculus. The fractional order model under consideration is nonlinear, it can be impossible to find the exact solution of this kind systems. For this reason, this part is devoted to investigate in detail the existence and uniqueness of the solution for the model (1.11) taking into consideration fixed point theory.

Let $\mathscr{P} = C(J) \times C(J)$ and $C(J)$ be a Banach space of continuous $\mathbb{R} \to \mathbb{R}$ valued functions on the interval J with the norm

$$\|(S, I, C)\| = \|S\| + \|I\| + \|C\| ,$$

where $\|S\| = sup\{|S(t)| : t \in J\}$, $\|I\| = sup\{|I(t)| : t \in J\}$, $\|C\| = sup\{|C(t)| : t \in N\}$. In order to simplify the Eq. (1.11), we edit this model in the following expressions:

$$\begin{aligned}
{}_0^{ABC}D_t^\eta (S(t)) &= K_1(t, S), \\
{}_0^{ABC}D_t^\eta (I(t)) &= K_2(t, I), \\
{}_0^{ABC}D_t^\eta (C(t)) &= K_3(t, C).
\end{aligned} \tag{1.12}$$

Using Theorem 3, the model (1.12) is of the following Volterra type integral equation with Atangana-Baleanu integral:

$$S(t)-S(0)=\frac{1-\eta}{F(\eta)}K_1(t,S)+\frac{\eta}{F(\eta)\Gamma(\eta)}\int_0^t (t-y)^{\eta-1}K_1(y,S)\,dy,$$

$$I(t)-I(0)=\frac{1-\eta}{F(\eta)}K_2(t,I)+\frac{\eta}{F(\eta)\Gamma(\eta)}\int_0^t (t-y)^{\eta-1}K_2(y,I)\,dy,$$

$$C(t)-C(0)=\frac{1-\eta}{F(\eta)}K_3(t,C)+\frac{\eta}{F(\eta)\Gamma(\eta)}\int_0^t (t-y)^{\eta-1}K_3(y,C)\,dy. \quad (1.13)$$

Theorem 4. *The kernel K_1 satisfies the Lipschitz condition and contraction if the below inequality holds:*

$$0\le b_1\varepsilon_2+b_2\varepsilon_3+\phi<1.$$

Proof. Let S and S_1 be two functions, then we get

$$\begin{aligned}\|K_1(t,S)-K_1(t,S_1)\| &= \|-b_1SI-b_2SC-\phi S+b_1S_1I+b_2S_1C+\phi S_1\|\\ &\le (b_1\|I(t)\|+b_2\|C(t)\|+\phi)\|S(t)-S_1(t)\|\\ &\le \gamma_1\|S(t)-S_1(t)\|. \end{aligned}\quad (1.14)$$

where $\gamma_1=b_1\varepsilon_2+b_2\varepsilon_3+\phi$ and $\|S(t)\|\le\varepsilon_1$, $\|I(t)\|\le\varepsilon_2$, $\|C(t)\|\le\varepsilon_3$. Thus, we get

$$\|K_1(t,S)-K_1(t,S_1)\|\le\gamma_1\|S(t)-S_1(t)\|. \quad (1.15)$$

As a consequence, Lipschitz condition satisfied for K_1 and because of $0\le b_1\varepsilon_2+b_2\varepsilon_3+\phi<1$ implies K_1 is also contraction.

Clearly, it can be shown that the other kernels K_2 and K_3 fulfil the Lipschitz condition and contraction.

Here, we consider the following iterative formula:

$$S_n(t)=\frac{1-\eta}{F(\eta)}K_1(t,S_{n-1})+\frac{\eta}{F(\eta)\Gamma(\eta)}\int_0^t (t-y)^{\eta-1}K_1(y,S_{n-1})\,dy,$$

$$I_n(t)=\frac{1-\eta}{F(\eta)}K_2(t,I_{n-1})+\frac{\eta}{F(\eta)\Gamma(\eta)}\int_0^t (t-y)^{\eta-1}K_2(y,I_{n-1})\,dy,$$

$$C_n(t)=\frac{1-\eta}{F(\eta)}K_3(t,C_{n-1})+\frac{\eta}{F(\eta)\Gamma(\eta)}\int_0^t (t-y)^{\eta-1}K_3(y,C_{n-1})\,dy. \quad (1.16)$$

with the initial conditions are

$$S_0(t) = S(0), I_0(t) = I(0), C_0(t) = C(0).$$

The difference between the successive terms is of the following form:

$$\Phi_n(t) = S_n(t) - S_{n-1}(t) = \frac{1-\eta}{F(\eta)}\left[K_1(t, S_{n-1}) - K_1(t, S_{n-2})\right]$$
$$+\frac{\eta}{F(\eta)\Gamma(\eta)}\int_0^t (t-y)^{\eta-1}\left[K_1(y, S_{n-1}) - K_1(y, S_{n-2})\right]dy,$$

$$\psi_n(t) = I_n(t) - I_{n-1}(t) = \frac{1-\eta}{F(\eta)}\left[K_2(t, I_{n-1}) - K_2(t, I_{n-2})\right]$$
$$+\frac{\eta}{F(\eta)\Gamma(\eta)}\int_0^t (t-y)^{\eta-1}\left[K_2(y, I_{n-1}) - K_2(y, I_{n-2})\right]dy,$$

$$\xi_n(t) = C_n(t) - C_{n-1}(t) = \frac{1-\eta}{F(\eta)}\left[K_3(t, C_{n-1}) - K_3(t, C_{n-2})\right]$$
$$+\frac{\eta}{F(\eta)\Gamma(\eta)}\int_0^t (t-y)^{\eta-1}\left[K_3(y, C_{n-1}) - K_3(y, C_{n-2})\right]dy, \quad (1.17)$$

In the light of the above calculations, it is explicit that

$$S_n(t) = \sum_{k=1}^{n}\Phi_k(t),$$
$$I_n(t) = \sum_{k=1}^{n}\psi_k(t),$$
$$C_n(t) = \sum_{k=1}^{n}\xi_k(t). \quad (1.18)$$

Performing the norm to both sides of the Eq. (1.17) and by using triangular identity, we find

$$\begin{aligned}\|\Phi_n(t)\| &= \|S_n(t) - S_{n-1}(t)\| \\ &\le \frac{1-\eta}{F(\eta)}\|\left[K_1(t, S_{n-1}) - K_1(t, S_{n-2})\right]\| \\ &+ \frac{\eta}{F(\eta)\Gamma(\eta)}\left\|\int_0^t (t-y)^{\eta-1}\left[K_1(y, S_{n-1}) - K_1(y, S_{n-2})\right]dy\right\| \quad (1.19)\end{aligned}$$

Because the kernel K_1 ensures Lipschitz condition as seen in Eq. (1.15), we assess

$$\begin{aligned} \|\Phi_n(t)\| &= \|S_n(t) - S_{n-1}(t)\| \\ &\leq \frac{1-\eta}{F(\eta)}\gamma_1 \|S_{n-1} - S_{n-2}\| \\ &+ \frac{\eta}{F(\eta)\Gamma(\eta)}\gamma_1 \int_0^t (t-y)^{\eta-1} \|S_{n-1} - S_{n-2}\| \, dy. \end{aligned} \tag{1.20}$$

and we own

$$\begin{aligned} \|\Phi_n(t)\| \leq \frac{1-\eta}{F(\eta)}\gamma_1 \left\|\Phi_{(n-1)}(t)\right\| \\ + \frac{\eta}{F(\eta)\Gamma(\eta)}\gamma_1 \int_0^t (t-y)^{\eta-1} \left\|\Phi_{(n-1)}(y)\right\| dy \end{aligned} \tag{1.21}$$

Using the same attitude we gain the followings:

$$\begin{aligned} \|\psi_n(t)\| &\leq \frac{1-\eta}{F(\eta)}\gamma_2 \left\|\psi_{(n-1)}(t)\right\| \\ &+ \frac{\eta}{F(\eta)\Gamma(\eta)}\gamma_2 \int_0^t (t-y)^{\eta-1} \left\|\psi_{(n-1)}(y)\right\| dy, \\ \|\xi_n(t)\| &\leq \frac{1-\eta}{F(\eta)}\gamma_3 \left\|\xi_{(n-1)}(t)\right\| \\ &+ \frac{\eta}{F(\eta)\Gamma(\eta)}\gamma_3 \int_0^t (t-y)^{\eta-1} \left\|\xi_{(n-1)}(y)\right\| dy. \end{aligned} \tag{1.22}$$

Considering the gained results, we state the below theorem.

Theorem 5. *If we can find t_0 such that*

$$\frac{1-\eta}{F(\eta)}\gamma_i + \frac{t_0^{\eta}}{F(\eta)\Gamma(\eta)}\gamma_i < 1 \text{ for } i = 1, 2, 3, \tag{1.23}$$

then the fractional model in the Eq. (1.11) has a solution.

Proof. With the aid of Eqs. (1.21) and (1.22), since the functions $S(t)$, $I(t)$ and $C(t)$ are bounded and carry out Lipschitz condition, we get the succeeding relations as below:

$$\begin{aligned}
\|\Phi_n(t)\| &\le \|S_n(0)\| \left[\frac{1-\eta}{F(\eta)}\gamma_1 + \frac{t^\eta}{F(\eta)\Gamma(\eta)}\gamma_1\right]^n, \\
\|\psi_n(t)\| &\le \|I_n(0)\| \left[\frac{1-\eta}{F(\eta)}\gamma_2 + \frac{t^\eta}{F(\eta)\Gamma(\eta)}\gamma_2\right]^n, \\
\|\xi_n(t)\| &\le \|C_n(0)\| \left[\frac{1-\eta}{F(\eta)}\gamma_3 + \frac{t^\eta}{F(\eta)\Gamma(\eta)}\gamma_3\right]^n.
\end{aligned} \tag{1.24}$$

Hence, we prove the existence and continuity of the aforementioned solutions. In order to show that the above functions are solutions of the model (1.11), we suppose

$$\begin{aligned}
S(t) - S(0) &= S_n(t) - e_n(t), \\
I(t) - I(0) &= I_n(t) - g_n(t), \\
C(t) - C(0) &= C_n(t) - h_n(t).
\end{aligned} \tag{1.25}$$

Next, we find

$$\begin{aligned}
\|e_n(t)\| = &\left\| \frac{1-\eta}{F(\eta)} \left[K_1(t,S) - K_1(t,S_{n-1})\right] \right. \\
&\left. + \frac{\eta}{F(\eta)\Gamma(\eta)} \int_0^t (t-y)^{\eta-1} \left[K_1(y,S) - K_1(y,S_{n-1})\right] dy \right\|
\end{aligned}$$

$$\begin{aligned}
&\le \frac{1-\eta}{F(\eta)} \|K_1(t,S) - K_1(t,S_{n-1})\| \\
&+ \frac{\eta}{F(\eta)\Gamma(\eta)} \int_0^t (t-y)^{\eta-1} \|K_1(y,S) - K_1(y,S_{n-1})\, dy\| \\
&\le \frac{1-\eta}{F(\eta)}\gamma_1 \|S - S_{n-1}\| + \frac{t^\eta}{F(\eta)\Gamma(\eta)}\gamma_1 \|S - S_{n-1}\|.
\end{aligned} \tag{1.26}$$

By continuing this process, we obtain

$$\|e_n(t)\| \le \left(\frac{1-\eta}{F(\eta)} + \frac{t_0^\eta}{F(\eta)\Gamma(\eta)}\right)^{n+1} \gamma_1^{n+1} a. \tag{1.27}$$

As n approaches to infinity, we find $\|e_n(t)\| \to 0$. Similarly, it can be seen $\|g_n(t)\| \to 0$, $\|h_n(t)\| \to 0$. Thus the proof is completed.

It is an important matter to prove the uniqueness for the solutions of the model (1.11). Let $S_1(t)$, $I_1(t)$ and $C_1(t)$ be another solutions, then

$$S(t) - S_1(t) = \frac{1-\eta}{F(\eta)}\left[K_1(t,S) - K_1(t,S_1)\right] + \frac{\eta}{F(\eta)\Gamma(\eta)}\int_0^t (t-y)^{\eta-1}\left[K_1(y,S) - K_1(y,S_1)\right]dy \quad (1.28)$$

Considering the fact that the kernel carries out the Lipschitz condition and implementing the norm (1.28), we get

$$\|S(t) - S_1(t)\| \le \frac{1-\eta}{F(\eta)}\gamma_1 \|S(t) - S_1(t)\| + \frac{t^\eta}{F(\eta)\Gamma(\eta)}\gamma_1 \|S(t) - S_1(t)\| \quad (1.29)$$

This gives rise to

$$\|S(t) - S_1(t)\|\left(1 - \frac{1-\eta}{F(\eta)}\gamma_1 - \frac{t^\eta}{F(\eta)\Gamma(\eta)}\gamma_1\right) \le 0. \quad (1.30)$$

If the inequality $\left(1 - \frac{1-\eta}{F(\eta)}\gamma_1 - \frac{t^\eta}{F(\eta)\Gamma(\eta)}\gamma_1\right) > 0$ holds, then $\|S(t) - S_1(t)\| = 0$. Thus, we find

$$S(t) = S_1(t).$$

Analogously, same results can be obtained for the other solutions $I(t)$ and $R(t)$.

4 Numerical Simulations and Discussion

In this section, benefiting from the above Euler approximation scheme, we give several numerical examples substantiating our theoretical outcomes. For this purpose, we choose the parameters $\lambda = 1$, $b_1 = 0.04$, $b_2 = 0.001$, $d_1 = 0.02$, $d_2 = 0.02$, $\phi = 0.1$ and initial conditions $S(0) = 3$, $I(0) = 1$, $C(0) = 5$ as given in [31].

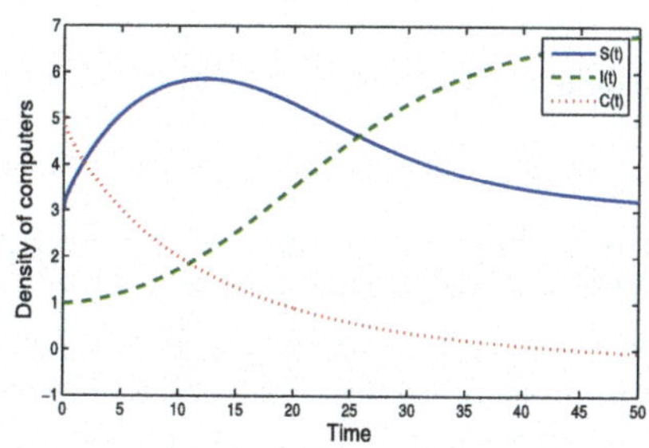

Fig. 1. Numerical simulations for the Eq. (1.11) at $\eta = 0.9$.

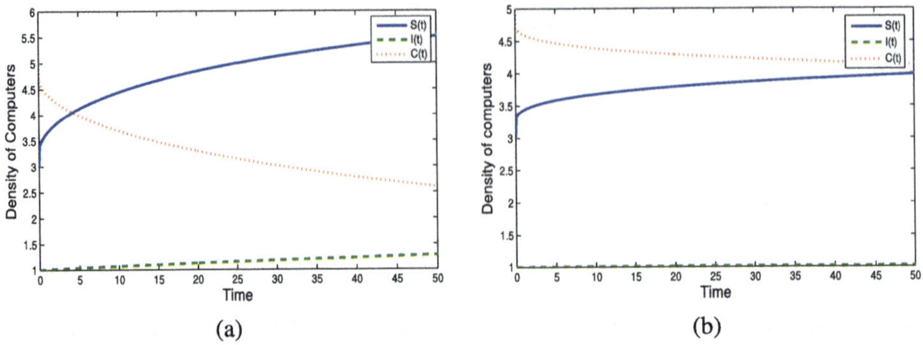

Fig. 2. Numerical simulations for the Eq. (1.11) at $\eta = 0.5$ and $\eta = 0.3$, respectively.

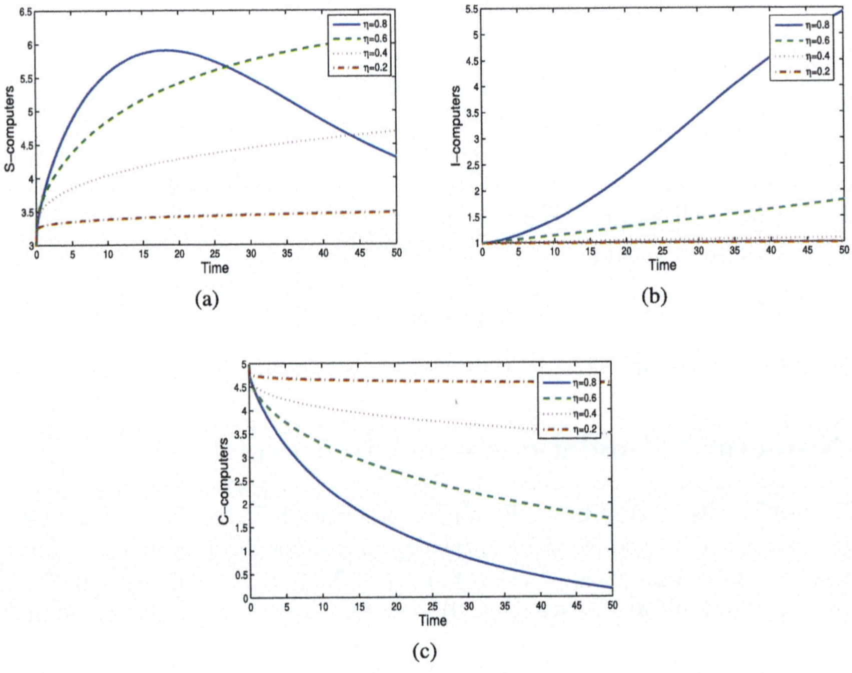

Fig. 3. Numerical simulations for the Eq. (1.11) at $\eta = 0.8$, $\eta = 0.6$, $\eta = 0.4$ and $\eta = 0.2$, respectively.

It is indicated in the figures that the number of S-computers, I-computers and C-computers rise or fall depending on the change in the fractional order η. It can be seen in Fig. 1 that the model remains under attack by infectious populations $I(t)$ for a long time, which is a critical behavior. Figure 2 together with the order the $\eta = 0.5$ and $\eta = 0.3$ demonstrate that $C(t)$ shows an increase while the numbers of exposed computers $S(t)$ and infective computers $I(t)$ decrease in numbers as the fractional order alpha decreases. Furthermore, Fig. 3 suggests that the model components' attitude can be perceived by the varying numerical data obtained from the fractional order η.

5 Concluding Remarks

Since blocking virus abundance plays a key role in computer security and the undesired outcomes of viruses are highly evident, the question how to suppress the avalanche of viruses has become a focus in both industrial and academic societies. In this paper, we aim to deeply examine a SICS model with the concept of ABC derivative assisting a memory effect.First of all, we propose a fractional SICS model based on the model given in [31], and then prove the existence and uniqueness conditions by benefiting from the fixed point theory. The simulation results are illustrated according to different values η, and briefly commented on. Due to the importance of averting infectious populations $I(t)$ in the cyber world, we observe that the number of $I(t)$ decreases when the fractional order η decreases; a great benefit of AB derivative. We expect our findings to be helpful for tackling the virus problem troubling the world of computing and computer users.

References

1. Murray, W.H.: The application of epidemiology to computer viruses. Comput. Secur. **7**(2), 130–150 (1988)
2. Mishra, B.K., Saini, D.K.: SEIRS epidemic model with delay for transmission of malicious objects in computer network. Appl. Math. Comput. **188**(2), 1476–1482 (2007)
3. Toutonji, O.A., Yoo, S.M., Park, M.: Stability analysis of VEISV propagation modeling for network worm attack. Appl. Math. Model. **36**(6), 2751–2761 (2012)
4. Yuan, H., Chen, G., Wu, J., Xiong, H.: Towards controlling virus propagation in information systems with point-to-group information sharing. Decis. Support Syst. **48**(1), 57–68 (2009)
5. Piqueira, J.R.C., Araujo, V.O.: A modified epidemiological model for computer viruses. Appl. Math. Comput. **213**(2), 355–360 (2009)
6. Han, X., Tan, Q.: Dynamical behavior of computer virus on Internet. Appl. Math. Comput. **217**(6), 2520–2526 (2010)
7. Ren, J., Yang, X., Yang, L.X., Xu, Y., Yang, F.: A delayed computer virus propagation model and its dynamics. Chaos, Solitons Fractals **45**(1), 74–79 (2012)
8. Zhu, Q., Yang, X., Yang, L.X., Zhang, C.: Optimal control of computer virus under a delayed model. Appl. Math. Comput. **218**(23), 11613–11619 (2012)
9. Chen, L., Carley, K.M.: The impact of countermeasure propagation on the prevalence of computer viruses. IEEE Trans. Syst. Man Cybern. B Cybern. **34**(2), 823–833 (2004)
10. Baleanu, D., Guvenc, Z.B., Machado, J.A.T.: New Trends in Nanotechnology and Fractional Calculus Applications. Springer, Dordrecht (2010)
11. Atangana, A., Alkahtani, B.T.: Analysis of non-homogenous heat model with new trend of derivative with fractional order. Chaos, Solitons Fractals **89**, 566–571 (2016)

12. Tyagi, S., Abbas, S., Hafayed, M.: Global Mittag-Leffler stability of complex valued fractional-order neural network with discrete and distributed delays. Rendiconti del Circolo Matematico di Palermo Series 2 **65**(3), 485–505 (2016)
13. Özdemir, N., Karadeniz, D., İskender, B.B.: Fractional optimal control problem of a distributed system in cylindrical coordinates. Phys. Lett. A **373**(2), 221–226 (2009)
14. Evirgen, F.: Analyze the optimal solutions of optimization problems by means of fractional gradient based system using VIM. Int. J. Optim. Control Theor. Appl. (IJOCTA) **6**(2), 75–83 (2016)
15. Avcı, D., Eroğlu, B.B., Özdemir, N.: Conformable heat equation on a radial symmetric plate. Therm. Sci. **21**(2), 819–826 (2017)
16. Özdemir, N., Yavuz, M.: Numerical solution of fractional black-scholes equation by using the multivariate pade approximation. Acta Phys. Pol A **132**, 1050–1053 (2016)
17. Hammouch, Z., Mekkaoui, T.: Circuit design and simulation for the fractional-order chaotic behavior in a new dynamical system. Complex Intell. Syst. **4**(4), 251–260 (2018)
18. Uçar, E., Özdemir, N., Altun, E.: Fractional order model of immune cells influenced by cancer cells. Math. Model. Nat. Phenom. **14**(3), 308 (2019)
19. Atangana, A., Baleanu, D.: New fractional derivatives with non-local and non-singular kernel: theory and applications to heat transfer model. Therm. Sci. **20**(2), 763–769 (2016)
20. Koca, I.: Analysis of rubella disease model with non-local and non-singular fractional derivatives. Int. J. Optim. Control Theor. Appl. (IJOCTA) **8**(1), 17–25 (2018)
21. Avcı, D., Yetim, A.: Analytical solutions to the advection-diffusion equation with the Atangana-Baleanu derivative over a finite domain. J. Balı kesir Univ. Inst. Sci. Technol. **20**(2), 382–395 (2018)
22. Gomez Aguilar, J.F.: Analytical and numerical solutions of a nonlinear alcoholism model via variable-order fractional differential equations. Physica A **494**, 52–75 (2018)
23. Baleanu, B., Fernandez, A.: On some new properties of fractional derivatives with Mittag-Leffler kernel. Commun. Nonlinear Sci. Numer. Simul. **59**, 444–462 (2018)
24. Yavuz, M., Özdemir, N., Baskonus, H.M.: Solutions of partial differential equations using the fractional operator involving Mittag-Leffler kernel. Eur. Phys. J. Plus **133**, 215 (2018)
25. Morales-Delgadoa, V.F., Gomez-Aguilar, J.F., Taneco-Hernandez, M.A., Escobar-Jimenezc, R.F., Olivares-Peregrino, V.H.: Mathematical modeling of the smoking dynamics using fractional differential equations with local and nonlocal kernel. J. Nonlinear Sci. Appl. **11**(8), 994–1014 (2018)
26. Fernandez, A., Baleanu, D., Srivastava, H.M.: Series representations for fractional-calculus operators involving generalised Mittag-Leffler functions. Commun. Nonlinear Sci. Numer. Simul. **67**, 517–527 (2019)
27. Uçar, S., Uçar, E., Özdemir, N., Hammouch, Z.: Mathematical analysis and numerical simulation for a smoking model with Atangana–Baleanu derivative. Chaos, Solitons Fractals **118**, 300–306 (2019)
28. Owolabi, K.M., Hammouch, Z.: Mathematical modeling and analysis of two-variable system with noninteger-order derivative. Chaos Interdisc. J. Nonlinear Sci. **29**, 013145 (2019)

29. Owolabi, K.M., Hammouch, Z.: Spatiotemporal patterns in the Belousov-Zhabotinskii reaction systems with Atangana-Baleanu fractional order derivative. Physica A. **523**, 1072–1090 (2019)
30. Jarad, F., Abdeljawad, T., Hammouch, Z.: On a class of ordinary differential equations in the frame of Atangana-Baleanu fractional derivative. Chaos, Solitons Fractals **117**, 16–20 (2018)
31. Zhu, Q., Yang, X., Yang, L.X., Zhang, X.: A mixing propagation model of computer viruses and countermeasures. Nonlinear Dyn. **73**(3), 1433–1441 (2013)
32. Baleanu, D., Jajarmi, A., Hajipour, M.: On the nonlinear dynamical systems within the generalized fractional derivatives with Mittag-Leffler kernel. Nonlinear Dyn. **94**(1), 397–414 (2018)

Some Novel Solutions of the Coupled Whitham-Broer-Kaup Equations

Hezha H. Abdulkareem[1,2(✉)], Hajar F. Ismael[1,2], Etibar Sadi Panakhov[2,3], and Hasan Bulut[2]

[1] Department of Mathematics, University of Zakho, Zakho, Iraq
hejazaxoy5@gmail.com
[2] Department of Mathematics, University of Firat, Elazig, Turkey
[3] Institute of Applied Mathematics, Bakun State University, Baku, Azerbaijan

Abstract. The shallow water equations provide a vast range of applications in the ocean, atmospheric modeling, and pneumatic computing, which can also be utilized to modeling flows in rivers and coastal areas. The Bernoulli sub-equation function method is utilized to build the analytic solutions of the (1+1) dimensional coupled Whitham-Broer-Kaup (WBK) equations. This partial differential equation model is translated into ordinary differential equations in order to construct new exponential prototype structures. As a result, the novel results are obtained and then plotted in 3D and 2D surfaces.

Keywords: Nonlinear Whitham-Broer-Kaup equation · Bernoulli sub-equation method · Exponential solution

1 Introduction

Many of the observed phenomena have been presented using nonlinear partial differential equations in engineering, applied mathematics and physics. NPDEs are extensively used in various scientific fields to describe complex phenomena, particularly in optical science, engineering, applied mathematics and physics. In research papers various numeric and analytic techniques have been used to find solutions of NLPDEs for example finite forward difference method [1,2], homotopy perturbation method [3], Adomian decomposition method [4,5], Adams-Bashforth-Moulton method [6], spectral methods [7], homotopy analysis method [8,9], shooting scheme [10–13], the sine-Gordon expansion method [14,15], the inverse scattering method [16], the Bernoulli sub-ODE function method [17,18], the modified auxiliary expansion method [19], the modified -expansion function method [20–22], the tan -expansion method [23,24], the extended sinh-Gordon expansion method [25,26] and the generalized exponential rational function method [27,28]. A number of articles have already resolved the shallow water equations numerically and analytically like the generally projective Riccati equation technique [29], a finite volume method [30], the variational iteration method

H. Dutta et al. (Eds.): CMES 2019, AISC 1111, pp. 200–208, 2020.
https://doi.org/10.1007/978-3-030-39112-6_14

[31,32], the Galerkin spectral method [33], second-order Runge-Kutta discontinuous Galerkin scheme [34], Bäcklund transformation with Lax pairs [35], the tanh-coth, Exp-function as well as Hirota's methods [36]. In this paper, we investigate the (1+1)–dimensional coupled Whitham-Broer-Kaup shallow water via the Bernoulli Sub-equation function method. Consider the nonlinear system of shallow water as follows:

$$u_t + uu_x + v_x + \beta u_{xx} = 0, \tag{1}$$

$$v_t + (uv)_x + \alpha u_{xxx} - \beta v_{xx} = 0, \tag{2}$$

in which $u(x,t)$ and $v(x,t)$ represents the velocity and the total depth, which could be utilized as a structure for water waves. If choosing $\alpha = 1$ and $\beta = 0$ Eqs. (1) and (2) will change to the approximate long wave equations in shallow water while, the same Eqs. will change to the modified Boussinesq equations if we choose $\alpha = 0$ and $\beta = 0.5$.

2 Structures of Bernoulli Sub-equation Function Method

The mainly modified steps of this technique are:

Let we have a nonlinear partial differential equation:

$$P(u_x, u_t, u_{xt}, u_{xx}, ...) = 0, \tag{3}$$

and defining the traveling wave transformation

$$u(x,t) = q(\eta), \eta = x + \gamma t, \tag{4}$$

where $\gamma \neq 0$. Applying Eq. (4) on Eq. (3) as a result, we get a nonlinear ordinary differential equation:

$$N(q, q', q'', ...) = 0. \tag{5}$$

Using a trial equation of solution as follows:

$$q(\eta) = \sum_{i=0}^{n} a_i F^i = a_0 + a_1 F + a_2 F^2 + ... + a_n F^n, \tag{6}$$

and

$$F' = bF + dF^M, b \neq 0, d \neq 0, M \in R - \{0,1,2\}. \tag{7}$$

here $F(\eta)$ is Bernoulli differential polynomial. Inserting Eq. (6) into Eq. (5) as well as using Eq. (7) produces:

$$\Omega(F(\eta)) = b_k F(\eta)^s + \cdots + b_1 F(\eta) + b_0 = 0, \tag{8}$$

via the balance principle,the connection of n and M will be evaluate.

By taking all the coefficients of $\Omega(F(\eta))$ to be zero, we get an algebraic equations system:

$$b_i = 0, \quad i = 0, \cdots, k, \tag{9}$$

solving Eq. (9), we will find the values of $a_0, a_1, ..., a_n$.

Step 4. Solving Bernoulli Eq. (7), two cases are observed depending on the values of b and d:

$$F(\eta) = \left[\frac{-d}{b} + \frac{E}{e^{b(M-1)\eta}}\right]^{\frac{1}{1-M}}, b \neq d, \tag{10}$$

$$F(\eta) = \left[\frac{(E-1) + (E+1)\tanh\left(\frac{b(1-M)\eta}{2}\right)}{1-\tanh\left(\frac{b(1-M)\eta}{2}\right)}\right]^{\frac{1}{1-M}}, b = d, E \in R. \tag{11}$$

Where E is the non-zero constant of integration, with the help of Mathematical packages, we gain the solutions to Eq. (5), using a complete polynomial discrimination system. Also, all the solutions gained in this method are plotted and the suitable parameter values on (1+1)-dimensional surfaces of solutions are taken into account.

3 Implementation of the BSEFM

Now we use the BSEFM on the shallow water equations to build novel solutions. Letting the traveling wave transformation, present the transformation $u(x,t) = U(\eta)$, and $v(x,t) = V(\eta), \eta = x + \gamma t$, in which γ is constant, the nonlinear system of shallow water is converted into a system of NLODEs [37,38]

$$\gamma U' + UU' + V' + \beta U'' = 0, \tag{12}$$

$$\gamma V' + (VU') + \alpha U''' - \beta V'' = 0. \tag{13}$$

Integrating Eqs. (12, 13) and letting a integration constant to zero, we get

$$\gamma U + \frac{U^2}{2} + V + \beta U' = 0, \tag{14}$$

$$\gamma V + VU + \alpha U'' - \beta V' = 0. \tag{15}$$

Taking derivative of Eq. (14) with respect to η and writing in terms of V', we get

$$V' = -\gamma U' - UU' - \beta U'', \tag{16}$$

putting Eq. (16) into Eq. (15), the NLODE can write as:

$$2\gamma^2 U + 3\gamma U^2 + U^3 - (\beta^2 + \alpha)U'' = 0. \tag{17}$$

Balancing U'' and U^3, the connection n and M yields,

$$M = n + 1.$$

Now by choosing the value of n, we get the value of M, Using these values in Eqs. (6–7) and then inserting resultant Eqs. into Eq. (17), we discuss the following cases:

Case 1. Using $n = 2$, $M = 3$ and then substituting them into Eq. (6), the following equations are obtained:

$$U = a_0 + a_1 F + a_2 F^2, \tag{18}$$

$$U^{'} = a_1 bF + a_1 dF^3 + 2a_2 bF^2 + 2a_2 dF^4, \tag{19}$$

$$U'' = a_1 b^2 F + 4a_2 b^2 F^2 + 4a_1 bdF^3 + 12a_2 bdF^4 + 3a_1 d^2 F^5 + 8a_2 d^2 F^6, \tag{20}$$

where $a_2 \neq 0$, $b \neq 0$, $d \neq 0$. Substituting Eqs. (18–20) into Eq. (17), a system of algebraic equations are found. Inserting Eqs. (10) or (11) into a system of algebraic equations, we can investigate the following solutions (Figs. 1, 2):

Case 1a. For $b \neq d$, $a_0 = -2, d = -\left(\frac{(a_2 b)}{2}\right), \alpha = \frac{1}{(4b^2)} - \beta^2, \gamma = 1$, we get

$$u(x,t) = -2 + \frac{a_2}{\frac{a_2}{2} + E\mathrm{e}^{-2b(t+x)}}, \tag{21}$$

$$v(x,t) = -\frac{4a_2 E\mathrm{e}^{2b(t+x)}\left(-1 + 2b\beta\right)}{\left(2E + a_2\mathrm{e}^{2b(t+x)}\right)^2}. \tag{22}$$

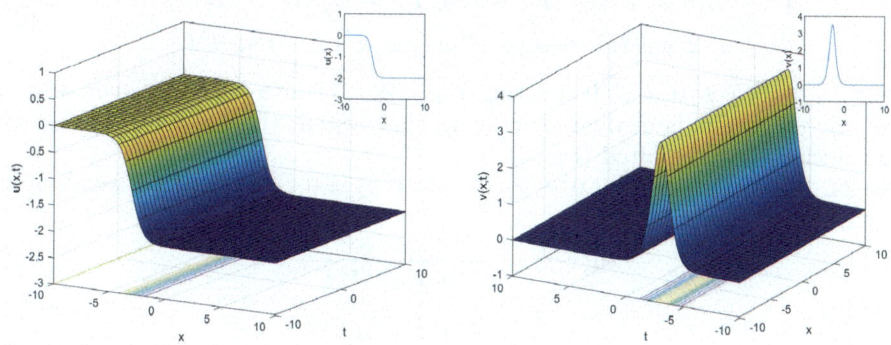

Fig. 1. 3D surfaces with its 2D figures of Eqs. (21) and (22) with values $M = 3, a_2 = 4, d = 2, b = -1, \beta = 3, E = 0.25$ and $t = 2$ for 2D surface.

Case 1b. $b = \frac{(2d)}{a_2}, \alpha = \frac{{a_2}^2}{(16d^2)} - \beta^2, \gamma = 1$, we get

$$u(x,t) = \frac{a_2}{-\frac{a_2}{2} + E\mathrm{e}^{-\frac{4d(t+x)}{a_2}}}, \tag{23}$$

$$v(x,t) = -\frac{4E\mathrm{e}^{\frac{4d(t+x)}{a_2}}\left(a_2 + 4d\beta\right)}{\left(-2E + a_2\mathrm{e}^{\frac{4d(t+x)}{a_2}}\right)^2}. \tag{24}$$

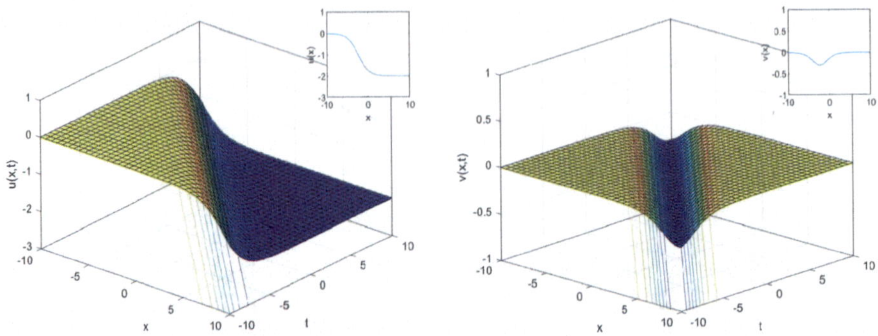

Fig. 2. 3D surfaces with its 2D figures of Eqs. (23) and (24) with values $M = 3, a_2 = 5, b = 2, d = 1, \beta = -2, E = -2$ and $t = 2$ for 2D surface.

Case 2. If taking $n = 3$ and $M = 4$ in Eq. (6), the following equations are found;

$$U = a_0 + a_1 F + a_2 F^2 + a_3 F^3, \tag{25}$$

$$U' = a_1 bF + 2a_2 bF^2 + 3a_3 bF^3 + a_1 dF^4 + 2a_2 dF^5 + 3a_3 dF^6, \tag{26}$$

$$\begin{aligned} U'' = {} & a_1 b^2 F + 4a_2 b^2 F^2 + 9a_3 b^2 F^3 + 5a_1 bdF^4 + 14a_2 bdF^5 \\ & + 27a_3 bdF^6 + 4a_1 d^2 F^7 + 10a_2 d^2 F^8 + 18a_3 d^2 F^9, \end{aligned} \tag{27}$$

where $a_3 \neq 0, b \neq 0, d \neq 0$. putting Eqs. (25–27) into Eq. (17), a system of algebraic equations is evaluated. Solving this system the following coefficients and solutions have resulted (Fig. 3):

Case 2a. When $a_0 = 0, a_1 = 0, a_3 = \frac{2d}{b}, \gamma = 1, a_2 = 0, \alpha = \frac{1}{9b^2} - \beta^2$, we obtain

$$u(x,t) = \frac{4}{-2 + Ee^{-3(t+x)}}, \tag{28}$$

$$v(x,t) = -\frac{4E(1+3\beta)e^{3(t+x)}}{\left(E - 2e^{3(t+x)}\right)^2}. \tag{29}$$

Case 2b. When $a_0 = 0, a_1 = 0, a_2 = 0, b = \frac{1}{3\sqrt{\alpha+\beta^2}}, d = \frac{a_3}{6\sqrt{\alpha+\beta^2}}, \gamma = 1$, we obtain Fig. 4

$$u(x,t) = \frac{a_3}{-2 + Ee^{-3(t+x)}}, \tag{30}$$

$$v(x,t) = -\frac{a_3 e^{3(t+x)}\left((-4+a_3)e^{3(t+x)} + E(2+6\beta)\right)}{2\left(E - 2e^{3(t+x)}\right)^2}. \tag{31}$$

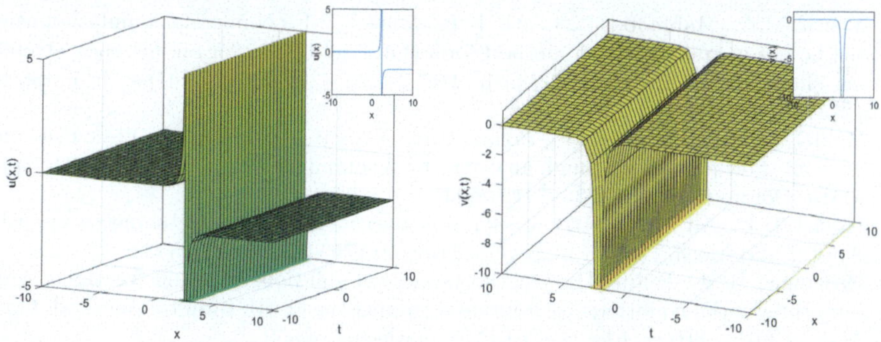

Fig. 3. 3D surfaces with its 2D figures of Eqs. (28) and (29) with values $M = 3, a_2 = 4, d = 2, b = -1, \beta = 3, E = 0.25$ and $t = 2$ for 2D surface.

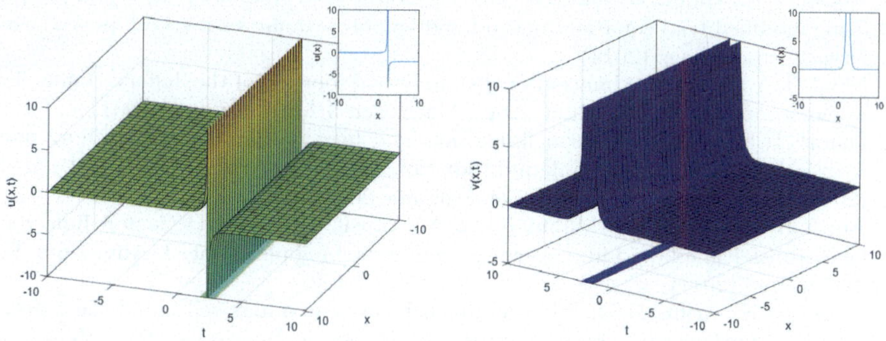

Fig. 4. 3D surfaces with its 2D figures of Eqs. (30) and (31) with values $M = 3, a_2 = 4, d = 2, b = -1, \beta = 3, E = 0.25$ and $t = 2$ for 2D surface.

4 Conclusion

Analytic solutions of Whitham-Broer-Kaup equations via utilizing the BSEFM are presented. In comparison with the results obtained at paper [30], Eqs. (21) and (23) are the same results, but the traveling wave solutions such as Eqs. (22), (24), (28), and (29) extracted via utilizing BSEFM are novel exponential functional solutions for Eqs. (1–2). Moreover, all solutions are inserted into Eqs. (1–2) and they verify the system.

References

1. Sulaiman, T.A., Bulut, H., Yokus, A., Baskonus, H.M.: On the exact and numerical solutions to the coupled Boussinesq equation arising in ocean engineering. Indian J. Phys. (2019). https://doi.org/10.1007/s12648-018-1322-1
2. Yokus, A., Baskonus, H.M., Sulaiman, T.A., Bulut, H.: Numerical simulation and solutions of the two-component second order KdV evolutionarysystem. Numer. Methods Partial Differ. Equ. (2018). https://doi.org/10.1002/num.22192

3. Yousif, M.A., Mahmood, B.A., Ali, K.K., Ismael, H.F.: Numerical simulation using the homotopy perturbation method for a thin liquid film over an unsteady stretching sheet. Int. J. Pure Appl. Math. **107**(2), 289–300 (2016). https://doi.org/10.12732/ijpam.v107i2.1
4. Bulut, H., Ergüt, M., Asil, V., Bokor, R.H.: Numerical solution of a viscous incompressible flow problem through an orifice by Adomian decomposition method. Appl. Math. Comput. **153**(3), 733–741 (2004)
5. Ismael, H.F., Ali, K.K.: MHD casson flow over an unsteady stretching sheet. Adv. Appl. Fluid Mech. (2017). https://doi.org/10.17654/FM020040533
6. Baskonus, H.M., Bulut, H.: On the numerical solutions of some fractional ordinary differential equations by fractional Adams-Bashforth-Moulton method. Open Math. (2015). https://doi.org/10.1515/math-2015-0052
7. Bueno-Orovio, A., Pérez-García, V.M., Fenton, F.H.: Spectral methods for partial differential equations in irregular domains: the spectral smoothed boundary method. SIAM J. Sci. Comput. **28**(3), 886–900 (2006)
8. Kocak, Z.F., Bulut, H. and Yel, G.: The solution of fractional wave equation by using modified trial equation method and homotopy analysis method, in AIP Conference Proceedings (2014)
9. Nofal, T.A.: An approximation of the analytical solution of the Jeffery-Hamel flow by homotopy analysis method. Appl. Math. Sci. **5**(53), 2603–2615 (2011)
10. Ismael, H.F.: Carreau-Casson fluids flow and heat transfer over stretching plate with internal heat source/sink and radiation. Int. J. Adv. Appl. Sci. J. **6**(2), 81–86 (2017). https://doi.org/10.1371/journal.pone.0002559
11. Ali, K.K., Ismael, H.F., Mahmood, B.A., Yousif, M.A.: MHD Casson fluid with heat transfer in a liquid film over unsteady stretching plate. Int. J. Adv. Appl. Sci. **4**(1), 55–58 (2017)
12. Ismael, H.F., Arifin, N.M.: Flow and heat transfer in a maxwell liquid sheet over a stretching surface with thermal radiation and viscous dissipation. JP J. Heat Mass Transf. **15**(4) (2018). https://doi.org/10.17654/HM015040847
13. Zeeshan, A., Ismael, H.F., Yousif, M.A., Mahmood, T., Rahman, S.U.: Simultaneous effects of slip and wall stretching/shrinking on radiative flow of magneto nanofluid through porous medium. J. Magn. **23**(4), 491–498 (2018). https://doi.org/10.4283/JMAG.2018.23.4.491
14. Baskonus, H.M., Sulaiman, T.A., Bulut, H.: On the novel wave behaviors to the coupled nonlinear Maccari's system with complex structure. Optik (Stuttg) (2017). https://doi.org/10.1016/j.ijleo.2016.10.135
15. Bulut, H., Sulaiman, T.A., Baskonus, H.M.: New solitary and optical wave structures to the Korteweg-de Vries equation with dual-power law nonlinearity. Opt. Quantum Electron. (2016). https://doi.org/10.1007/s11082-016-0831-4
16. Vakhnenko, V.O., Parkes, E.J., Morrison, A.J.: A Bäcklund transformation and the inverse scattering transform method for the generalised Vakhnenko equation, Chaos. Solitons Fractals (2003). https://doi.org/10.1016/S0960-0779(02)00483-6
17. Baskonus, H.M., Bulut, H.: An effective schema for solving some nonlinear partial differential equation arising in nonlinear physics. Open Phys. (2015). https://doi.org/10.1515/phys-2015-0035
18. Baskonus, H.M., Bulut, H.: Exponential prototype structures for (2+1)-dimensional Boiti-Leon-Pempinelli systems in mathematical physics. Waves Random Complex Media (2016). https://doi.org/10.1080/17455030.2015.1132860
19. Wei, G., Ismael, H.F., Bulut, H., Baskonus, H.M.: Instability modulation for the (2+1)-dimension paraxial wave equation and its new optical soliton solutions in Kerr media. Phys. Scr. (2019). https://doi.org/10.1088/1402-4896/ab4a50

20. Ilhan, O.A., Bulut, H., Sulaiman, T.A., Baskonus, H.M.: Dynamic of solitary wave solutions in some nonlinear pseudoparabolic models and Dodd-Bullough-Mikhailov equation. Indian J. Phys. (2018). https://doi.org/10.1007/s12648-018-1187-3
21. Cattani, C., Sulaiman, T.A., Baskonus, H.M., Bulut, H.: Solitons in an inhomogeneous Murnaghan's rod. Eur. Phys. J. Plus (2018). https://doi.org/10.1140/epjp/i2018-12085-y
22. Houwe, A., Hammouch, Z., Bienvenue, D., Nestor, S. and Betchewe, G.: Nonlinear Schrödingers equations with cubic nonlinearity: M-derivative soliton solutions by $\exp(-\Phi(\xi))$-expansion method (2019)
23. Manafian, J., Aghdaei, M.F.: Abundant soliton solutions for the coupled Schrödinger-Boussinesq system via an analytical method. Eur. Phys. J. Plus (2016). https://doi.org/10.1140/epjp/i2016-16097-3
24. Hammouch, Z., Mekkaoui, T., Agarwal, P.: Optical solitons for the Calogero-Bogoyavlenskii-Schiff equation in (2 + 1) dimensions with time-fractional conformable derivative. Eur. Phys. J. Plus (2018). https://doi.org/10.1140/epjp/i2018-12096-8
25. Cattani, C., Sulaiman, T.A., Baskonus, H.M., Bulut, H.: On the soliton solutions to the Nizhnik-Novikov-Veselov and the Drinfel'd-Sokolov systems. Opt. Quantum Electron. (2018). https://doi.org/10.1007/s11082-018-1406-3
26. Bulut, H., Sulaiman, T.A., Baskonus, H.M.: Dark, bright optical and other solitons with conformable space-time fractional second-order spatiotemporal dispersion. Optik (Stuttg). (2018). https://doi.org/10.1016/j.ijleo.2018.02.086
27. Osman, M.S., Ghanbari, B.: New optical solitary wave solutions of Fokas-Lenells equation in presence of perturbation terms by a novel approach. Optik (Stuttg). (2018). https://doi.org/10.1016/j.ijleo.2018.08.007
28. Ghanbari, B., Kuo, C.-K.: New exact wave solutions of the variable-coefficient (1+ 1)-dimensional Benjamin-Bona-Mahony and (2+ 1)-dimensional asymmetric Nizhnik-Novikov-Veselov equations via the generalized exponential rational function method. Eur. Phys. J. Plus **134**(7), 334 (2019)
29. Chen, Y., Li, B., Zhang, H.: New exact travelling wave solutions for the shallow long wave approximate equations. Appl. Math. Comput. **160**(1), 77–88 (2005)
30. Benkhaldoun, F., Elmahi, I., Seaïd, M.: A new finite volume method for flux-gradient and source-term balancing in shallow water equations. Comput. Methods Appl. Mech. Eng. **199**(49–52), 3324–3335 (2010)
31. Imani, A.A., Ganji, D.D., Rokni, H.B., Latifizadeh, H., Hesameddini, E., Rafiee, M.H.: Approximate traveling wave solution for shallow water wave equation. Appl. Math. Model. **36**(4), 1550–1557 (2012)
32. Rafei, M., Daniali, H.: Application of the variational iteration method to the Whitham-Broer-Kaup equations. Comput. Math. Appl. **54**(7–8), 1079–1085 (2007)
33. Kröger, T., Lukáčová-Medvid'ová, M.: An evolution Galerkin scheme for the shallow water magnetohydrodynamic equations in two space dimensions. J. Comput. Phys. **206**(1), 122–149 (2005)
34. Kesserwani, G., Ghostine, R., Vazquez, J., Ghenaim, A., Mosé, R.: Application of a second-order Runge-Kutta discontinuous Galerkin scheme for the shallow water equations with source terms. Int. J. Numer. Methods Fluids **56**(7), 805–821 (2008)
35. Shang, Y.: Böcklund transformation, Lax pairs and explicit exact solutions for the shallow water waves equation. Appl. Math. Comput. **187**(2), 1286–1297 (2007)
36. Wazwaz, A.-M.: Solitary wave solutions of the generalized shallow water wave (GSWW) equation by Hirota's method, tanh-coth method and Exp-function method. Appl. Math. Comput. **202**(1), 275–286 (2008)

37. Kumar, M., Tiwari, A.K., Kumar, R.: More of coupled solutions Whitham–Broer–Kaup equations. Proc. Natl. Acad. Sci. India Sect. A Phys. Sci. **89**(4), 747–755 (2019)
38. Xie, F., Yan, Z., Zhang, H.: Explicit and exact traveling wave solutions of Whitham-Broer-Kaup shallow water equations. Phys. Lett. A **285**(1–2), 76–80 (2001)

Numerical Solution of the Homogeneous Telegraph Equation by Using Galerkin Finite Element Method

Dursun Irk[1(✉)] and Emre Kirli[2]

[1] Department of Mathematics and Computer Science, University of Eskişehir Osmangazi, Eskişehir, Turkey
dirk@ogu.edu.tr
[2] Department of Mathematics, University of İstanbul Bilgi, İstanbul, Turkey
emre.kirli@bilgi.edu.tr

Abstract. In this study, high order numerical solution of one dimensional homogeneous Telegraph equation is presented using quadratic B-spline Galerkin finite element method. In the method, second and fourth order single step methods are used for the time integration. Second order single step method is also known as Crank Nicolson method. The numerical example is studied to illustrate the accuracy and the efficiency of the method.

Keywords: Quadratic B-spline · Galerkin method · Homogeneous Telegraph equation

1 Introduction

One dimensional homogeneous Telegraph equation arises in the study of propagation of electrical signals in a cable of transmission line and wave phenomena [1, 2]. Various numerical techniques have been developed and compared for solving the one dimensional Telegraph equation (see [3, 4] and referenced in). In the next section, after the time discretization of the homogeneous Telegraph equation is performed by using higher accurate finite difference method, a finite element space discretization is used to obtain a system of algebraic equation. In the numerical experiment section, proposed methods are tested for the test problem and a summary of main findings of the work is presented in the last section. While the numerical solutions of partial differential equations are investigated, finite difference or finite element method is frequently used. While the time discretization of such equations is done, generally Crank Nicolson method having second order accuracy is used. The aim of this study is to see how the results change when a method with an accuracy of 4 is used instead of the Crank Nicolson method for the numerical solution of the homogeneous Telegraph equation.

We consider the following one dimensional homogeneous Telegraph equation

$$u_{tt} + 2\alpha u_t + \beta^2 u - u_{xx} = 0 \tag{1}$$

H. Dutta et al. (Eds.): CMES 2019, AISC 1111, pp. 209–217, 2020.
https://doi.org/10.1007/978-3-030-39112-6_15

with the boundary conditions

$$u(a,\, t) = g_1(t),\;\; u(b,\, t) = g_2(t),\;\; t \in (0,\, T] \tag{2}$$

and initial conditions

$$\begin{aligned} u(x,\, 0) &= f_1(x),\;\; a \le x \le b \\ u_t(x,\, 0) &= f_2(x),\;\; a \le x \le b \end{aligned} \tag{3}$$

in a restricted solution domain over an space/time interval $[a,\, b] \times [0,\, T]$.

2 Application of the Method

The homogeneous Telegraph equation can be converted to the following system by standard change of variables

$$\begin{aligned} u_t &= v, \\ v_t &= u_{xx} - 2\alpha v - \beta^2 u. \end{aligned} \tag{4}$$

The following one step method will be used for time discretization of the Eq. (4)

$$\begin{aligned} u^{n+1} &= u^n + \theta_1 u_t^{n+1} + \theta_2 u_t^n + \theta_3 u_{tt}^{n+1} + \theta_4 u_{tt}^n, \\ v^{n+1} &= v^n + \theta_1 v_t^{n+1} + \theta_2 v_t^n + \theta_3 v_{tt}^{n+1} + \theta_4 v_{tt}^n. \end{aligned} \tag{5}$$

When

$$\theta_1 = \theta_2 = \frac{\Delta t}{2},\; \theta_3 = \theta_4 = 0,$$

the method (5) is of order 2 known as Crank-Nicolson method (M1). When

$$\theta_1 = \theta_2 = \frac{\Delta t}{2},\;\; \theta_3 = -\frac{(\Delta t)^2}{12},\;\; \theta_4 = \frac{(\Delta t)^2}{12},$$

the method (5) is of order 4 (M2). Using (4) in (5), we have

$$\left(1 + \theta_3\beta^2\right) u^{n+1} + (2\alpha\theta_3 - \theta_1)\, v^{n+1} - \theta_3 (u_{xx})^{n+1} = \left(1 - \theta_4\beta^2\right) u^n + (-2\alpha\theta_4 + \theta_2)\, v^n + \theta_4 (u_{xx})^n \tag{6}$$

and

$$\begin{aligned} &\left(\beta^2\theta_1 - 2\alpha\beta^2\theta_3\right) u^{n+1} + \left(1 + 2\alpha\theta_1 + \beta^2\theta_3 - 4\alpha^2\theta_3\right) v^{n+1} + (-\theta_1 + 2\alpha\theta_3)\, u_{xx}^{n+1} - \theta_3 v_{xx}^{n+1} = \\ &\left(-\beta^2\theta_2 + 2\alpha\beta^2\theta_4\right) u^n + \left(1 - 2\alpha\theta_2 - \beta^2\theta_4 + 4\alpha^2\theta_4\right) v^n + (\theta_2 - 2\alpha\theta_4)\, u_{xx}^n + \theta_4 v_{xx}^n. \end{aligned} \tag{7}$$

When making calculations, the space-time plane $[a, b] \times [0, T]$ will be discretized by grids with Δt and h. Thus,

$$\begin{aligned} u(x_m,\, t_n) &= u_m^n, \\ v(x_m,\, t_n) &= v_m^n, \end{aligned} \quad m = 0,\, 1,\, \ldots,\, N; \quad n = 0,\, 1,\, 2,\, \ldots$$

where $x_m = a+mh$, $t_n = n\Delta t$, will be used for the exact solution at the points (x_m, t_n). U_m^n and V_m^n will be used for the approximate solutions. The space interval $[a, b]$ will be divided into equal length N sub-interval as

$$a = x_0 < x_1 < \ldots < x_{N-1} < x_N = b.$$

Then the quadratic B-spline functions are defined at these knots as

$$T_m(x) = \frac{1}{h^2}\begin{cases} (x_{m+2}-x)^2 - 3(x_{m+1}-x)^2 \\ + 3(x_m - x)^2, & x_{m-1} \le x < x_m \\ (x_{m+2}-x)^2 - 3(x_{m+1}-x)^2, & x_m \le x < x_{m+1} \\ (x_{m+2}-x)^2, & x_{m+1} \le x < x_{m+2} \\ 0, & \text{otherwise.} \end{cases} \tag{8}$$

Over the problem domain, the approximate solutions $U(x, t)$ and $V(x, t)$ to the exact solutions $u(x, t)$ and $v(x, t)$ can be written as a combination of the quadratic B-splines

$$U(x, t) = \sum_{j=-1}^{N} T_j(x)\delta_j(t),$$
$$V(x, t) = \sum_{j=-1}^{N} T_j(x)\rho_j(t) \tag{9}$$

where δ_j and ρ_j are time dependent unknown parameters. Since each quadratic B-splines covers 3 intervals, each element $[x_m, x_{m+1}]$ is covered by three splines. Therefore over the element $[x_m, x_{m+1}]$, an approximation to the exact solutions $u(x, t)$ and $v(x, t)$ in terms of quadratic B-splines can be written as

$$U(x, t) = \sum_{j=m-1}^{m+1} T_j(x)\delta_j(t)$$
$$V(x, t) = \sum_{j=m-1}^{m+1} T_j(x)\rho_j(t). \tag{10}$$

Using quadratic trigonometric B-spline function (8) and the trial solution (10), the values of $U_m = U(x_m, t)$, $V_m = V(x_m, t)$ and $U_m' = U(x_m, t)$, $V_m' = V(x_m, t)$ are obtained as follows

$$U_m = \delta_{m-1} + \delta_m,$$
$$U_m' = \frac{2}{h}(\delta_m - \delta_{m-1}),$$
$$V_m = \rho_{m-1} + \rho_m,$$
$$V_m' = \frac{2}{h}(\rho_m - \rho_{m-1}). \tag{11}$$

Applying Galerkin method to the Eqs. (6–7) with weight function $W(x)$ and then integrating by parts lead to the equation:

$$\begin{aligned}&\int_a^b\left[\left(1+\theta_3\beta^2\right)W(x)U^{n+1}+\theta_3 W_x(x)U_x^{n+1}+\left(2\alpha\theta_3-\theta_1\right)W(x)V^{n+1}\right]dx\\&-\theta_3 W(x)U_x^{n+1}\Big|_a^b=\int_a^b\left[\left(1-\theta_4\beta^2\right)W(x)u^n+\left(-2\alpha\theta_4+\theta_2\right)W(x)v^n-\right.\\&\left.\theta_4 W_x(x)U_x^n\right]dx+\theta_4 W(x)U_x^n\Big|_a^b\end{aligned} \tag{12}$$

and

$$\begin{aligned}&\int_a^b\left[\left(\beta^2\theta_1-2\alpha\beta^2\theta_3\right)W(x)U^{n+1}-(-\theta_1+2\alpha\theta_3)W_x(x)U_x^{n+1}+\right.\\&\left.\left(1+2\alpha\theta_1+\beta^2\theta_3-4\alpha^2\theta_3\right)W(x)V^{n+1}+\theta_3 W_x(x)V_x^{n+1}\right]dx+\\&(-\theta_1+2\alpha\theta_3)W(x)U_x^{n+1}\Big|_a^b-\theta_3 W(x)V_x^{n+1}\Big|_a^b=\int_a^b\left[(-\beta^2\theta_2+2\alpha\beta^2\theta_4)W(x)u^n+\right.\\&\left.(1-2\alpha\theta_2-\beta^2\theta_4+4\alpha^2\theta_4)W(x)V^n-(\theta_2-2\alpha\theta_4)W_x(x)U_x^n-\theta_4 W_x(x)V_x^n\right]dx\\&+(\theta_2-2\alpha\theta_4)W(x)U_x^n\Big|_a^b+\theta_4 W(x)V_x^n\Big|_a^b.\end{aligned} \tag{13}$$

If the weight function $W(x)$ is taken as quadratic B-spline shape function T_m and using the expression (10) in the Eqs. (12–13), a fully discrete approximation is obtained over the element $[x_m,\ x_{m+1}]$ as

$$\begin{aligned}&\sum_{j=m-1}^{m+1}\left\{\int_{x_m}^{x_{m+1}}\left[\left(1+\theta_3\beta^2\right)T_iT_j\delta_j^{n+1}+\theta_3T_i'T_j'\delta_j^{n+1}+\left(2\alpha\theta_3-\theta_1\right)T_iT_j\rho_j^{n+1}\right]dx-\theta_3T_iT_j'\Big|_{x_m}^{x_{m+1}}\delta_j^{n+1}\right\}\\&-\sum_{j=m-1}^{m+1}\left\{\int_{x_m}^{x_{m+1}}\left[\left(1-\theta_4\beta^2\right)T_iT_j\delta_j^n+\left(-2\alpha\theta_4+\theta_2\right)T_iT_j\rho_j^n-\theta_4T_i'T_j'\delta_j^n\right]dx+\theta_4T_iT_j'\Big|_{x_m}^{x_{m+1}}\delta_j^n\right\}\end{aligned} \tag{14}$$

and

$$\begin{aligned}&\sum_{j=m-1}^{m+1}\int_{x_m}^{x_{m+1}}\left[\left(\beta^2\theta_1-2\alpha\beta^2\theta_3\right)T_iT_j\delta_j^{n+1}-(-\theta_1+2\alpha\theta_3)T_i'T_j'\delta_j^{n+1}+\right.\\&\left.\left(1+2\alpha\theta_1+\beta^2\theta_3-4\alpha^2\theta_3\right)T_iT_j\rho_j^{n+1}+\theta_3T_i'T_j'\rho_j^{n+1}\right]dx+(-\theta_1+2\alpha\theta_3)T_iT_j'\Big|_{x_m}^{x_{m+1}}\delta_j^{n+1}\\&-\theta_3\,T_iT_j'\Big|_{x_m}^{x_{m+1}}\rho_j^{n+1}\\&-\sum_{j=m-1}^{m+1}\int_{x_m}^{x_{m+1}}\left[(-\beta^2\theta_2+2\alpha\beta^2\theta_4)T_iT_j\delta_j^n+(1-2\alpha\theta_2-\beta^2\theta_4+4\alpha^2\theta_4)T_iT_j\rho_j^n-\right.\\&\left.(\theta_2-2\alpha\theta_4)T_i'T_j'\delta_j^n-\theta_4T_i'T_j'\rho_j^n\right]dx+(\theta_2-2\alpha\theta_4)T_iT_j'\Big|_{x_m}^{x_{m+1}}\delta_j^n+\theta_4\,T_iT_j'\Big|_{x_m}^{x_{m+1}}\rho_j^n.\end{aligned} \tag{15}$$

A typical finite interval $[x_m,\ x_{m+1}]$ is mapped to the interval $[0,\ h]$ by a local coordinate transformation defined by $\xi = x - x_m$. Therefore (14–15) by using quadratic B-spline shape functions in terms of ξ over the element $[0,\ h]$ can be written as

$$\begin{aligned}&\sum_{j=m-1}^{m+1}\left\{\int_0^h\left[\left(1+\theta_3\beta^2\right)T_iT_j\delta_j^{n+1}+\theta_3T_i'T_j'\delta_j^{n+1}+\left(2\alpha\theta_3-\theta_1\right)T_iT_j\rho_j^{n+1}\right]d\xi-\theta_3T_iT_j'\Big|_0^h\delta_j^{n+1}\right\}\\&-\sum_{j=m-1}^{m+1}\left\{\int_0^h\left[\left(1-\theta_4\beta^2\right)T_iT_j\delta_j^n+\left(-2\alpha\theta_4+\theta_2\right)T_iT_j\rho_j^n-\theta_4T_i'T_j'\delta_j^n\right]d\xi+\theta_4T_iT_j'\Big|_0^h\delta_j^n\right\}\end{aligned} \tag{16}$$

and

$$\sum_{j=m-1}^{m+1}\int_0^h\Big[\left(\beta^2\theta_1-2\alpha\beta^2\theta_3\right)T_iT_j\delta_j^{n+1}-(-\theta_1+2\alpha\theta_3)T_i'T_j'\delta_j^{n+1}+$$
$$\left(1+2\alpha\theta_1+\beta^2\theta_3-4\alpha^2\theta_3\right)T_iT_j\rho_j^{n+1}+\theta_3T_i'T_j'\rho_j^{n+1}\Big]d\xi+(-\theta_1+2\alpha\theta_3)T_iT_j'\Big|_0^h\delta_j^{n+1}$$
$$-\theta_3\,T_iT_j'\Big|_0^h\rho_j^{n+1}$$
$$-\sum_{j=m-1}^{m+1}\int_0^h\Big[(-\beta^2\theta_2+2\alpha\beta^2\theta_4)T_iT_j\delta_j^n+(1-2\alpha\theta_2-\beta^2\theta_4+4\alpha^2\theta_4)T_iT_j\rho_j^n-$$
$$(\theta_2-2\alpha\theta_4)T_i'T_j'\delta_j^n-\theta_4T_i'T_j'\rho_j^n\Big]d\xi+(\theta_2-2\alpha\theta_4)T_iT_j'\Big|_0^h\delta_j^n+\theta_4\,T_iT_j'\Big|_0^h\rho_j^n. \tag{17}$$

(16–17) can be written in the matrices form as

$$\begin{aligned}&\left[\left(1+\theta_3\beta^2\right)\boldsymbol{A}^e\left(\boldsymbol{\delta}^e\right)^{n+1}-\theta_3\boldsymbol{B}^e\left(\boldsymbol{\delta}^e\right)^{n+1}+(2\alpha\theta_3-\theta_1)\boldsymbol{A}^e(\boldsymbol{\rho}^e)^{n+1}-\theta_3\boldsymbol{C}^e\left(\boldsymbol{\delta}^e\right)^{n+1}\right]-\\&\left[\left(1-\theta_4\beta^2\right)\boldsymbol{A}^e\left(\boldsymbol{\delta}^e\right)^n+\theta_4\boldsymbol{B}^e\left(\boldsymbol{\delta}^e\right)^n+(-2\alpha\theta_4+\theta_2)\boldsymbol{A}^e(\boldsymbol{\rho}^e)^n+\theta_4\boldsymbol{C}^e\left(\boldsymbol{\delta}^e\right)^n\right]\end{aligned} \tag{18}$$

and

$$\begin{aligned}&\Big[\left(\beta^2\theta_1-2\alpha\beta^2\theta_3\right)\boldsymbol{A}^e\left(\boldsymbol{\delta}^e\right)^{n+1}+(-\theta_1+2\alpha\theta_3)\boldsymbol{B}^e\left(\boldsymbol{\delta}^e\right)^{n+1}+\\&\left(1+2\alpha\theta_1+\beta^2\theta_3-4\alpha^2\theta_3\right)\boldsymbol{A}^e(\boldsymbol{\rho}^e)^{n+1}-\theta_3\boldsymbol{B}^e(\boldsymbol{\rho}^e)^{n+1}\Big]+\\&(-\theta_1+2\alpha\theta_3)\boldsymbol{C}^e\left(\boldsymbol{\delta}^e\right)^{n+1}+\theta_3\boldsymbol{C}^e(\boldsymbol{\rho}^e)^{n+1}-\\&\left[(-\beta^2\theta_2+2\alpha\beta^2\theta_4)\boldsymbol{A}^e\left(\boldsymbol{\delta}^e\right)^n+(1-2\alpha\theta_2-\beta^2\theta_4+4\alpha^2\theta_4)\boldsymbol{A}^e(\boldsymbol{\rho}^e)^n+\right.\\&\left.(\theta_2-2\alpha\theta_4)\boldsymbol{B}^e\left(\boldsymbol{\delta}^e\right)^n+\theta_4\boldsymbol{B}^e(\boldsymbol{\rho}^e)^n+(\theta_2-2\alpha\theta_4)\boldsymbol{C}^e\left(\boldsymbol{\delta}^e\right)^n+\theta_4\boldsymbol{C}^e(\boldsymbol{\rho}^e)^n\right]\end{aligned} \tag{19}$$

where

$$A_{ij}^e=\int_{x_m}^{x_{m+1}}T_iT_jdx,\quad B_{ij}^e=\int_{x_m}^{x_{m+1}}T_iT_j'dx-\int_{x_m}^{x_{m+1}}T_i'T_j'dx,\quad C_{ij}^e=T_iT_j'\Big|_0^h,$$
$$\boldsymbol{\delta}^e=(\delta_{m-1},\,\delta_m,\,\delta_{m+1})^T,\quad \boldsymbol{\rho}^e=(\rho_{m-1},\,\rho_m,\,\rho_{m+1})^T.$$

Combining contributions from all elements lead to the linear matrix equation

$$\begin{aligned}&\left[\left(1+\theta_3\beta^2\right)\boldsymbol{A}\boldsymbol{\delta}^{n+1}-\theta_3\boldsymbol{B}\boldsymbol{\delta}^{n+1}+(2\alpha\theta_3-\theta_1)\boldsymbol{A}\boldsymbol{\rho}^{n+1}-\theta_3\boldsymbol{C}\boldsymbol{\delta}^{n+1}\right]=\\&\left[\left(1-\theta_4\beta^2\right)\boldsymbol{A}\boldsymbol{\delta}^n+\theta_4\boldsymbol{B}\boldsymbol{\delta}^n+(-2\alpha\theta_4+\theta_2)\boldsymbol{A}\boldsymbol{\rho}^n+\theta_4\boldsymbol{C}\boldsymbol{\delta}^n\right]\end{aligned} \tag{20}$$

and

$$\begin{aligned}&\left(\beta^2\theta_1-2\alpha\beta^2\theta_3\right)\boldsymbol{A}\boldsymbol{\delta}^{n+1}+(-\theta_1+2\alpha\theta_3)\boldsymbol{B}\boldsymbol{\delta}^{n+1}+\left(1+2\alpha\theta_1+\beta^2\theta_3-4\alpha^2\theta_3\right)\boldsymbol{A}\boldsymbol{\rho}^{n+1}\\&-\theta_3\boldsymbol{B}\boldsymbol{\rho}^{n+1}+(-\theta_1+2\alpha\theta_3)\boldsymbol{C}\boldsymbol{\delta}^{n+1}+\theta_3\boldsymbol{C}\boldsymbol{\rho}^{n+1}=(-\beta^2\theta_2+2\alpha\beta^2\theta_4)\boldsymbol{A}\boldsymbol{\delta}^n\\&+(1-2\alpha\theta_2-\beta^2\theta_4+4\alpha^2\theta_4)\boldsymbol{A}\boldsymbol{\rho}^n+(\theta_2-2\alpha\theta_4)\boldsymbol{B}\boldsymbol{\delta}^n+\theta_4\boldsymbol{B}\boldsymbol{\rho}^n+(\theta_2-2\alpha\theta_4)\boldsymbol{C}\boldsymbol{\delta}^n+\theta_4\boldsymbol{C}\boldsymbol{\rho}^n\end{aligned} \tag{21}$$

where global element parameters

$$\boldsymbol{\delta}=(\delta_{-1},\,\delta_0,\,\ldots,\,\delta_{N-1},\,\delta_N)^T,$$

$$\boldsymbol{\rho} = (\rho_{-1}, \rho_0, \ldots, \rho_{N-1}, \rho_N)^T.$$

The system of Eqs. (20–21) consists of $(2N + 4)$ equations of $(2N + 4)$ unknown parameters

$$(\delta_{-1}, \delta_0, \ldots, \delta_{N-1}, \delta_N, \rho_{-1}, \rho_0, \ldots, \rho_{N-1}, \rho_N).$$

Once the first two and last two equations have been deleted in the system of Eqs. (20–21), unknowns

$$\delta_{-1}^{n+1}, \rho_{-1}^{n+1}, \delta_N^{n+1}, \rho_N^{n+1}$$

can be eliminated from the system using the following boundary conditions

$$U(a, x) = g_1(t), \quad U(b, x) = g_2(t)$$
$$V(a, x) = \frac{dg_1(t)}{dt}, \quad V(b, x) = \frac{dg_2(t)}{dt}.$$

Thus, $(2N)$ unknowns and $(2N)$ equations obtained from the system of Eqs. (20–21) can be solved easily with the Matlab package program.

Once the initial vector

$$\mathbf{d}^0 = (\delta_{-1}^0, \ldots, \delta_{N-1}^0, \delta_N^0, \rho_{-1}^0, \rho_0^0, \ldots, \rho_{N-1}^0, \rho_N^0)$$

is found using the initial and boundary conditions, the unknown vector

$$\mathbf{d}^1 = (\delta_{-1}^1, \ldots, \delta_{N-1}^1, \delta_N^1, \rho_{-1}^1, \rho_0^1, \ldots, \rho_{N-1}^1, \rho_N^1)$$

can be found using the (20–21). Therefore unknown vector $\mathbf{d}^{n+1}$, $(n = 0, 1, \ldots)$ can be found repeatedly by solving the recurrence relation (20–21) using previous unknown vector $\mathbf{d}^n$.

3 Test Problem

The error norm

$$L_\infty = \max_m |u_m - U_m|, \tag{22}$$

and the order of convergence

$$\text{order} = \frac{\log\left|\frac{u - U_{\Delta t_m}}{u - U_{\Delta t_{m+1}}}\right|}{\log\left|\frac{\Delta t_m}{\Delta t_{m+1}}\right|}, \tag{23}$$

will be calculated.

The homogeneous Telegraph equation has the exact solution [5]

$$u(x,\,t) = e^{x-t} \tag{24}$$

with the boundary conditions

$$u(a,\,t) = e^{a-t},\ u(b,\,t) = e^{b-t} \quad ,\, t \in (0,\,T] \tag{25}$$

and initial conditions

$$\begin{aligned} u(x,\,0) &= e^{x} \\ u_t(x,\,0) &= v(x,\,0) = -e^{x}. \end{aligned} \tag{26}$$

The numerical simulation is accomplished with $\alpha = 0.5$ and $\beta = 1$ by the terminating time $t = 5$ over the domain [0, 4]. The program is run until time t = 5 and the analytical solutions at various times are shown in the Fig. 1.

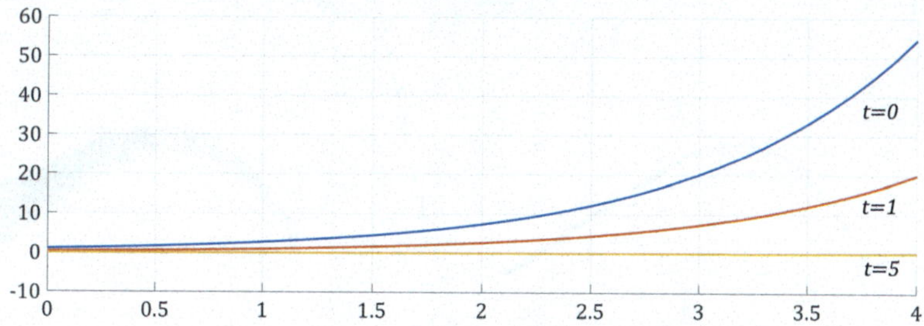

Fig. 1. Analytical solutions at different times

The error norm L_∞ and rate of convergence for the presented two methods are listed in Table 1. According to the Table 1 that M2 gives better result than M1 and the error norm decreases as the time step decreases for the both proposed methods and the convergence rate is almost 4 for the M2, which is the accuracy of the proposed method for the time discretization of the Telegraph equation.

Absolute error (difference between the exact and numerical solutions) distribution at $t = 5$ is also depicted in Fig. 2 for each proposed methods. When the figures are examined, it can be seen that maximum errors do not occur at the end points of the range in both methods. Therefore, it can be said that there is no error due to the application of boundary conditions.

Table 1. L_∞ and rate of convergence at time $t = 5$ for $h = 0.001$.

	M1	M1	M2	M2
Δt	L_∞	Order	L_∞	Order
1	2.57×10^{-1}	3.19	1.41×10^{-3}	4.33
0.5	2.82×10^{-2}	2.98	6.76×10^{-5}	4.00
0.2	1.84×10^{-3}	1.66	1.73×10^{-6}	4.30
0.1	5.84×10^{-4}	1.82	8.80×10^{-8}	3.94
0.05	1.65×10^{-4}	2.19	5.75×10^{-9}	3.73
0.02	2.21×10^{-5}		1.88×10^{-10}	

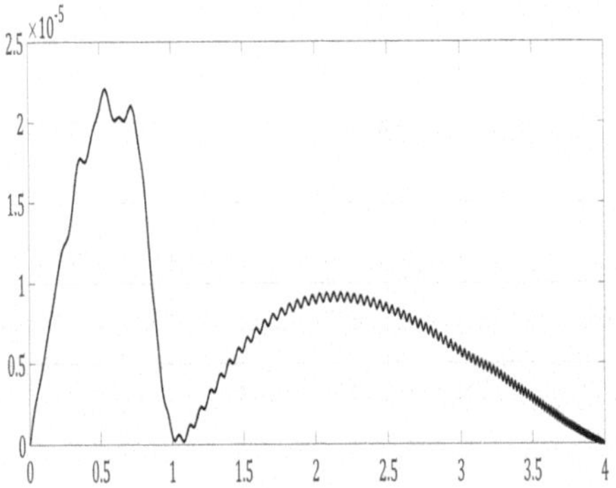

M1

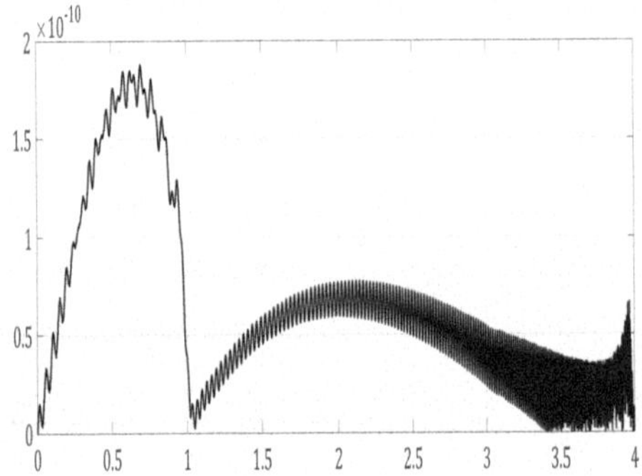

M2

Fig. 2. Absolute errors when $h = 0.001$ and $\Delta t = 0.02$.

4 Conclusion

The high-order Galerkin finite-element method based on Taylor series expansion for the time discretization and quadratic B-spline functions for the space discretization was proposed to solve numerically the homogeneous Telegraph equation. The test problem was simulated well with the proposed all algorithms. As a result, according to the results obtained from the proposed methods, the M2 method has been found to give very good results for approximate solution of the homogeneous Telegraph equation.

Acknowledgements. This work was supported by the Scientific Research Council of Eskişehir Osmangazi University under Project No. 2018-2090.

References

1. Pascal, H.: Pressure wave propagation in a fluid flowing through a porous medium problems related to interpretation of Stoneley's wave attenuation in acoustical well logging. Int. J. Eng. Sci. **24**(9), 1553–1570 (1986)

2. Bohme, G.: Non-Newtonian Fluid Mechanics. North-Holland, New York (1987)
3. Nazir, T., Abbas, M., Yaseen, M.: Numerical solution of second-order hyperbolic Telegraph equation via new cubic trigonometric B-splines approach. Cogent Math. **4**, 1382061 (2017)
4. Dehghan, M., Shokri, A.: A Numerical Method for Solving the Hyperbolic Telegraph Equation. Numer. Methods Part. Differ. Equ. **24**(4), 1080–1093 (2008)
5. Momani, S.: Analytic and approximate solutions of the space-and time-fractional telegraph equations. Appl. Math. Comput. **170**(2), 1126–1134 (2005)

A Class of Exact Solutions for Unsteady MHD Natural Convection Flow of a Viscous Fluid over a Moving Inclined Plate with Exponential Heating, Constant Concentration and Chemical Reaction

Azhar Ali Zafar[1], M. Bilal Riaz[2,3], and Zakia Hammouch[4(✉)]

[1] GC University, Lahore 54000, Pakistan
azharalizafar@gmail.com
[2] University of Management and Technology, Lahore 54000, Pakistan
bilalsehole@gmail.com
[3] Institute for Groundwater Sciences, University of the Free State, Bloemfontein, South Africa
[4] FST Errachidia, Moulay Ismail University of Meknes, Meknes, Morocco
z.hammouch@fste.umi.ac.ma

Abstract. The purpose of this article is to study analytically the hydromagnetic natural convection flow of an electrically conducting, incompressible viscous fluid over a moving infinite inclined plate. Moreover, the dynamic of fluid is studied under the influence of exponential heating and constant concentration. Porous effects are taken into consideration and in order to investigate the influence of the transverse magnetic field, two cases when the transverse magnetic field is held fixed to the fluid or to the plate are considered. The Laplace transform technique is used to obtain exact solutions for such motions. The dimensionless Latin symbols velocity, and also the corresponding skin friction, is presented as sum of mechanical, thermal and concentration components. Finally, for illustration, as well as for a check of results, some special cases with applications in engineering are considered and influence of the system parameters is graphically brought to light.

Keywords: Magnetohydrodynamic · Inclined plate · Natural convection · Exponential heating · Chemical reaction

Nomenclature

Latin Symbols

B	Magnetic field strength
C	Dimensional concentration in the fluid

H. Dutta et al. (Eds.): CMES 2019, AISC 1111, pp. 218–232, 2020.
https://doi.org/10.1007/978-3-030-39112-6_16

C'_w	Concentration of the fluid near the plate
C'_∞	Concentration of the fluid far away from the plate
T'_w	Constant temperature of the plate
T'_∞	Free stream temperature
u	Velocity of the fluid
c_p	Specific heat at constant pressure
u_0	Characteristic velocity of the plate
D	Chemical molecular diffusivity
Gc	Mass Grashof number
Gr	Thermal Grashof number
g	Acceleration due to gravity
K	Permeability of porous medium
k_R	Rosseland mean attenuation coefficient
M	Magnetic field parameter
N	Ratio of the buoyancy forces
Nr	Radiation conduction
$\Pr$	Prandtl number
$\Pr_{eff}$	Effective Prandtl number
q	The transform parameter
q_r	radioactive heat flux
R	Radiative parameter
Sc	Schmidt number

Greek Symbols

β_T	Volumetric coefficient of thermal expansion
β_C	Volumetric coefficient of expansion with concentration
ν	Kinematic coefficient of viscosity
τ	Skin friction
γ	Inclination angle from the vertical direction
σ	Electric conductivity
μ	Coefficient of viscosity
ρ	Density
κ	Thermal conductivity of the fluid

1 Introduction

From the past few years, the study of magnetohydrodynamic (MHD) natural convection flow of electrically conducting fluids with heat and mass transfer has gained a special attention due to their multiple applications in meteorology, electrical power generation, solar physics, geophysics and chemical engineering. The study of MHD natural convection flow over a moving inclined plate has many useful consequences so attracted the attention of many researchers. For example, the flow characteristics for the natural convection boundary layer flow over a flat plate with arbitrary inclination also depend on the angle of inclination and on the distance from the leading edge [1]. Uddin and Kumar [2] noticed that as the angle of the plate from vertical direction increases the value of friction factor and heat transfer coefficient decreases while studying the unsteady free convection in a fluid flow past an infinite inclined plate immersed in a porous medium has been considered for viscous dissipative heat. Moreover, Palani [3] studied the convection effects on flow past an inclined plate with variable surface temperatures in water at 4°, Singh and Makinde [4] investigated the MHD free convection flow with Newtonian heating in the presence of exponentially decaying volumetric heat source along the inclined plane. The thermal radiation effect on an unsteady MHD flow past an inclined, porous, heated plate in the presence of chemical reaction and viscous dissipation is investigated by Barik et al. [5]. Chen [6] found that increasing the angle of inclination decreases the effect of buoyancy force while investigating the natural flow over a permeable inclined surface with variable wall temperature and concentration. MHD natural convection flow with Newtonian heating and mass diffusion was analytical solved by Vieru et al. [7], when the plate applies an arbitrary time-dependent shear stress to the fluid. Fetecau et al. [8] investigated the slip effects on the radiative MHD free convection flow over a moving plate with mass diffusion and heat source. Recently, a general study of such flow with radiative effects, heat source and shear stress on the boundary has been developed by Fetecau et al. [9].

However, in all these studies as well as in many other which have been previously published, the magnetic lines of force are fixed to the fluid. There are many interesting papers [10–16] in which exact solutions are obtained for hydromagnetic free convection flows through porous media with heat and mass transfer, but they also correspond to the case when the magnetic field lines of force are fixed to the fluid.

The first exact solutions for free convection flows when the magnetic lines of force are fixed to the fluid or to the plate seem to be those obtained by Tokis [17] corresponding to motions induced by uniform, constantly accelerating or decaying oscillatory translations of the plate. More recently, Narahari and Debnath [18] developed an interesting study of unsteady MHD free convection flow with constant heat flux and heat source when the magnetic lines of force are fixed to the fluid or to the plate.

In this article, we present a general study of MHD natural convection flow over a moving inclined plate that is embedded in a porous medium with exponential heating, constant concentration and chemical reaction. However, our purpose is not only to extend some previous results by including porous effects and mass transfer, but we also want to provide new results both for general and oscillating motions of the inclined plate. It is worth pointing out the fact that "the fluid velocity does not remain zero at infinity if the magnetic field is fixed to the plate" [19,20]. Moreover, to investigate the contribution of mechanical, thermal and concentration influence on the fluid velocity as well as on skin friction, we will present these as a sum of mechanical, thermal and concentration components.

2 Statement of the Problem

By choosing a suitable cartesian coordinate system $Oxy'z$, let us consider the unsteady free convection flow of an electrically conducting incompressible viscous fluid over a non-conducting infinite inclined plate making an angle γ with the verticle and in the presence of a uniform magnetic field of strength B. The magnetic field is applied perpendicular to the plate and its magnetic lines of force are fixed to the fluid or to the plate. Initially, the plate and the fluid are at rest at the constant temperature T'_∞ and species concentration C'_∞. After the time $t' = 0^+$, the plate begins to slide in its plane against the gravitational field with the velocity $u_0 g'(t')$ and its temperature is maintained at the value $T'_\infty + T'_w \left(1 - ae^{-b't'}\right)$ where a, b' and T'_w are constants. Moreover, is a constant velocity, a piecewise continuous function with the condition that $g'(0) = 0$. The plate is also maintained at a constant concentration C'_w.

We assume that all physical properties are constant except the density variation with temperature in the body force and the induced magnetic field is negligible in comparison with the applied magnetic field B. Furthermore, neglecting viscous dissipation and Joule heating and taken into consideration porous and radiative effects, the chemical reaction between the fluid and species concentration. Under the usual Boussinesq's approximation, our problem reduces to the following set of partial differential equations [18,19].

$$\begin{cases} \dfrac{\partial u'}{\partial t'} = \nu \dfrac{\partial^2 u}{\partial y'^2} + g\beta_T \left(T' - T'_\infty\right)\cos\gamma + g\beta_c \left(C' - C'_\infty\right)\cos\gamma - \dfrac{\nu}{K'}u' - \dfrac{\sigma B_0^2}{\rho}\left(u' - lu_0 g'(t')\right), \\ \rho c_p \dfrac{\partial T'}{\partial t'} = k\dfrac{\partial^2 T'}{\partial y'^2} - \dfrac{\partial q_r}{\partial y'}, \\ \dfrac{\partial C'}{\partial t'} = D\dfrac{\partial^2 C'}{\partial y'^2} - R'\left(C' - C'_\infty\right), \\ y', t' > 0, \end{cases} \tag{1}$$

with the initial and boundary conditions

$$\begin{cases} u'(y', 0) = 0, T'(y', 0) = T'_\infty, \ C'(y', 0) = C'_\infty, y' \geq 0, \\ u'(0, t') = u_0 g'(t'), T'(0, t') = T'_\infty + T'_w \left(1 - ae^{-b't'}\right), \ C'(0, t') = C'_w, t' \geq 0, \\ u'(y', t') \to 0, T'(y', t') \to T'_\infty, \ C'(y', t') \to C'_\infty, \ y' \to \infty, \end{cases} \tag{2}$$

Into above equations, the unknown functions $u'(y', t')$, $T'(y', t')$ and $C'(y', t')$ are the velocity, the temperature and the species concentration while ν, g, β_T, β_C, K', σ, ρ, c_p, k, D, R' and q_r are defined in the nomenclature. The parameter l is 0 when the magnetic field is fixed relative to the fluid (MFFRF) and 1 (one) when the magnetic field is fixed relative to the plate (MFFRP).

By adopting the Rosseland diffusion approximation for an optically thick fluid (see Seth et al. [21] or Narahari and Dutta [22])

$$q_r = -\frac{4}{3}\frac{\sigma}{k_R}\frac{\partial T'^4}{\partial y'}, \tag{3}$$

assuming the temperature difference between the fluid temperature and the free stream temperature to be small enough, the energy equation $(2.2)_1$ can be written in the form [23,24].

$$\Pr_{eff}\frac{\partial T'(y',t')}{\partial t'} = \frac{\partial^2 T'(y',t')}{\partial y'^2};\ y',t' > 0, \tag{4}$$

where $\Pr_{eff} = \dfrac{\Pr}{1+N_r}$, $\Pr = \dfrac{\mu c_p}{k}$ and $N_r = \dfrac{16}{3}\dfrac{\sigma}{kk_R}T_\infty^3$.
Introducing the following dimensionless variables, functions and parameters

$$\begin{cases} y = \dfrac{u_0}{\nu}y',\ t = \dfrac{u_0^2}{\nu}t',\ u = \dfrac{u'}{u_0}, \\ T = \dfrac{T' - T'_\infty}{T'_w},\ C = \dfrac{C' - C'_\infty}{C'_w - C'_\infty},\ b = \dfrac{\nu}{u_0^2}b',\ K = \left(\dfrac{\nu}{u_0}\right)^2\dfrac{1}{K'}, \\ R = \dfrac{\nu}{u_0^2}R',\ g(t) = g'\left(\dfrac{\nu}{u_0^2}t'\right), \end{cases} \tag{5}$$

and choosing the characteristic velocity u_0 to be equal with $\sqrt[3]{\nu g\beta_T T_w}$, our problem reduce to the following dimensionless partial differential equations

$$\begin{cases} \dfrac{\partial u(y,t)}{\partial t} = \dfrac{\partial^2 u(y,t)}{\partial y'^2} + T(y,t)\cos\gamma + NC(y,t)\cos\gamma - Ku(y,t) - M(u(y,t) - lg(t)), \\ \Pr_{eff}\dfrac{\partial T(y,t)}{\partial t} = \dfrac{\partial^2 T(y,t)}{\partial y^2}, \\ \dfrac{\partial C(y,t)}{\partial t} = \dfrac{1}{Sc}\dfrac{\partial^2 C(y,t)}{\partial y^2} - RC(y,t), \\ y,t > 0, \end{cases} \tag{6}$$

with the initial and boundary conditions

$$\begin{cases} u(y,0) = 0, T(y,0) = 0,\ C(y,0) = 0, y \geq 0, \\ u(0,t) = g(t), T(0,t) = 1 - ae^{-b't'},\ C(0,t) = 1, t \geq 0, \\ u(y,t) \to 0, T(y,t) \to 0,\ C(y,t) \to 0, y' \to \infty, \end{cases} \tag{7}$$

Into above relations, K is the inverse permeability parameter of the porous medium, R is the dimensionless chemical reaction parameter while the ratio of the buoyancy forces N, the magnetic parameter M and Schmidt number Sc are defined by

$$N = \frac{\beta_C (C_w - C_\infty)}{\beta_T T_w}, \; M = \frac{\sigma B^2}{\rho} \frac{\nu}{V^2}, \; Sc = \frac{\nu}{D}.$$

where Pr_{eff} and Sc are transport parameters regarding the thermal and mass diffusivity and N represents the relative contribution of the mass transport rate on the free convection flow. Moreover, depending upon β_C, N can be also positive or negative because β_T is always positive [20] and $N = 0$ for the case when the buoyancy force effect from mass diffusion is absent.

3 Solution of the Problem

As the temperature and concentration fields corresponding to this problem can be easily obtained from previous works (see [[20], Eq. (20)], respectively [[25], Eq. (15)]). Our prime interest is to find the fluid velocity, however in order to determine it using the Laplace transform technique, we need the Laplace transforms of $T(y,t)$ and $C(y,t)$, namely

$$\bar{T}(y,q) = \left(\frac{1}{q} - \frac{a}{q+b}\right) e^{-y\sqrt{\mathrm{Pr}_{eff}\, q}}, \; \bar{C}(y,q) = \frac{1}{q} e^{-y\sqrt{Sc(q+R)}}, \tag{8}$$

obtained from [20] and [25].

Applying the Laplace transform to Eq. (2.9) and using the corresponding initial and boundary conditions, we find the differential equation

$$\begin{aligned} q\bar{u}(y,q) = & \frac{\partial^2 \bar{u}(y,q)}{\partial y^2} + \bar{T}(y,q)\cos\gamma + N\bar{C}(y,q)\cos\gamma - K\bar{u}(y,q) \\ & - M(\bar{u}(y,q) - lG(q)), \; y,t > 0, \end{aligned} \tag{9}$$

with the boundary conditions

$$\bar{u}(0,q) = G(q), \bar{u}(y,q) \to 0, \text{ as } y \to \infty, \tag{10}$$

where $\bar{u}(y,q)$ and $G(q)$ denote the Laplace transforms of $u(y,t)$, respectively $g(t)$. Introducing Eqs. (3.1) into (3.2), we get

$$\begin{aligned} & \frac{\partial^2 \bar{u}(y,q)}{\partial y^2} - (q+H)\bar{u}(y,q) = -lMG(q) - \left(\frac{1}{q} - \frac{a}{q+b}\right) e^{-y\sqrt{\mathrm{Pr}_{eff}\, q}} \cos\gamma \\ & - N\frac{1}{q} e^{-y\sqrt{Sc(q+R)}} \cos\gamma, \end{aligned} \tag{11}$$

where $H = M + K$.

The solution of the ordinary differential equation (3.4) with the boundary conditions (3.3) is

$$\bar{u}(y,q) = G(q)\, e^{-y\sqrt{q+H}} + \varepsilon M \frac{G(q)}{q+H}\left(1 - e^{-y\sqrt{q+H}}\right) + \frac{(1-a)\,q+b}{q\,(q+b)\,[(1-\mathrm{Pr}_{eff})\,q+H]}\left(e^{-y\sqrt{\mathrm{Pr}_{eff}\,q}} - e^{-y\sqrt{q+H}}\right)\cos\gamma + \frac{N}{q\,[(1-Sc)\,q-(ScR-H)]}\left(e^{-y\sqrt{Sc(q+R)}} - e^{-y\sqrt{q+H}}\right)\cos\gamma. \quad (12)$$

Next, introducing the relations

$$\frac{(1-a)\,q+b}{q\,(q+b)\,[(1-\mathrm{Pr}_{eff})\,q+H]} = \frac{1}{1-\mathrm{Pr}_{eff}}\left[\frac{1}{E}\frac{1}{q} - \frac{a}{E-b}\frac{1}{q+b} + \frac{(1-a)\,E-b}{E\,(b-E)}\frac{1}{q+E}\right],$$

$$\frac{1}{q\,[(1-Sc)\,q-(ScR-H)]} = \frac{1}{H-ScR}\left(\frac{1}{q} - \frac{1}{q+F}\right),$$

$$E = \frac{H}{1-\mathrm{Pr}_{eff}}, \quad F = \frac{ScR-H}{Sc-1},$$

into Eq. (3.5), applying the inverse Laplace transform and using the convolution theorem and Eqs. (A1) and (A2) from Appendix, we can present the velocity field under the form

$$u(y,t) = u_m(y,t) + u_T(y,t) + u_C(y,t), \quad (13)$$

where

$$u_m(y,t) = \frac{y}{2\sqrt{\pi}} \int_0^t \frac{g(t-s)}{s\sqrt{s}} \exp\left(\frac{-y^2}{4s} - Hs\right) ds + lM \int_0^t g(t-s)\, e^{-Hs}\, \mathrm{erf}\left(\frac{y}{2\sqrt{s}}\right) ds, \quad (14)$$

$$u_T(y,t) = \frac{1}{1-\mathrm{Pr}_{eff}}\left[\frac{1}{E}\left[\Psi\left(y\sqrt{\mathrm{Pr}_{eff}}, t, 0, 0\right) - \Psi(y,t,H,0)\right] + \frac{a}{b-E}\left[\Psi\left(y\sqrt{\mathrm{Pr}_{eff}}, t, 0, -b\right) - \Psi(y,t,H,-b)\right] + \frac{(1-a)\,E-b}{E\,(b-E)}\left[\Psi\left(y\sqrt{\mathrm{Pr}_{eff}}, t, 0, -E\right) - \Psi(y,t,H,-E)\right]\right]\cos\gamma, \quad (15)$$

$$u_C(y,t)=\frac{N}{H-ScR}\left[\Psi\left(y\sqrt{Sc},t,R,0\right)-\Psi(y,t,H,0)\right.$$
$$\left.-\Psi\left(y\sqrt{Sc},t,R,-F\right)+\Psi(y,t,H,-F)\right]\cos\gamma, \tag{16}$$

are its mechanical, thermal and concentration components and the function $\Psi(y,t,a,b)$ is defined in Appendix.

It is not difficult to show that $u(y,t)$, given by Eqs. (3.6)–(3.9), satisfies the imposed initial and boundary conditions. In order to verify the boundary condition $(2.12)_1$, for instance, we rewrite $u_m(y,t)$ in the equivalent form

$$u_m(y,t)=\frac{2}{\sqrt{\pi}}\int_{\frac{y}{2\sqrt{t}}}^{\infty}g\left(t-\frac{y^2}{4s^2}\right)\exp\left(-s^2-\frac{Hy^2}{4s^2}\right)ds$$
$$+lM\int_0^t g(t-s)\,e^{-Hs}\,\mathrm{erf}\left(\frac{y}{2\sqrt{s}}\right)ds, \tag{17}$$

As regards the limit of velocity as $y\to\infty$, it results that

$$\lim_{y\to\infty}u_m(y,t)=\begin{cases}0 & if \quad l=0\\ M\int_0^t g(t-s)\,e^{-Hs}ds & if \quad l=1\end{cases} \tag{18}$$

Consequently, in the case when the MFFRP, the fluid does not remain at rest far away from the plate.

From physical point of view, it is also important to determine the skin friction or shear on the plate. Introducing Eq. (3.5)

$$\tau=-\left.\frac{\partial u(y,t)}{\partial y}\right|_{y=0}=-L^{-1}\left\{\left.\frac{\partial\bar{u}(y,q)}{\partial y}\right|_{y=0}\right\}, \tag{19}$$

we find that (see also Eqs. (A3)–(A5) from Appendix)

$$\tau=\tau_m+\tau_T+\tau_C, \tag{20}$$

where

$$\tau_m=\int_0^t g'(t-s)\left[\sqrt{H}\,\mathrm{erf}\left(\sqrt{Hs}\right)+\frac{e^{-Hs}}{\sqrt{\pi s}}\right]ds$$
$$-l\frac{M}{\sqrt{H}}\int_0^t g'(t-s)\,\mathrm{erf}\left(\sqrt{Hs}\right)ds, \tag{21}$$

$$\tau_T=\frac{1}{1-\mathrm{Pr}_{eff}}\left\{\left[\frac{1}{E\sqrt{\pi t}}+\frac{a}{b-E}\phi(t;0,b)+\frac{(1-a)E-b}{E(b-E)}\phi(t;0,E)\right]\right.$$
$$\left.-\frac{1}{E}\phi(t;H,0)-\frac{a}{b-E}\phi(t;H,b)-\frac{(1-a)E-b}{E(b-E)}\phi(t;H,E)\right\}\cos\gamma, \tag{22}$$

$$\tau_C = \frac{N}{ScR - H} \{\phi(t; H, F) - \phi(t; H, 0) + \sqrt{Sc}\left[\phi(t; R, 0) - \phi(t; R, F)\right]\Big\} \cos\gamma, \tag{23}$$

are the mechanical, thermal and concentration components of the skin friction and the function $\phi(t; a, b)$ is defined in the Appendix.

By taking $K = 0$ and $\gamma = 0$ into Eqs. (3.6) and (3.13), we recover the corresponding results of [[19], Eqs. (20) and (27)].

As the concentration and thermal parts of velocity and skin friction are independent of $g(t)$, so in the following section we will discuss the special cases regarding to the mechanical part of velocity and skin friction.

4 Special Cases

In the following, in order to get some physical insight of present results and for validation of the obtained results with possible engineering applications, we consider the following cases.

4.1 Case $g(t) = H(t)$ (Uniform Motion of the Plate)

let us take $K = 0$, $\gamma = 0$, $g(t) = H(t)$ (the Heaviside unit step function) in our relations (3.7) and (3.14) and use Eqs. (A6) and (A7) from Appendix, we get

$$u_m(y,t) = (1 - l)\,\Psi(y, t; 0, M) + l\left[1 - \exp(-Mt)\,\mathrm{erf}\left(\frac{y}{2\sqrt{t}}\right)\right], \tag{24}$$

and

$$\tau_m(t) = \left[\sqrt{H}\,\mathrm{erf}(Ht) + \frac{e^{-Ht}}{\sqrt{\pi t}}\right] H(t) - l\frac{M}{\sqrt{H}} H(t)\,\mathrm{erf}\left(\sqrt{Ht}\right). \tag{25}$$

As it was to be expected, the corresponding results are identical to those obtained by Tokis [[17], Eqs. (12) and (13a)] and Narahari and Debnath [[18], Eqs. (11a), (13)] with $a_0 = 0$ and in the absence of thermal and concentration effects and $\gamma = 0$.

4.2 Case $g(t) = H(t)\,t^\alpha$ (Variably Accelerating Plate)

Thermal and concentration components of velocity do not depend on the plate motion. However, the heat and mass transfer can influence the fluid motion and we have to know if their influence is significant or it can be neglected in some motions with possible engineering applications. Taking, $g(t) = H(t)\,t^\alpha$ with $\alpha > 0$, the Eqs. (3.7) and (3.14) take the forms

$$u_m(y,t) = \frac{y}{2\sqrt{\pi}} \int_0^t \frac{(t-s)^\alpha}{s\sqrt{s}} \exp\left(-\frac{y^2}{4s} - Hs\right) ds + lM \int_0^t (t-s)^\alpha e^{-Hs}\,\mathrm{erf}\left(\frac{y}{2\sqrt{s}}\right) ds, \tag{26}$$

$$\tau_m(t) = \alpha \int_0^t (t-s)^{\alpha-1} \left[\sqrt{H}\,\mathrm{erf}\left(\sqrt{Hs}\right) + \frac{e^{-Hs}}{\sqrt{\pi s}}\right] ds$$
$$-l\alpha \frac{M}{\sqrt{H}} \int_0^t (t-s)^{\alpha-1}\,\mathrm{erf}\left(\sqrt{Hs}\right) ds;\ \alpha > 0, \tag{27}$$

which corresponds to motions induced by a slowly, constantly or highly accelerating plate.

4.3 Case $g(t) = H(t)\cos(\omega t)$ or $H(t)\sin(\omega t)$ (Oscillating Plate)

Introducing into Eqs. (3.7) and (3.14) and using the fact that $H'(t) = \delta(t)$ and

$$\int_0^t \delta(t-s)\,g(s)\,ds = \int_0^t \delta(s)\,g(t-s)\,ds = g(t), \tag{28}$$

where $\delta(.)$ is the Dirac delta function, we find that

$$u_{cm}(y,t) = \frac{y}{2\sqrt{\pi}} \int_0^t \frac{\cos[\omega(t-s)]}{s\sqrt{s}} \exp\left(-\frac{y^2}{4s} - Hs\right) ds$$
$$+lM \int_0^t \cos[\omega(t-s)]\, e^{-Hs}\mathrm{erf}\left(\frac{y}{2\sqrt{s}}\right) ds, \tag{29}$$

$$u_{sm}(y,t) = \frac{y}{2\sqrt{\pi}} \int_0^t \frac{\sin[\omega(t-s)]}{s\sqrt{s}} \exp\left(-\frac{y^2}{4s} - Hs\right) ds$$
$$+lM \int_0^t \sin[\omega(t-s)]\, e^{-Hs}\mathrm{erf}\left(\frac{y}{2\sqrt{s}}\right) ds, \tag{30}$$

$$\tau_{cm} = H(t)\left\{\sqrt{H}\,\mathrm{erf}\left(\sqrt{Ht}\right) + \frac{e^{-Ht}}{\sqrt{\pi t}} - l\frac{M}{\sqrt{H}}\,\mathrm{erf}\left(\sqrt{Ht}\right)\right\}$$
$$-\omega \int_0^t \sin[\omega(t-s)] \left[\sqrt{H}\,\mathrm{erf}\left(\sqrt{Hs}\right) + \frac{e^{-Hs}}{\sqrt{\pi s}}\right] ds$$
$$+l\omega \frac{M}{\sqrt{H}} \int_0^t \sin[\omega(t-s)]\,\mathrm{erf}\left(\sqrt{Hs}\right) ds, \tag{31}$$

$$\tau_{sm} = \omega \int_0^t \cos[\omega(t-s)] \left[\sqrt{H}\,\mathrm{erf}\left(\sqrt{Hs}\right) + \frac{e^{-Hs}}{\sqrt{\pi s}}\right] ds$$
$$-l\omega \frac{M}{\sqrt{H}} \int_0^t \cos[\omega(t-s)]\,\mathrm{erf}\left(\sqrt{Hs}\right) ds. \tag{32}$$

As expected, for $\omega = 0$, the solutions (4.3.2) and (4.3.4) reduce to those given by Eqs. (4.1.1) and (4.1.2) corresponding to the motion with uniform velocity on the boundary.

Indeed, assigning to $g(.)$ suitable forms, we can determine exact solutions for any motion with technical relevance of this type. Consequently, the problem under debate is completely solved.

5 Numerical Results and Discussion

In this paper, exact general solutions are determined for dimensionless velocity and skin friction corresponding to the MHD natural convection flow over a moving inclined plate with exponential heating, constant concentration and chemical reaction. Radiative and porous effects are taken into consideration and the magnetic field is fixed to the fluid or to the plate. In order to get some physical insight of obtained results and to avoid repetition, three special cases are considered. Figures 1 and 2 present the profiles of the dimensionless velocity $u(y,t)$, respectively mechanical component of velocity $u_m(y,t)$ against y at different times for a slowly accelerating motion of the plate. As expected, both velocities are increasing functions of time. Furthermore, the velocities corresponding to (MFFRP) are appreciably large as compared with (MFFRF). In all cases, the velocities smoothly decrease from maximum values on the boundary to asymptotical values for increasing y. However, as it is clearly seen from these figures, the asymptotic values of both velocities are not zero at infinity if the magnetic field is fixed to the plate.

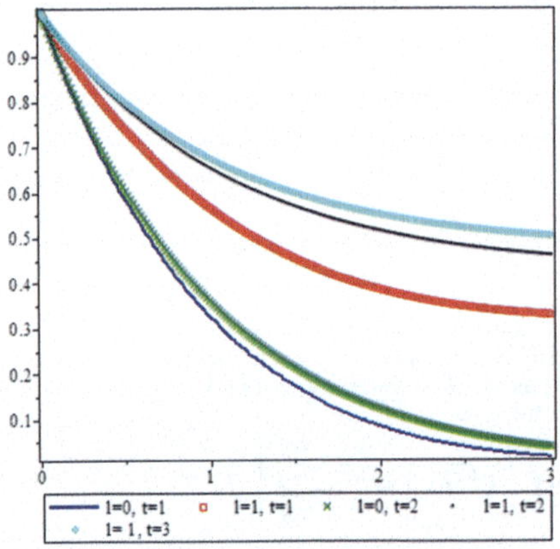

(a)

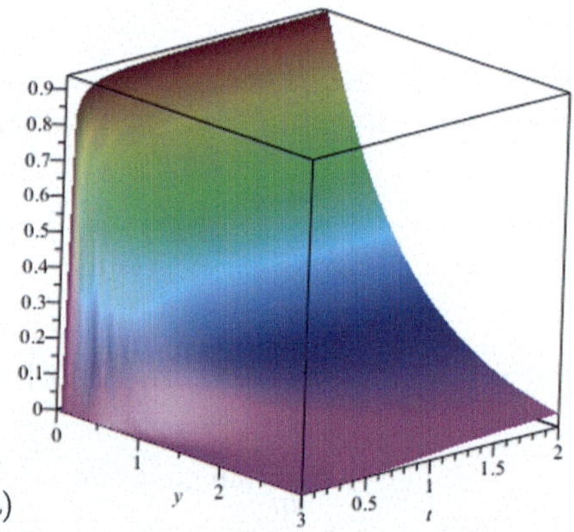

(b)

Fig. 1. (a) Profiles of $u_m(y,t)$ against y for $\gamma = \frac{\pi}{6}$ and different values of t, (b) 3D plot of $u_m(y,t)$.

In Fig. 3 we have plotted velocities $u_m(y,t)$, $u_m(y,t) + u_C(y,t)$ and $u_m(y,t) + u_C(y,t) + u_T(y,t)$ versus y to investigate the contributions of mechanical, thermal and concentration components of velocity on the fluid motion. It is observed that contributions of mechanical, thermal and concentration components of velocity on the fluid motion are significant and they cannot be neglected. In all diagrams $a = 0.75$, $b = 0.15$, $\alpha = 0.5$, $\Pr = 0.7$, $M = 0.5$, $\Pr_{eff} = 0.5$, $Sc = 0.6$, $R = 0.7$, $N = 0.5$, $K = 0.3$.

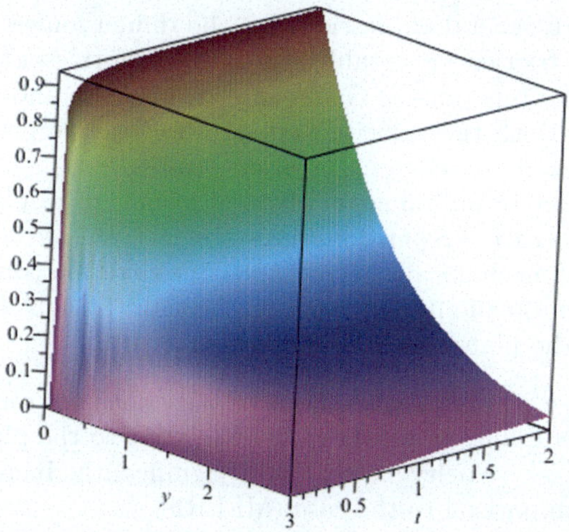

Fig. 2. Profiles of $u\,(y,t)$ against y for $\gamma = \frac{\pi}{6}$ and different values of t.

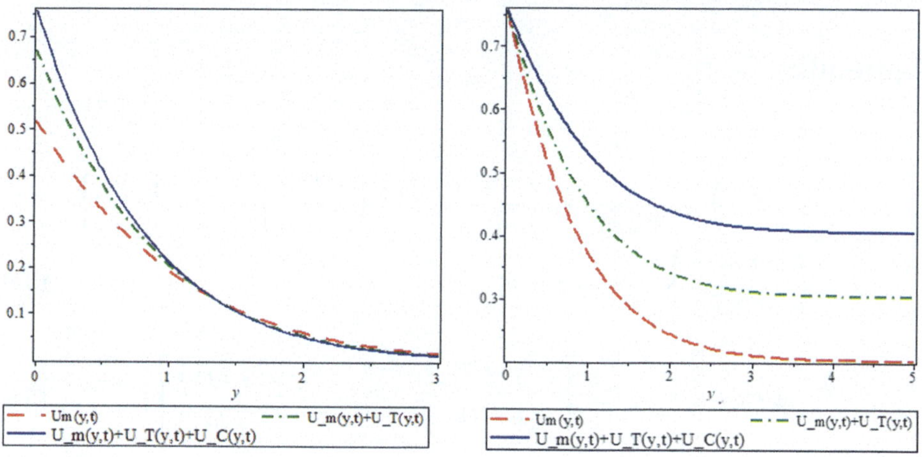

Fig. 3. Profiles of $u_{\alpha m}\,(y,t)$, $u_{\alpha m}\,(y,t) + u_T\,(y,t)$ and $u_{\alpha m}\,(y,t) + u_T\,(y,t) + u_C\,(y,t)$ against y at $\gamma = \frac{\pi}{6}$, $t = 1.5$. and $l = 0$ when $g(t) = H(t).t^{\alpha}$.

6 Conclusions

Hydromagnetic natural convection flow of an electrically conducting, incompressible viscous fluid over a moving infinite inclined plate with exponentially heating, constant concentration and chemical reaction is analytically and graphically studied. Viscous dissipation and Joule heating are neglected but the porous and radiative effects are taken into consideration. The plate is moving with arbitrary time-dependent velocity in its plane while the transverse magnetic field is fixed to the fluid or to the plate and our interest is focused on the fluid motion.

Consequently, exact general expressions for the dimensionless velocity and the corresponding skin friction are established in simple forms in terms of error and complementary error functions of Gauss and the problem under consideration is completely solved. Both the velocity and skin friction are presented as sum of their mechanical, thermal and concentration components.

However, in order to obtain some physical insight of results that have been obtained as well as to avoid repetition, three special cases are considered. Finally, the contributions of mechanical, thermal and concentration components of velocity and skin friction on the fluid motion are brought to light for a slowly accelerating motion of the plate. The main conclusions are:

- Contrary to our expectations, the fluid velocity does not remain zero at infinity if the magnetic lines of force are fixed relative to the plate.
- The dimensionless velocity of the fluid significantly increases in the case (MFFRP) in comparison to the case (MFFRF).
- Contributions of mechanical, thermal and concentration components of velocity and skin friction on the fluid motion are significant and they cannot be neglected.

Appendix

$$L^{-1}\left\{e^{-y\sqrt{q}}\right\} = \frac{y}{2t\sqrt{\pi t}}\exp\left(\frac{-y^2}{4t}\right),\; L^{-1}\left\{\frac{e^{-y\sqrt{q}}}{q}\right\} = \operatorname{erf}c\left\{\frac{y}{2\sqrt{t}}\right\},$$
$$L^{-1}\left\{\frac{e^{-y\sqrt{q+a}}}{q-b}\right\} = \Psi\left(y,t;a,b\right), \tag{33}$$

$$\Psi\left(y,t;a,b\right) = \frac{e^{bt}}{2}\begin{bmatrix} e^{-y\sqrt{a+b}}\operatorname{erf}c\left(\frac{y}{2\sqrt{t}} - \sqrt{(a+b)\,t}\right) \\ +e^{y\sqrt{a+b}}\operatorname{erf}c\left(\frac{y}{2\sqrt{t}} + \sqrt{(a+b)\,t}\right)\end{bmatrix}. \tag{34}$$

$$L^{-1}\left\{qF\left(q\right)\right\} = f'\left(t\right) + \delta\left(t\right)f\left(0\right)\; if\; L^{-1}\left\{F\left(q\right)\right\} = f\left(t\right)\;(\delta\left(.\right)\text{ is the Dirac delta function}). \tag{35}$$

$$L^{-1}\left\{\frac{1}{(q+b)\sqrt{q+a}}\right\} = \frac{e^{-bt}}{\sqrt{a-b}}\operatorname{erf}\left(\sqrt{(a-b)\,t}\right), L^{-1}\left\{\frac{1}{\sqrt{q}}\right\} = \frac{1}{\sqrt{\pi t}}. \tag{36}$$

$$L^{-1}\left\{\frac{\sqrt{q+a}}{q+b}\right\} = \frac{e^{-at}}{\sqrt{\pi t}} + \frac{1}{\sqrt{a-b}}e^{-bt}\operatorname{erf}\left(\sqrt{a-bt}\right) = \phi\left(t;a,b\right), \tag{37}$$

$$\int_0^t \frac{1}{\sqrt{s}}\exp\left(-\frac{y^2}{4s} - as\right)ds = \frac{\sqrt{\pi}}{2\sqrt{a}}\left\{e^{-y\sqrt{a}}\operatorname{erf}c\left(\frac{y}{2\sqrt{t}} - \sqrt{at}\right)\right.$$
$$\left. - e^{y\sqrt{a}}\operatorname{erf}c\left(\frac{y}{2\sqrt{t}} + \sqrt{at}\right)\right\}, \tag{38}$$

$$\int_0^t \frac{1}{s\sqrt{s}} \exp\left(-\frac{y^2}{4s} - as\right) ds = \frac{\sqrt{\pi}}{y} \left\{ e^{-y\sqrt{a}} \operatorname{erf} c\left(\frac{y}{2\sqrt{t}} - \sqrt{at}\right) - e^{y\sqrt{a}} \operatorname{erf} c\left(\frac{y}{2\sqrt{t}} + \sqrt{at}\right) \right\}, \quad (39)$$

$$\int_0^\infty e^{-p^2 s^2 - \frac{q^2}{s^2}} \cos\left(a^2 s^2 + \frac{b^2}{s^2}\right) ds = \frac{\sqrt{\pi}}{2^4 \sqrt{p^4 + a^4}} e^{-2c\cos(\alpha+\beta)} \cos\left[\alpha + 2c \sin\left(\alpha + \beta\right)\right], \quad (40)$$

$$\int_0^\infty e^{-p^2 s^2 - \frac{q^2}{s^2}} \sin\left(a^2 s^2 + \frac{b^2}{s^2}\right) ds = \frac{\sqrt{\pi}}{2^4 \sqrt{p^4 + a^4}} e^{-2c\cos(\alpha+\beta)} \sin\left[\alpha + 2c \sin\left(\alpha + \beta\right)\right], \quad (41)$$

where $\alpha = \frac{1}{2} arctg\left(\frac{a^2}{p^2}\right)$, $\beta = \frac{1}{2} arctg\left(\frac{b^2}{q^2}\right)$ and $c = \sqrt[4]{(p^4 + a^4)(q^4 + b^4)}$.

References

1. Umemura, A., Law, C.K.: Natural convection boundary layer flow over a heated plate with arbitrary inclination. J. Fluid Mech. **219**(5), 71–84 (1990)
2. Uddin, Z., Kumar, M.: Unsteady free convection in a fluid past an inclined plate immersed in a porous medium. Comput. Model New Tech. **14**(3), 41–47 (2010)
3. Palani, G.: Convection effects on flow past an inclined plate with variable surface temperatures in water at 4 C. Ann. Fac. Eng. Hunedora **IV**(1), 75–82 (2008)
4. Singh, G., Makinde, O.D.: Computational dynamics of MHD free convection flow along an inclined plate with Newtonian heating in the presence of volumetric heat generation. Chem. Eng. Commun. **199**(9), 1144–1154 (2012)
5. Barik, R.N., Dash, G.C., Rath, P.K.: Thermal radiation effect on an unsteady MHD flow past inclined porous heated plate in the presence of chemical reaction and viscous-dissipation. Appl. Math. Comput. **226**, 423–434 (2014)
6. Chen, C.H.: Heat and mass transfer in MHD flow by natural convection from a permeable inclined surface with variable wall temperature and concentration. Acta Mech. **172**, 219–235 (2004)
7. Vieru, D., Fetecau, C., Fetecau, C., Nigar, N.: Magnetohydrodynamic natural convection flow with Newtonian heating and mass diffusion over an infinite plate that applies shear stress to a viscous fluid. Zeitschrift fur Naturforschung A **69**, 714–724 (2014)
8. Fetecau, C., Vieru, D., Fetecau, C., Pop, I.: Slip effects on the unsteady radiative MHD free convection flow over a moving plate with mass diffusion and heat source. Eur. Phys. J. Plus **130**(6), 1–13 (2015)
9. Fetecau C., Shahraz A., Pop I., Fetecau, C.: Unsteady general solutions for MHD natural convection flow with radiative effects, heat source and shear stress on the boundary (2016). https://doi.org/10.1108/HFF-02-2016-0069

10. Reddy, B.P.: Effects of thermal diffusion and viscous dissipation on unsteady MHD free convection flow past a vertical porous plate under oscillatory suction velocity with heat sink. Int. J. Appl. Mech. Eng. **19**(2), 303–320 (2014)
11. Abid, H., Ismail, Z., Khan, I., Hussein, A.G., Shafie, S.: Unsteady boundary layer MHD free convection flow in a porous medium with constant mass diffusion and Newtonian heating. Eur. Phys. J. Plus **129**(46), 1–16 (2014)
12. Khan, A., Khan, I., Ali, F., Shafie, S.: Effects of wall shear stress on MHD conjugate flow over an inclined plate in a porous medium with ramped wall temperature. Math. Probl. Eng. **2014**, 15 (2014). Article ID 861708
13. Sreekala, L., Reddy, E.K.: Steady MHD Couette flow of an incompressible viscous fluid through a porous medium between two infinite parallel plates under effect of inclined magnetic field. Int. J. Eng. Sci. **3**, 1837 (2014)
14. Marneni, N., Tippa, S., Pendyala, R., Nayan, M.Y.: Ramped temperature effect on unsteady MHD natural convection flow past an infinite inclined plate in the presence of radiation, heat source and chemical reaction. Recent Adv. Appl. Theor. Mech. **7**, 126–137 (2013)
15. Zhang, C., Zheng, L., Zhang, X., Chen, G.: MHD flow and radiation heat transfer of nanofluids in porous media with variable surface heat flux and chemical reaction. Appl. Math. Model. **39**, 165–181 (2015)
16. Seth, G.S., Hussain, S.M., Sarkar, S.: Hydromagnetic natural convection flow with heat and mass transfer of a chemically reacting and heat absorbing fluid past an accelerated moving vertical plate with ramped temperature and ramped surface concentration through a porous medium. J. Egypt. Math. Soc. **23**, 197–207 (2015)
17. Tokis, J.N.: A class of exact solutions of the unsteady magnetohydrodynamic free-convection flows. Astrophys. Space Sci. **112**, 413–422 (1985)
18. Narahari, M., Debnath, L.: Unsteady magnetohydrodynamic free convection flow past an accelerated vertical plate with constant heat flux and heat generation or absorption. Zeitschrift für Angewandte Mathematik und Mechanik **93**(1), 38–49 (2013)
19. Shah, N.A., Zafar, A.A., Fetecau, C.: Hydromagnetic natural convection flow over a moving verticle plate with exponential heating, constant concentration and chemical reaction? (Sent for publication)
20. Shah, N.A., Zafar, A.A., Akhtar, S.: General solution for MHD free convection flow over a vertical plate with ramped wall temperature and chemical reaction. (Sent for publication)
21. Seth, G.S., Ansari, M.S., Nandkeolyar, R.: MHD natural convection flow with radiative heat transfer past an impulsively moving plate with ramped wall temperature. Heat Mass Transfer **47**, 555–561 (2011)
22. Narahari, N., Dutta, B.K.: Effects of thermal radiation and mass diffusion on free convection flow near a vertical plate with Newtonian heating. Chem. Eng. Commun. **199**(5), 628–643 (2012)
23. Fetecau, C., Rana, M., Fetecau, C.: Radiative and porous effects on free convection flow near a vertical plate that applies shear stress to the fluid. Zeitschrift fur Naturforschung A **68a**, 130–138 (2013)
24. Magyari, E., Pantokratoras, A.: Note on the effect of thermal radiation in the linearized Rosseland approximation on the heat transfer characteristics of various boundary layer flows. Int. Commun. Heat Mass Transfer **38**(5), 554–556 (2011)
25. Rubbab, Q., Vieru, D., Fetecau, C., Fetecau, C.: Natural convection flow near a vertical plate that applies a shear stress to a viscous fluid. PLoS ONE **8**(11), 1–7 (2013)

Analytical Solutions to the Coupled Boussinesq–Burgers Equations via Sine-Gordon Expansion Method

Karmina K. Ali[1,2(✉)], Resat Yilmazer[2], and Hasan Bulut[2]

[1] Faculty of Scicences, Department of Mathematics, University of Zakho, Zakho, Iraq
karmina.ali@uoz.edu.krd
[2] Faculty of Scicences, Department of Mathematics, Firat University, Elazig, Turkey

Abstract. In the current study, we investigate the coupled Boussinesq-Burgers equations through sine-Gordon expansion method. BBEs arises in the research of fluid flow and describes the spreading of shallow water waves. A traveling wave transformation has been applied to turn the governing equation into a nonlinear ordinary differential equation. As a result, we produce some novel analytical solutions, such as topological, non-topological, and kink-type soliton solutions. Furthermore, 2D, 3D and contour surfaces are also plotted for all obtaining solutions.

Keywords: Analytical solutions · Coupled Boussinesq-Burgers equations · Sine-Gordon expansion method

1 Introduction

Nonlinear evolution equations (NLEEs) have been used as models to explain many physical phenomena in fluid mechanics, plasma waves, solid state physics, chemical physics, etc. [1–5]. It is very important to explore the attitudes of the models that occur in ocean dynamics because of the important roles they play in our daily activities [6–8]. NLEE solutions not only represent the problems identified, but also provide more insight into the physical aspects of the problems in the relevant area [9,10]. Recently, distinct computational and numerical methods have been use to handle these type of nonlinear models, such as the Hirota's bilinear method [11], Adomian decomposition method [12], Bernoulli sub equation function method [13], Homotopy perturbation method [14], and many more [15–19].

The intention of this research is to explore the coupled Boussinesq-Burgers equations [1–10] using sin-Gordon expansion method [20–22].

The coupled BBEs is given by

$$u_t + 2uu_x - \frac{1}{2}v_x = 0, \tag{1}$$

H. Dutta et al. (Eds.): CMES 2019, AISC 1111, pp. 233–240, 2020.
https://doi.org/10.1007/978-3-030-39112-6_17

$$v_t + 2(uv)_x - \frac{1}{2}u_{xxx} = 0, \tag{2}$$

Eqs. (1) and (2) appear in the fluid flow survey and explain the proliferation of severe shallow waves, where $u = u(x,t)$ denoted the horizontal velocity $v = v(x,t)$ donated the height of the water surface above the horizontal level at the bottom. Recently, different computational approaches have been used to investigate the BBEs such as Exp-function method [1], nonlinear transformation [2], Bäcklund transformation [3], The modified $\exp(-\phi(\xi))$-expansion function method [4], Darboux transformation [5], Homotopy perturbation method [6], the extended homogeneous balance [7], Jacobi elliptic function method [8], generalized algebraic method [9], the extended homogeneous balance method [10].

2 The SGEM

In the current section, the basic concepts of sine-Gordon expansion method is presented.

Consider the sine-Gordon equation

$$v_{xx} - v_{tt} = n^2 \sin(v), \tag{3}$$

where $v = v(x,t)$ and $n \in R$.
Using the wave transformation $v = v(x,t) = G(\gamma)$, $\gamma = p(x - bt)$ into Eq. (3), we get a NODE

$$G'' = \frac{n^2}{p^2(1-b^2)} \sin(G), \tag{4}$$

where $G = G(\gamma)$, γ is the amplitude of wave and b is the wave speed.

After some simplification, Eq. (4) gives:

$$\left[\left(\frac{G}{2}\right)'\right]^2 = \frac{n^2}{p^2(1-b^2)} \sin^2\left(\frac{G}{2}\right) + E. \tag{5}$$

Setting $E = 0$, $z(\gamma) = \frac{G}{2}$ and $d^2 = \frac{n^2}{p^2(1-b^2)}$ into Eq. (5), we obtain

$$z'(\gamma) = d\sin(z), \tag{6}$$

assume that $d = 1$, Eq. (6) turn into:

$$z'(\gamma) = \sin(z). \tag{7}$$

The following two significant equations have been gained from Eq. (7)

$$\sin(z) = \sin(z(\gamma)) = \left.\frac{2qe^{\gamma}}{q^2e^{2\gamma}+1}\right|_{q=1} = sech(\gamma), \tag{8}$$

$$\cos(z) = \sin(z(\gamma)) = \left.\frac{q^2e^{2\gamma}-1}{q^2e^{2\gamma}+1}\right|_{q=1} = \tanh(\gamma), \tag{9}$$

where q is an integration constant.

For the solution of the following nonlinear partial differential equation (NPDE);

$$W\left(p, p_x, p_t, \ldots\right), \tag{10}$$

we consider

$$G\left(\gamma\right)=\sum_{i=1}^{\tau}\tanh^{i-1}\left(\gamma\right)\left[B_i sech\left(\gamma\right)+A_i\tanh\left(\gamma\right)\right]+A_0, \tag{11}$$

Eq. (11) can be rewritten according to Eqs. (8) and (9) as follows:

$$G\left(z\right)=\sum_{i=1}^{\tau}\cos^{i-1}\left(z\right)\left[B_i\sin\left(z\right)+A_i\cos\left(z\right)\right]+A_0. \tag{12}$$

Using the balance principle to identify the value of τ, by considering the highest power of nonlinear term and the highest derivative in the obtained NODE. Equating the coefficients of the same power of $\sin^i\left(w\right), \cos^i\left(w\right)$ to zero, a system of equations can be obtained, these system can be solved by supporting one of the computer programs, gives the values of $A_0,\ A_i,\ B_i,\ p$ and b. Eventually, inserting the finding values into Eq. (12), we get the novel traveling wave solutions to the Eq. (10).

3 Implementation of SGEM

In the current section, the application of SGEM to the BBEs equation is presented.

Consider the coupled Boussinesq-Burgers equations (Eqs. (1) and (2)) stated in Sect. 1.

Inserting the wave transformation

$$u\left(x,t\right)=G\left(\gamma\right),\ v\left(x,t\right)=V\left(\gamma\right)\ ,\ \gamma=x-wt, \tag{13}$$

into Eqs. (1) and (2), the following NODE can be obtain:

$$-wG'+2GG'-\frac{1}{2}V'=0, \tag{14}$$

$$-wV'+2GV'+2VG'-\frac{1}{2}G'''=0, \tag{15}$$

integrating Eq. (14) respect to γ.

Let the constant of integration to be zero, one can obtains:

$$-wG+G^2-\frac{1}{2}V=0, \tag{16}$$

from Eq. (16), we have

$$V=2G^2-2wG, \tag{17}$$

substituting Eq. (17) into Eq. (15), and integrate once with respect to γ, we get

$$G'' - 8G^3 + 12wG^2 - 4w^2G = 0. \tag{18}$$

Applying the balance principle between G^3 and G'' in Eq. (18), yield $\tau = 1$, with $\tau = 1$ Eq. (12) take the form

$$G(z) = B_1 \sin(z) + A_1 \cos(z) + A_0. \tag{19}$$

where

$$G'' = -2A_1 \cos(z) \sin(z)^2 - B_1 \sin(z)^3 + B_1 \sin(z) \left(1 - \sin(z)^2\right), \tag{20}$$

$$\begin{aligned} G^2 = {} & {A_0}^2 + 2A_0A_1 \cos(z) + {A_1}^2 \cos(z)^2 + 2A_0B_1 \sin(z) \\ & + 2A_1B_1 \cos(z) \sin(z) + {B_1}^2 \sin(z)^2, \end{aligned} \tag{21}$$

and

$$\begin{aligned} G^3 = {} & {A_0}^3 + 3{A_0}^2A_1 \cos(z) + 3A_0{A_1}^2 \cos(z)^2 + 3{A_1}^2B_1 \cos(z)^2 \sin(z) \\ & + 3{A_0}^2B_1 \sin(z) + 6A_0A_1B_1 \cos(z) \sin(z) + 3A_0{B_1}^2 \sin(z)^2 \\ & + 3A_1{B_1}^2 \cos(z) \sin(z)^2 + {B_1}^3 \sin(z)^3 + {A_1}^3 \cos(z)^3. \end{aligned} \tag{22}$$

Substituting Eqs. (19), (20), (21), and (22) into Eq. (18), we get

$$\begin{aligned} & -8{A_0}^3 + 12{A_0}^2w - 4A_0w^2 - 24{A_0}^2A_1 \cos(z) + 24A_0A_1w \cos(z) \\ & -8{A_1}^3 \cos(z)^3 - 4A_1w^2 \cos(z) - 24A_0{A_1}^2 \cos(z)^2 + 12{A_1}^2w \cos(z)^2 \\ & +B_1 \sin(z) - 8{B_1}^3 \sin(z)^3 24{A_0}^2B_1 \sin(z) + 24A_0B_1w \sin(z) \\ & -4B_1w^2 \sin(z) - 48A_0A_1B_1 \cos(z) \sin(z) + 24A_1B_1w \cos(z) \sin(z) \\ & -24{A_1}^2B_1 \cos(z)^2 \sin(z) - 24A_0{B_1}^2 \sin(z)^2 + 12{B_1}^2w \sin(z)^2 \\ & -2A_1 \cos(z) \sin(z)^2 - 24A_1{B_1}^2 \cos(z) \sin(z)^2 - 2B_1 \sin(z)^3 = 0. \end{aligned} \tag{23}$$

According to Eq. (23), we get the following equations:
constant: $-8{A_0}^3 - 24A_0{A_1}^2 + 12{A_0}^2w + 12{A_1}^2w - 4A_0w^2 = 0.$
$\cos(z) : -24{A_0}^2A_1 - 8{A_1}^3 + 24A_0A_1w - 4A_1w^2 = 0.$
$\sin(z) : -B_1 - 24{A_0}^2B_1 - 8{B_1}^3 + 24A_0B_1w - 4B_1w^2 = 0.$
$\cos(z)\sin(z) : -48A_0A_1B_1 + 24A_1B_1w = 0.$
$\cos^2(z)\sin(z) : 2B_1 - 24{A_1}^2B_1 + 8{B_1}^3 = 0.$
$\sin^2(z) : 24A_0{A_1}^2 - 24A_0{B_1}^2 - 12{A_1}^2w + 12{B_1}^2w = 0.$
$\cos(z)\sin^2(z) : -2A_1 + 8{A_1}^3 - 24A_1{B_1}^2 = 0.$
Solving the above system, gives the following families of solutions:

Family 1. When

$$A_0 = -\frac{1}{4},\ A_1 = -\frac{1}{4},\ B_1 = \frac{i}{4},\ w = -\frac{1}{2}, \tag{24}$$

we get

$$\begin{aligned} u_1(x,t) &= -\tfrac{1}{4} + \tfrac{1}{4} i sech\left(\tfrac{t}{2} + x\right) - \tfrac{1}{4} \tanh\left(\tfrac{t}{2} + x\right), \\ v_1(x,t) &= -\tfrac{1}{4} + \tfrac{1}{4} i sech\left(\tfrac{t}{2} + x\right) - \tfrac{1}{4} \tanh\left(\tfrac{t}{2} + x\right) \\ & \quad + 2\left(-\tfrac{1}{4} + \tfrac{1}{4} i sech\left(\tfrac{t}{2} + x\right) - \tfrac{1}{4} \tanh\left(\tfrac{t}{2} + x\right)\right)^2. \end{aligned} \tag{25}$$

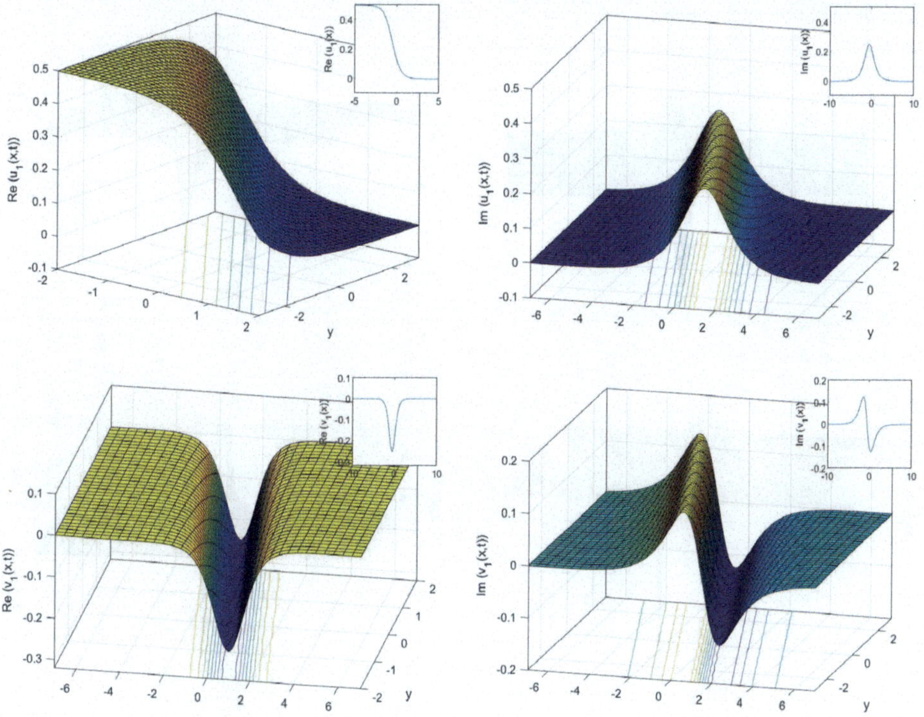

Fig. 1. 2D, 3D and contour surfaces of Eq. (25).

Family 2. when

$$A_0 = \frac{i}{2\sqrt{2}},\ A_1 = 0,\ B_1 = -\frac{i}{2},\ w = \frac{i}{\sqrt{2}}, \tag{26}$$

we get

$$\begin{aligned} u_2(x,t) &= \tfrac{i}{2\sqrt{2}} - \tfrac{1}{2} i \sec\left(\tfrac{t}{\sqrt{2}} + ix\right), \\ v_2(x,t) &= -i\sqrt{2}\left(\tfrac{i}{2\sqrt{2}} - \tfrac{1}{2} i \sec\left(\tfrac{t}{\sqrt{2}} + ix\right)\right) \\ &\quad + 2\left(\tfrac{i}{2\sqrt{2}} - \tfrac{1}{2} i \sec\left(\tfrac{t}{\sqrt{2}} + ix\right)\right)^2. \end{aligned} \tag{27}$$

Family 3. when

$$A_0 = \frac{1}{2},\ A_1 = -\frac{1}{2},\ B1 = 0,\ w = 1, \tag{28}$$

we get

$$\begin{aligned} u_3(x,t) &= \tfrac{1}{2} + \tfrac{1}{2}\tanh(t - x), \\ v_3(x,t) &= -2\left(\tfrac{1}{2} + \tfrac{1}{2}\tanh(t - x)\right) \\ &\quad + 2\left(\tfrac{1}{2} + \tfrac{1}{2}\tanh(t - x)\right)^2. \end{aligned} \tag{29}$$

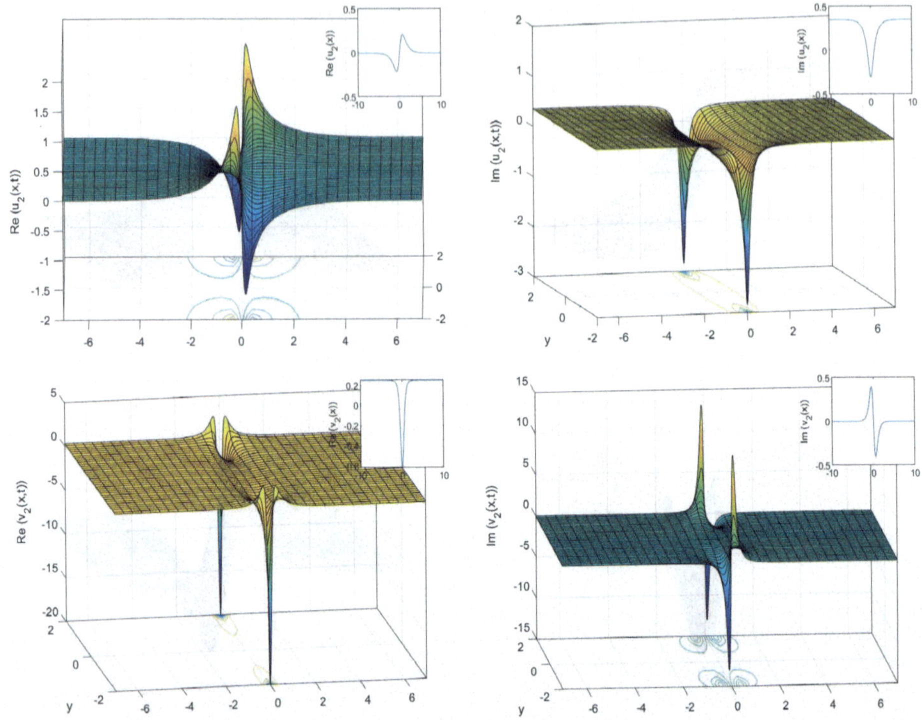

Fig. 2. 2D, 3D and contour surfaces of Eq. (27).

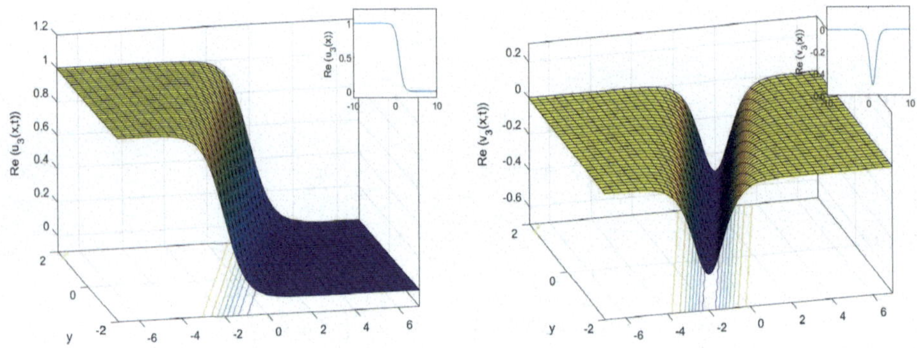

Fig. 3. 2D, 3D and contour surfaces of Eq. (29).

4 Conclusion

In this paper, with aid of sine-Gordon expansion method the coupled Boussinesq-Burgers equation have been investigated. We successfully constructed different types of exact solutions such as topological, non-topological, and kink-type soliton solutions. 2D, 3D and contour surfaces are plotted to explain the physical structure of the obtaining solutions (Figs. 1, 2 and 3).

References

1. Ravi, L.K., Saharay, S., Sahoo, S.: New exact solutions of coupled Boussinesq-Burgers equations by Exp-function method. J. Ocean Eng. Sci. **2**, 34–46 (2017)
2. Li, X., Li, B., Chen, J., Wang, M.: Exact solutions to the Boussinesq-Burgers equations. J. Appl. Math. Phys. **5**, 1720–1724 (2017)
3. Wang, P., Tian, B., Liu, W.J., Lü, X., Jiang, Y.: Lax pair, Bäcklund transformation and multi-soliton solutions for the Boussinesq-Burgers equations from shallow water waves. Appl. Math. Comput. **218**(5), 1726–1734 (2011)
4. Sulaiman, T.A., Bulut, H., Yokus, A., Baskonus, H.M.: On the exact and numerical solutions to the coupled Boussinesq equation arising in ocean engineering. Indian J. Phys. **93**(5), 647–656 (2019)
5. Li, X., Chen, A.: Darboux transformation and multi-soliton solutions of Boussinesq-Burgers equation. Phys. Lett. A **342**(5–6), 413–420 (2005)
6. Gupta, A.K., Ray, S.S.: Comparison between homotopy perturbation method and optimal homotopy asymptotic method for the soliton solutions of Boussinesq-Burger equations. Comput. Fluids **103**, 34–41 (2014)
7. Mohammed, K.: Exact traveling wave solutions of the Boussinesq-Burgers equation. Math. Comput. Model. **49**, 666–671 (2009)
8. Rady, A.A., Khalfallah, M.: On soliton solutions for Boussinesq-Burgers equations. Commun. Nonlinear Sci. Numer. Simul. **15**(4), 886–894 (2010)
9. Gao, L., Xu, W., Tang, Y., Meng, G.: New families of travelling wave solutions for Boussinesq-Burgers equation and (3+1)-dimensional Kadomtsev-Petviashvili equation. Phys. Lett. A **366**(4–5), 411–421 (2007)
10. Zheng-Yan, W., Ai-Hua, C.: Explicit solutions of Boussinesq-Burgers equation. Chin. Phys. **16**(5), 1233 (2007)
11. Ilhan, O.A., Manafian, J., Shahriari, M.: Lump wave solutions and the interaction phenomenon for a variable-coefficient Kadomtsev-Petviashvili equation. Comput. Math. Appl. **78**(8), 2429–2448 (2019)
12. Ismael, H.F., Ali, K.K.: MHD Casson Flow Over an Unsteady Stretching Sheet **20**(4), 533–541 (2017)
13. Baskonus, H.M.: Complex soliton solutions to the Gilson-Pickering model. Axioms **8**(1), 1–14 (2019)
14. Yousif, M.A., Mahmood, B.A., Ali, K.K., Ismael, H.F.: Numerical simulation using the homotopy perturbation method for a thin liquid film over an unsteady stretching sheet. Int. J. Pure Appl. Math. **107**(2), 289–300 (2016)
15. Ghanbari, B., Bekir, A., Saeed, R.K.: Oblique optical solutions of mitigating internet bottleneck with quadratic-cubic nonlinearity. Int. J. Modern Phys. B **33**(20), 1950224 (2019)
16. Ali, K.K., Varol, A.: Weissenberg and Williamson MHD flow over a stretching surface with thermal radiation and chemical reaction. JP J. Heat Mass Transf. **18**(1), 15 (2019)
17. Sulaiman, T.A., Aktürk, T., Bulut, H., Baskonus, H.M.: Investigation of various soliton solutions to the Heisenberg ferromagnetic spin chain equation. J. Electromagn. Waves Appl. **32**(9), 1093–1105 (2018)

18. Ali, K.K., Ismael, H.F., Mahmood, B.A., Yousif, M.A.: MHD casson fluid with heat transfer in a liquid film over unsteady stretching plate. Int. J. Adv. Appl. Sci. **4**(1), 55–58 (2017)
19. Gao, W., Ismael, H.F., Bulut, H., Baskonus, H.M.: Instability modulation for the (2+ 1)-dimension paraxial wave equation and its new optical soliton solutions in Kerr media. Phys. Scr. (2019). https://doi.org/10.1088/1402-4896/ab4a50
20. Baskonus, H.M.: The sine-Gordon expansion method to the Davey–Stewartson equation with power-law nonlinearity. In: 1st International Symposium on Computational Mathematics and Engineering Sciences, Errichidia/Morocco (2016)
21. Baskonus, H.M., Bulut, H.: New wave behaviors of the system of equations for the ion sound and Langmuir waves. Waves Random Complex Media **26**(4), 613–625 (2016)
22. Baskonus, H.M., Bulut, H.: New complex exact travelling wave solutions for the generalized-Zakharov equation with complex structures. Int. J. Optim. Control Theor. Appl. (IJOCTA) **6**(2), 141–150 (2016)

On a Functional Equation Arising from Subcontrary Mean and Its Pertinences

B. V. Senthil Kumar[1](✉), Hemen Dutta[2], and Khalifa Al-Shaqsi[1]

[1] Department of Information Technology, Nizwa College of Technology, Nizwa 611, Oman
{senthilkumar,khalifa.alshaqsi}@nct.edu.om, hemen_dutta08@rediffmail.com
[2] Department of Mathematics, Gauhati University, Guwahati 781 014, Assam, India

Abstract. Modelling equations involving functions is a powerful tool in many physical problems which do not require derivatives of function. The study of solution, stability and application of functional equations is an emerging field in the present scenario of research in abstract and applied mathematics. The purpose of this study is to deal with a new functional equation arising from subcontrary mean (harmonic mean) and its various fundamental stabilities relevant to Ulam's ideology of stability and also its pertinences in different fields such as physics, finance, geometry and in other sciences. We illustrate a numerical example to relate the equation dealt in this study with the fuel economy in automobiles.

Keywords: Arithmetic mean · Harmonic mean · Functional equation · Ulam stability

2010 Mathematics Subject Classication: 39B82 · 39B72

1 Introduction and Preliminaries

The research work on approximating functional, differential and integral inequalities is a hot topic in analysis. The historical background of the stability of mathematical equations is available in the literaure ([5,6,9,11,20]). There are many published papers and textbooks on various types of functional equations, their solutions and stability results; one can refer to [1,7,8,10,12,13,15,19]. The Ulam's approximation of several functional and differential equations are dealt via invariant point technique in ([2–4]). There are many interesting applications of functional equations, especially multiplicative inverse functional equations in various fields [16–18].

In this study, a new multiplicative inverse functional equation arising from subcontrary mean (harmonic mean) of the form

$$m_j\left(\frac{2pq}{p+q}\right) = \frac{1}{2}\left[m_j(p) + m_j(q)\right] \tag{1.1}$$

H. Dutta et al. (Eds.): CMES 2019, AISC 1111, pp. 241–247, 2020.
https://doi.org/10.1007/978-3-030-39112-6_18

is proposed. One can easily verify that the multiplicative inverse function $m_j(p) = \frac{1}{p}$ is a solution of Eq. (1.1). The fundamental stabilities of Eq. (1.1) are solved using fixed point alternative theorem and some applications of Eq. (1.1) are presented. Also, Eq. (1.1) is interpreted with the measurements of fuel economy in automobiles.

2 Stability Results of Subcontrary Mean Functional Equation (1.1)

In this fragmant, let us presume that $q \neq p$, for all $p, q \in \mathbb{R}^\star$. Then we proceed to obtain fundamental stability results of Eq. (1.1) involving a common control mapping as upper bound, a positive fixed constant and sum of powers of norms as upper bounds in the setting of real numbers excluding zero. For the purpose of uncomplicated computation, let us symbolize the operator D as below:

$$Dm_j(p,q) = m_j\left(\frac{2}{\frac{1}{p}+\frac{1}{q}}\right) - \frac{1}{2}\left[m_j(p) + m_j(q)\right].$$

Theorem 2.1. *Consider a function* $m_j : \mathbb{R}^\star \longrightarrow \mathbb{R}$ *with the condition that* $m_j(p)$ *tends to* 0 *when* p *tends to* ∞. *Also, assume that the function* m_j *satisfies the following inequality*

$$|Dm_j(p,q)| \leq \upsilon\left(\frac{1}{p}, \frac{1}{q}\right) \tag{2.1}$$

for all $p, q \in \mathbb{R}^\star$, *where* $\upsilon : \mathbb{R}^\star \times \mathbb{R}^\star \longrightarrow [0,\infty)$ *is a given function. Suppose there persists* $L < 1$ *such that the mapping* $p \mapsto \Upsilon(p) = \upsilon\left(\frac{1}{p}, 0\right)$ *has the property* $\Upsilon\left(\frac{p}{2}\right) \leq 2L\Upsilon(p)$ *for all* $p \in \mathbb{R}^\star$. *If the mapping* υ *has the property*

$$\lim_{n\to\infty} 2^{-n}\upsilon\left(2^{-n}p, 2^{-n}q\right) = 0, \tag{2.2}$$

for all $p, q \in \mathbb{R}^\star$, *then a unique multiplicative inverse mapping* $M_j : \mathbb{R}^\star \longrightarrow \mathbb{R}$ *exists such that*

$$|m_j(p) - M_j(p)| \leq \frac{1}{1-L}\Upsilon(p) \tag{2.3}$$

for all $p \in \mathbb{R}^\star$.

Proof. Let us define a set S as follows: $T = \{\phi : \mathbb{R}^\star \longrightarrow \mathbb{R}, \text{where } \phi \text{ is a mapping}\}$. Assume ρ be the generalized metric on T which is described as:

$$\rho(\psi,\phi) = \rho_\Upsilon(\psi,\phi) = \inf\{\lambda > 0 : |\psi(p) - \phi(p)| \leq \lambda\Upsilon(p), \text{ for all } p \in \mathbb{R}^\star\}. \tag{2.4}$$

From the above definition of ρ shows that the set T is complete space. Now, define a mapping $\mu : T \longrightarrow T$ by

$$\mu\phi(p) = \frac{1}{2}\phi\left(\frac{p}{2}\right) \qquad (p \in \mathbb{R}^\star) \tag{2.5}$$

for all $\phi \in T$. Next, let us show that μ is a strictly contractive function on the set T. For given $\psi, \phi \in T$, suppose $0 \leq \lambda_{\psi\phi} \leq \infty$ is an arbitrary constant with $\rho(\psi, \phi) \leq \lambda_{\psi\phi}$. Therefore, we have

$$\begin{aligned}
\rho(\psi, \phi) < \lambda_{\psi\phi} &\Longrightarrow |\psi(p) - \phi(p)| \leq \lambda_{\psi\phi}\Upsilon(p), \quad (\forall p \in \mathbb{R}^\star) \\
&\Longrightarrow \left|\frac{1}{2}\psi\left(\frac{p}{2}\right) - \frac{1}{2}\phi\left(\frac{p}{2}\right)\right| \leq \frac{1}{2}\lambda_{\psi\phi}\Upsilon\left(\frac{p}{2}\right), \quad (\forall p \in \mathbb{R}^\star) \\
&\Longrightarrow \left|\frac{1}{2}\psi\left(\frac{p}{2}\right) - \frac{1}{2}\phi\left(\frac{p}{2}\right)\right| \leq L\lambda_{\psi\phi}\Upsilon(p), \quad (\forall p \in \mathbb{R}^\star) \\
&\Longrightarrow \rho(\mu\psi, \mu\phi) \leq L\lambda_{\psi\phi}.
\end{aligned}$$

The above inequality implies that $\rho(\mu\psi, \mu\phi) \leq L\rho(\psi, \phi)$ for all $\psi, \phi \in T$, which inturn indicates that μ is a strictly contractive mapping of T, with the Lipschitz constant L. Now, plugging (p, q) by $(p, 0)$ in (2.1), we get

$$\left|\frac{1}{2}m_j\left(\frac{p}{2}\right) - m_j(p)\right| \leq \upsilon\left(\frac{1}{p}, 0\right) = \Upsilon(p)$$

for all $p \in \mathbb{R}^\star$. Hence (2.4) produces that $\rho(\mu r_j, r_j) \leq 1$. So, by employing the fixed point alternative Theorem, there exists a function $M_j : \mathbb{R}^\star \longrightarrow \mathbb{R}$ satisfying the following:

(1) M_j is a fixed point of ρ, that is

$$M_j\left(\frac{p}{2}\right) = 2M_j(p) \tag{2.6}$$

for all $p \in \mathbb{R}^\star$. The mapping M_j is the distinctive invariant point of μ in the set $\Gamma = \{\phi \in T : \rho(M_j, m_j) < \infty\}$. This implies that M_j is the unique mapping satisfying (2.6) such that there exists $0 < \lambda < \infty$ satisfying $|M_j(p) - m_j(p)| \leq \lambda\Upsilon(p),\ \forall p \in \mathbb{R}^\star$.

(2) $\rho\left(\mu^n m_j, M_j\right) \to 0$ as $n \to \infty$. Thus, we have

$$\lim_{n\to\infty} 2^{-n} m_j\left(2^{-n}p\right) = M_j(p) \tag{2.7}$$

for all $p \in \mathbb{R}^\star$.

(3) $d(M_j, m_j) \leq \frac{1}{1-L}\rho(M_j, \mu m_j)$, which implies $\rho(M_j, m_j) \leq \frac{1}{1-L}$.

So, the inequality (2.3) holds. On the other hand, from (2.1), (2.2) and (2.7), we have

$$\begin{aligned}
\left|DM_j(p, q)\right| &= \lim_{n\to\infty} 2^{-n}\left|m_j\left(\frac{2}{\frac{1}{2^{-n}p} + \frac{1}{2^{-n}q}}\right) - \frac{1}{2}\left[2^{-n}m_j(2^{-n}p) + 2^{-n}m_j(2^{-n}q)\right]\right| \\
&\leq \lim_{n\to\infty} 2^{-n}\upsilon\left(2^{-n}p, 2^{-n}q\right) = 0
\end{aligned}$$

for all $p, q \in \mathbb{R}^\star$, which shows that M_j is a solution of the Eq. (1.1) and hence $M_j : \mathbb{R}^\star \longrightarrow \mathbb{R}$ is a multiplicative inverse function. Now, we exhibit that M_j is the distinctive multiplicative inverse mapping satisfying (1.1) and (2.3). Let us suppose that $M_j' : \mathbb{R}^\star \longrightarrow \mathbb{R}$ be one more multiplicative inverse function satisfying (1.1) and (2.3). Since M_j' is an invariant point of μ and $\rho(m_j, M_j') < \infty$, we have $M_j' \in T^\star = \{\psi \in T | \rho(m_j, \psi) < \infty\}$. From the invariant point alternative theorem and since both M_j and M_j' are invariant points of μ, we have $M_j = M_j'$. Therefore, M_j is unique which completes the proof of Theorem 2.1. □

The upcoming theorem is the dual of Theorem 2.1. The proof is obtained by similar arguments as in Theorem 2.1 and so, for completeness, we present only the statement.

Theorem 2.2. *Suppose that the mapping $m_j : \mathbb{R}^\star \longrightarrow \mathbb{R}$ satisfies the condition $m_j(\infty) = 0$ and the inequality (2.1), where $\upsilon : \mathbb{R}^\star \times \mathbb{R}^\star \longrightarrow [0, \infty)$ is a specified function. Suppose there is an $L < 1$ exists such that the function $p \mapsto \Upsilon(p) = 2\upsilon\left(\frac{2}{p}, 0\right)$ has the property $\Upsilon(2p) \leq \frac{1}{2}L\Upsilon(p)$, for all $p \in \mathbb{R}^\star$. If the mapping υ has the property $\lim_{n\to\infty} 2^n\upsilon(2^n p, 2^n q) = 0$, for all $p, q \in \mathbb{R}^\star$, then a unique multiplicative inverse function $M_j : \mathbb{R}^\star \longrightarrow \mathbb{R}$ exists such that $|m_j(p) - M_j(p)| \leq \frac{1}{1-L}\Upsilon(p)$ for all $p \in \mathbb{R}^\star$.*

The following corollary is the investigation of various stabilities of Eq. (1.1) pertinent to UHS and UHR stability. The proof directly follow from the above theorems.

Corollary 2.3. *Let $m_j : \mathbb{R}^\star \longrightarrow \mathbb{R}$ be a function. Let there exists a constant η (not depending on p, q)≥ 0 and real numbers $\ell \neq -1$ and $\delta \geq 0$ such that the functional inequality*

$$|Dm_j(p,q)| \leq \begin{cases} \frac{\eta}{2} \\ \delta\left(\left|\frac{1}{p}\right|^\ell + \left|\frac{1}{q}\right|^\ell\right) \end{cases}$$

holds for all $p, q \in \mathbb{R}$. Then a distinctive multiplicative inverse function $M_j : \mathbb{R}^\star \longrightarrow \mathbb{R}$ exists which satisfies (1.1) and

$$|m_j(\alpha) - M_j(\alpha)| \leq \begin{cases} \eta \\ \frac{2^{\ell+1}\delta}{|2^{\ell+1}-1|}\left|\frac{1}{p}\right|^\ell, & \ell \neq -1 \end{cases}$$

for all $p \in \mathbb{R}^\star$.

Proof. The proof is obtained by considering $\upsilon\left(\frac{1}{p}, \frac{1}{q}\right) = \frac{\eta}{2}, \delta\left(\left|\frac{1}{p}\right|^\ell + \left|\frac{1}{q}\right|^\ell\right)$ for all $p, q \in \mathbb{R}^\star$, and then selecting $L = \frac{1}{2}$ in Theorems 2.1 and 2.2. □

3 Applications of Equation (1.1)

We summarize here various applications of Eq. (1.1) in many other fields where harmonic mean is involved.

- In chemistry, the density of an alloy is the harmonic mean of densitiies of its constituents. Hence Eq. (1.1) can be used to estimate the density of the alloy.
- In an electrical circuit, the effective resistance of two resistors connected in parallel is the harmonic mean of the resistance of the parallel resistor. Hence to calculate the effective resistance of an electric circuit, we can apply Eq. (1.1).
- In finance, the price-earning ratio is the harmonic mean of data points since it gives equal weight to each data point. Thus we can estimate price-earning ratio using Eq. (1.1).
- In geometry, consider an incircle in a triangle. Then the radius of the incircle is equal to the one-third of the harmonic mean of altitudes of the triangle. In this situation, Eq. (1.1) can be utilized.
- In computer science, to evaluate algorithms and systems especially in machine learning and information retrieval, the harmonic mean of the precision and the recall is employed as an accumulated performance score. Hence to evaluate algorithms and systems, we can use Eq. (1.1).

4 Interpretation of Equation (1.1)

We wind up this study with an interpretation of Eq. (1.1) with the standard measurements of fuel economy in automobiles.

There are two standard measurements used for fuel economy in automobiles. They are miles per gallon and litres per 100 km. It is clear that the dimensions of these quantitites are reciprocal to each other. Hence, the calculation of average value of the fuel economy of a car in one measurement implies the harmonic mean of the other. In other words, the average value of fuel economy expressed in litres per 100 km to miles per gallon will produce the harmonic mean of the fuel economy expressed in miles per gallon.

4.1 Numerical Example

Suppose there are two cars with fuel economy 10 L/100 km and 20 L/100 km, respectively. Let $m_j(p)$ and $m_j(q)$ denote fuel economy of the cars. Since $m_j(p) = \frac{1}{p}$, the value of the right hand side of Eq. (1.1) is

$$\frac{1}{2}[m_j(p) + m_j(q)] = \frac{1}{2}\left[\frac{1}{10} + \frac{1}{20}\right] = 0.075 \quad \text{L/100 km}.$$

Now,

$$p = \frac{1}{10} \quad \text{L/100 km} = 2824.81 \quad \text{miles/gal},$$
$$q = \frac{1}{20} \quad \text{L/100 km} = 5649.81 \quad \text{miles/gal}.$$

Then, $\frac{2pq}{p+q} = 3766.41 \quad \text{miles/gal}.$

But, $0.075\,\text{L/100 km} = 3766.41 \quad \text{miles/gal}.$

Thus, the arithmetic mean of $\frac{1}{p}$ and $\frac{1}{q}$ is obtained by the mapping of harmonic mean of p and q by the solution of Eq. (1.1).

Acknowledgment. The first and third authors are supported by The Research Council, Oman (Under Project proposal ID: BFP/RGP/CBS/18/099).

References

1. Bodaghi, A., Senthil Kumar, B.V.: Estimation of inexact reciprocal-quintic and reciprocal-sextic functional equations. Mathematica **49**(82), 3–14 (2017). No. 1–2
2. Cădariu, L., Radu, V.: Fixed ponts and the stability of Jensen's functional equation. J. Inequ. Pure Appl. Math. **4**(1), 4 (2003). Article 4
3. Cădariu, L., Radu, V.: On the stability of the Cauchy functional equation: a fixed point apporach. Grazer Math. Ber. **346**, 43–52 (2006)
4. Cădariu, L., Radu, V.: Fixed points and stability for functional equations in probabilistic metric and random normed spaces. Fixed Point Theory Appl. **2009**, 18 (2009). Article ID 589143
5. Găvruta, P.: A generalization of the Hyers-Ulam-Rassias stability of approximately additive mapppings. J. Math. Anal. Appl. **184**, 431–436 (1994)
6. Hyers, D.H.: On the stability of the linear functional equation. Proc. Nat. Acad. Sci. U.S.A. **27**, 222–224 (1941)
7. Kim, S.O., Senthil Kumar, B.V., Bodaghi, A.: Stability and non-stability of the reciprocal-cubic and reciprocal-quartic functional equations in non-Archimedean fields. Adv. Differ. Equ. **77**, 1–12 (2017)
8. Pinelas, S., Arunkumar, M., Sathya, E.: Hyers type stability of a radical reciprocal quadratic functional equation originating from 3 dimensional Pythagorean means. Int. J. Math. Appl. **5**(4), 45–52 (2017)
9. Rassias, J.M.: On approximately of approximately linear mappings by linear mappings. J. Funct. Anal. **46**, 126–130 (1982). USA
10. Rassias, J.M., Thandapani, E., Ravi, K., Senthil Kumar, B.V.: Functional Equations and Inequalities: Solutions and Stability Results. Word Scientific Publishing Company, Singapore (2017)
11. Rassias, T.M.: On the stability of the linear mapping in Banach spaces. Proc. Amer. Math. Soc. **72**, 297–300 (1978)
12. Ravi, K., Rassias, J.M., Senthil Kumar, B.V.: Ulam stability of generalized reciprocal functional equation in several variables. Int. J. App. Math. Stat. **19**(D10), 1–19 (2010)
13. Ravi, K., Rassias, J.M., Senthil Kumar, B.V.: Ulam stability of reciprocal difference and adjoint functional equations. Aust. J. Math. Anal. Appl. **8**(1), 1–18 (2011). Article 13
14. Ravi, K., Senthil Kumar, B.V.: Ulam-Gavruta-Rassias stability of Rassias reciprocal functional equation. Global J. Appl. Math. Sci. **3**(1–2), 57–79 (2010)
15. Ravi, K., Thandapani, E., Senthil Kumar, B.V.: Stability of reciprocal type functional equations. PanAm. Math. J. **21**(1), 59–70 (2011)
16. Senthil Kumar, B.V., Kumar, A., Suresh, G.: Functional equations related to spatial filtering in image enhancement. Int. J. Control Theory Appl. **9**(28), 555–564 (2016)
17. Senthil Kumar, B.V., Dutta, H.: Non-Archimedean stability of a generalized reciprocal-quadratic functional equation in several variables by direct and fixed point methods. Filomat **32**(9), 3199–3209 (2018)
18. Senthil Kumar, B.V., Dutta, H.: Fuzzy stability of a rational functional equation and its relevance to system design. Int. J. Gen. Syst. **48**(2), 157–169 (2019)

19. Senthil Kumar, B.V., Dutta, H.: Approximation of multiplicative inverse undecic and duodecic functional equations. Math. Meth. Appl. Sci. **42**, 1073–1081 (2019)
20. Ulam, S.M.: Problems in Modern Mathematics. Wiley-Interscience, New York (1964). Chapter VI

Inequalities of Bullen's Type for Logarithmically Convexity with Numerical Applications

Havva Kavurmacı-Önalan[1], Ahmet Ocak Akdemir[2(✉)], and Hemen Dutta[3]

[1] Faculty of Education, Department of Mathematics Education, Van Yüzüncü Yıl University, Van, Turkey
havvaonalan@yyu.edu.tr

[2] Faculty of Science and Arts, Department of Mathematics, Ağrı İbrahim Çeçen University, Ağrı, Turkey
ahmetakdemir@agri.edu.tr

[3] Department of Mathematics, Gauhati University, Guwahati, India
hemen_dutta08@rediffmail.com

Abstract. In this study, we consider a familier inequality of Hermite-Hadamard inequality that is well known as Bullen's inequality in the literature. We remind an integral identity that derives Bullen's type integral inequalities. By using this integral identity, we have established new Bullen's type inequalities for functions whose second derivatives in absolute value are logarithmically convex. So, new error bounds for averaged midpoint-trapezoid quadrature rules are obtained and applications in numerical integration are given.

Keywords: Power mean inequality · Logarithmically convex · Averaged midpoint-trapezoid formula

1991 Mathematics Subject Classification: 26D10 · 26D15

1 Introduction

The following double inequality offers estimations of the mean value of a convex mapping $f : [a, b] \to \mathbb{R}$,

$$f\left(\frac{a+b}{2}\right) \le \frac{1}{b-a}\int_a^b f(x)\,dx \le \frac{f(a)+f(b)}{2}.$$

This famous inequality called as Hermite-Hadamard inequality that appears firstly in the studies of Hermite [5] and Hadamard [11]. During the last few years, many researchers focused their attention on the study, generalizations and similar results of the above inequality, see the papers [7–9, 13, 15–17]. Many researchers used new lemmas to obtain Hermite-Hadamard type inequalities for different kinds of convexity.

H. Dutta et al. (Eds.): CMES 2019, AISC 1111, pp. 248–255, 2020.
https://doi.org/10.1007/978-3-030-39112-6_19

Here we give another inequality for convex functions in the literature:

Suppose that $f : I \subset \mathbb{R} \to \mathbb{R}$ is a convex function on the subset of real numbers, then the following inequality

$$\frac{1}{b-a}\int_a^b f(x)\,dx \leq \frac{1}{2}\left[f\left(\frac{a+b}{2}\right)+\frac{f(a)+f(b)}{2}\right]$$

is known as Bullen's inequality for convex functions [17], p. 39.

In [6], Niculescu mentioned log −convex functions as following:

A positive function f is called log −convex on a real interval $I = [a, b]$, if for all $x, y \in [a, b]$ and $\lambda \in [0, 1]$,

$$f(\lambda x + (1-\lambda) y) \leq f^{\lambda}(x) f^{1-\lambda}(y).$$

If f is a positive log −concave function, then the inequality is reversed. Equivalently, a function f is log −convex on I if f is positive and $\log f$ is convex on I. Several researchers spent efforts to provide new refinements, more general results and different versions of extensions for log −convexity. A brief historical background can be found in [1–3, 10, 18].

Recently, finding new results in error analysis by using various quadrature formulas is a popular topic. Approaching error analysis in terms of inequalities gave a new dimension to numerical analysis. The averaged midpoint-trapezoid quadrature rule is discussed in [4, 12, 14, 19–22].

The aim of this paper is to obtain new Bullen's type inequalities for functions whose second derivatives are log −convex by using Lemma 2 which is obtained by Sarıkaya and Aktan in [13]. Then we give some applications in numerical integration.

2 Main Results

Now we begin with the following lemma, for $\lambda = \frac{1}{2}$, [13]:

Lemma 1. *Suppose that I be a subset of real numbers where $a, b \in I$ with $a < b$. $f : I \to \mathbb{R}$ is a twice differentiable mapping such that f'' is integrable. Then the following identity holds:*

$$\frac{1}{2}\left[f\left(\frac{a+b}{2}\right)+\frac{f(a)+f(b)}{2}\right]-\frac{1}{b-a}\int_a^b f(x)\,dx = (b-a)^2\int_0^1 k(t)\,f''(ta+(1-t)b)\,dt$$

where

$$k(t) = \begin{cases} \frac{1}{2}t\left(\frac{1}{2}-t\right), & 0 \leq t \leq \frac{1}{2} \\ \frac{1}{2}(1-t)\left(t-\frac{1}{2}\right), & \frac{1}{2} \leq t \leq 1. \end{cases}$$

We give our first theorem and its proof as following:

Theorem 1. *Suppose that I be a subset of real numbers where $a, b \in I$ with $a < b$. $f : I \to \mathbb{R}$ is a twice differentiable mapping such that f'' is integrable. If $|f''|^q$ is log-convex on I, $q \geq 1$, the following inequality holds:*

$$\left| \frac{1}{2} \left[f\left(\frac{a+b}{2}\right) + \frac{f(a) + f(b)}{2} \right] - \frac{1}{b-a} \int_a^b f(x)\, dx \right|$$

$$\leq \frac{|f''(b)| (b-a)^2}{4} \left(\frac{1}{24}\right)^{1-\frac{1}{q}}$$

$$\times \left\{ \left(\frac{(\sqrt{\eta}+1)\ln\eta - 4\sqrt{\eta} + 4}{\ln^3 \eta} \right)^{\frac{1}{q}} + \left(\frac{(\sqrt{\eta}+\eta)\ln\eta - 4\eta + 4\sqrt{\eta}}{\ln^3 \eta} \right)^{\frac{1}{q}} \right\}$$

where η is defined as $\eta = \frac{|f''(a)|^q}{|f''(b)|^q} \neq 1$.

Proof. Starting with the identity that is given in Lemma 1 and by applying welll-known power mean integral inequality, we have

$$\left| \frac{1}{2} \left[f\left(\frac{a+b}{2}\right) + \frac{f(a) + f(b)}{2} \right] - \frac{1}{b-a} \int_a^b f(x)\, dx \right|$$

$$\leq \frac{(b-a)^2}{4} \left\{ \int_0^{\frac{1}{2}} |t(1-2t)| \left|f''(ta + (1-t)b)\right| dt \right.$$

$$\left. + \int_{\frac{1}{2}}^1 |(1-t)(2t-1)| \left|f''(ta + (1-t)b)\right| dt \right\}$$

$$\leq \frac{(b-a)^2}{4} \left\{ \left(\int_0^{\frac{1}{2}} |t(1-2t)|\, dt \right)^{1-\frac{1}{q}} \left(\int_0^{\frac{1}{2}} |t(1-2t)| \left|f''(ta + (1-t)b)\right|^q dt \right)^{\frac{1}{q}} \right.$$

$$\left. + \left(\int_{\frac{1}{2}}^1 |(1-t)(2t-1)|\, dt \right)^{1-\frac{1}{q}} \left(\int_{\frac{1}{2}}^1 |(1-t)(2t-1)| \left|f''(ta + (1-t)b)\right|^q dt \right)^{\frac{1}{q}} \right\}.$$

By taking into account that

$$\int_0^{\frac{1}{2}} |t(1-2t)|\, dt = \int_{\frac{1}{2}}^1 |(1-t)(2t-1)|\, dt = \frac{1}{24}$$

and since $|f''|^q$ is log-convex on I, we have

$$\int_0^{\frac{1}{2}} |t(1-2t)| \left|f''(ta + (1-t)b)\right|^q dt \leq \int_0^{\frac{1}{2}} \left(t - 2t^2\right) |f''(a)|^{qt} |f''(b)|^{q(1-t)} dt$$

$$= |f''(b)|^q \frac{(\sqrt{\eta}+1)\ln\eta - 4\sqrt{\eta} + 4}{\ln^3 \eta}$$

and

$$\int_{\frac{1}{2}}^{1}(3t-2t^2-1)\left|f''(ta+(1-t)b)\right|^q dt \le \int_{\frac{1}{2}}^{1}(3t-2t^2-1)\left|f''(a)\right|^{qt}\left|f''(b)\right|^{q(1-t)} dt$$

$$=\left|f''(b)\right|^q \frac{(\sqrt{\eta}+\eta)\ln\eta-4\eta+4\sqrt{\eta}}{\ln^3\eta}$$

where $\eta=\frac{|f''(a)|^q}{|f''(b)|^q}$, the proof is completed.

Corollary 1. *If we choose $q=1$ in Theorem 1, we get*

$$\left|\frac{1}{2}\left[f\left(\frac{a+b}{2}\right)+\frac{f(a)+f(b)}{2}\right]-\frac{1}{b-a}\int_a^b f(x)\,dx\right|$$

$$\le\frac{|f''(b)|(b-a)^2}{4}\left[\frac{(\mu+2\sqrt{\mu}+1)\ln\mu-4\mu+4}{\ln^3\mu}\right]$$

where μ is defined as $\mu=\frac{|f''(a)|}{|f''(b)|}$.

Theorem 2. *Suppose that I be a subset of real numbers where $a,b\in I$ with $a<b$. $f:I\to\mathbb{R}$ is a twice differentiable mapping such that f'' is integrable. If $|f''|^q$ is log-convex on I, $q>1$, the following inequality holds:*

$$\left|\frac{1}{2}\left[f\left(\frac{a+b}{2}\right)+\frac{f(a)+f(b)}{2}\right]-\frac{1}{b-a}\int_a^b f(x)\,dx\right|$$

$$\le|f''(b)|(b-a)^2\left[\left(\frac{1}{2}\right)^{\frac{4q-3}{q-1}}\frac{(q-1)^2}{(2q-1)(3q-1)}\right]^{1-\frac{1}{q}}$$

$$\times\left\{\left(\frac{2+\sqrt{\eta}(-2+\ln\eta)}{4\ln^2\eta}\right)^{\frac{1}{q}}+\left(\frac{2\eta-\sqrt{\eta}(2+\ln\eta)}{4\ln^2\eta}\right)^{\frac{1}{q}}\right\}$$

where η is defined in Theorem 1.

Proof. Using Lemma 1, the property of the modulus and Hölder integral inequality we have

$$\left|\frac{1}{2}\left[f\left(\frac{a+b}{2}\right)+\frac{f(a)+f(b)}{2}\right]-\frac{1}{b-a}\int_a^b f(x)\,dx\right|$$

$$\le(b-a)^2\left\{\int_0^{\frac{1}{2}}\left|\frac{t}{2}\left(\frac{1}{2}-t\right)\right|\left|f''(ta+(1-t)b)\right|dt\right.$$

$$\left.+\int_{\frac{1}{2}}^{1}\left|\left(\frac{1-t}{2}\right)\left(t-\frac{1}{2}\right)\right|\left|f''(ta+(1-t)b)\right|dt\right\}$$

$$\leq (b-a)^2 \left\{ \left(\int_0^{\frac{1}{2}} \left| \frac{t}{2} \left(\frac{1}{2} - t \right)^{\frac{q}{q-1}} \right| dt \right)^{1-\frac{1}{q}} \left(\int_0^{\frac{1}{2}} \frac{t}{2} \left| f''(ta+(1-t)b) \right|^q dt \right)^{\frac{1}{q}} \right.$$
$$\left. + \left(\int_{\frac{1}{2}}^1 \left| \left(\frac{1-t}{2} \right) \left(t - \frac{1}{2} \right)^{\frac{q}{q-1}} \right| dt \right)^{1-\frac{1}{q}} \left(\int_{\frac{1}{2}}^1 \frac{1-t}{2} \left| f''(ta+(1-t)b) \right|^q dt \right)^{\frac{1}{q}} \right\}.$$

By taking into account that

$$\int_0^{\frac{1}{2}} \left| \frac{t}{2} \left(\frac{1}{2} - t \right)^{\frac{q}{q-1}} \right| dt = \int_{\frac{1}{2}}^1 \left| \left(\frac{1-t}{2} \right) \left(t - \frac{1}{2} \right)^{\frac{q}{q-1}} \right| dt = \left(\frac{1}{2} \right)^{\frac{4q-3}{q-1}} \frac{(q-1)^2}{(2q-1)(3q-1)}$$

and since $|f''|^q$ is log-convex on I, we have

$$\int_0^{\frac{1}{2}} \frac{t}{2} \left| f''(ta+(1-t)b) \right|^q dt \leq \int_0^{\frac{1}{2}} \frac{t}{2} \left| f''(a) \right|^{qt} \left| f''(b) \right|^{q(1-t)} dt$$
$$= \left| f''(b) \right|^q \left(\frac{2+\sqrt{\eta}(-2+\ln\eta)}{4\ln^2\eta} \right)$$

and

$$\int_{\frac{1}{2}}^1 \frac{1-t}{2} \left| f''(ta+(1-t)b) \right|^q dt \leq \int_{\frac{1}{2}}^1 \frac{1-t}{2} \left| f''(a) \right|^{qt} \left| f''(b) \right|^{q(1-t)} dt$$
$$= \left| f''(b) \right|^q \left(\frac{2\eta-\sqrt{\eta}(2+\ln\eta)}{4\ln^2\eta} \right)$$

where η is defined in Theorem 1, the proof is completed.

3 Applications for Numerical Integration

Let $\pi = \{a = x_0 < x_1 < ... < x_n = b\}$ be a partition of the interval $[a,b]$, $h_i = x_{i+1} - x_i$, for $i = 0, 1, 2, ..., n-1$ and consider the averaged midpoint-trapezoid quadrature formula

$$\int_a^b f(x)\,dx = A_{MT}(\pi, f) + R_{MT}(\pi, f),$$

where

$$A_{MT}(\pi, f) = \frac{1}{4} \sum_{i=0}^{n-1} h_i \left[f(x_i) + 2f\left(\frac{x_i + x_{i+1}}{2} \right) + f(x_{i+1}) \right]$$

Here, the term $R_{MT}(\pi, f)$ denotes the associated approximation error.

The following results holds.

Proposition 1. *Let $I \subset \mathbb{R}$ be an open interval, $a, b \in I$ with $a < b$. $f : I \to \mathbb{R}$ is a twice differentiable mapping such that f'' is integrable. If $|f''|^q$ is log-convex on I, $q \geq 1$, then, for partition π of $[a, b]$, the following inequality holds:*

$$|R_{MT}(\pi, f)| \leq \left(\frac{1}{24}\right)^{1-\frac{1}{q}} \sum_{i=0}^{n-1} \frac{(h_i)^3 |f''(x_{i+1})|}{4}$$

$$\times \left\{ \left(\frac{(\sqrt{\eta_i} + 1) \ln \eta_i - 4\sqrt{\eta_i} + 4}{(\ln \eta_i)^3} \right)^{\frac{1}{q}} + \left(\frac{(\sqrt{\eta_i} + \eta_i) \ln \eta_i - 4\eta_i + 4\sqrt{\eta_i}}{(\ln \eta_i)^3} \right)^{\frac{1}{q}} \right\}$$

where η_i is defined as $\eta_i = \frac{|f''(x_i)|^q}{|f''(x_{i+1})|^q}$.

Proof. We apply Theorem 1 to the intervals $[x_i, x_{i+1}]$, $i = 0, 1, ..., n-1$ and sum. Then the triangle inequality gives the proof.

Corollary 2. *Let $I \subset \mathbb{R}$ be an open interval, $a, b \in I$ with $a < b$. $f : I \to \mathbb{R}$ is a twice differentiable mapping such that f'' is integrable. Assume that $|f''|$ is a log-convex on I. Then, for partition π of $[a, b]$, the following inequality holds:*

$$|R_{MT}(\pi, f)| \leq \sum_{i=0}^{n-1} \frac{|f''(x_{i+1})| (h_i)^3}{4} \left[\frac{(\mu_i + 2\sqrt{\mu_i} + 1) \ln \mu_i - 4\mu_i + 4}{(\ln \mu_i)^3} \right]$$

where μ_i is defined as $\mu_i = \frac{|f''(x_i)|}{|f''(x_{i+1})|}$.

Proof. We apply Proposition 1 for $q = 1$ to the intervals $[x_i, x_{i+1}]$, $i = 0, 1, ..., n-1$ and sum. Then the triangle inequality gives the proof.

Proposition 2. *Let $I \subset \mathbb{R}$ be an open interval, $a, b \in I$ with $a < b$. $f : I \to \mathbb{R}$ is a twice differentiable mapping such that f'' is integrable. If $|f''|^q$ is log-convex on I, $q > 1$, then, for partition π of $[a, b]$, the following inequality holds:*

$$|R_{MT}(\pi, f)| \leq \sum_{i=0}^{n-1} |f''(x_{i+1})| (h_i)^3 \left[\left(\frac{1}{2}\right)^{\frac{4q-3}{q-1}} \frac{(q-1)^2}{(2q-1)(3q-1)} \right]^{1-\frac{1}{q}}$$

$$\times \left\{ \left(\frac{2 + \sqrt{\eta_i}(-2 + \ln \eta_i)}{4 \ln^2 \eta_i} \right)^{\frac{1}{q}} + \left(\frac{2\eta_i - \sqrt{\eta_i}(2 + \ln \eta_i)}{4 \ln^2 \eta_i} \right)^{\frac{1}{q}} \right\}$$

where η_i is defined in Proposition 1.

Proof. We apply Theorem 2 to the intervals $[x_i, x_{i+1}]$, $i = 0, 1, ..., n-1$ and sum. Then the triangle inequality gives the proof.

4 Conclusion

This paper includes some new integral inequalities of Bullen's type for differentiable logarithmically convex functions. The results have been applied for numerical integration in terms of remainder term. The main results have performed by using Hölder inequality and its variants. Interested researchers can improved our results and new bounds can be presented by using different kinds of convex functions.

References

1. Fink, A.M.: Hadamard's inequality for log −concave functions. Math. Comput. Model. **32**(5–6), 625–629 (2000)
2. Pachpatte, B.G.: A note on integral inequalities involving two log −convex functions. Math. Inequal. Appl. **7**(4), 511–515 (2004)
3. Pachpatte, B.G.: A note on Hadamard type integral inequalities involving several log −convex functions. Tamkang J. Math. **36**(1), 43–47 (2005)
4. Pearce, C.E.M., Pečarić, J.: Inequalities for differentiable mappings with application to special means and quadrature formulae. Appl. Math. Lett. **13**(2), 51–55 (2000)
5. Hermite, C.: Sur deux limites d'une intégrale définie. Mathesis **3**, 82 (1883)
6. Niculescu, C.P.: The Hermite-Hadamard inequality for log − convex functions. Nonlinear Anal. **75**, 662–669 (2012)
7. Set, E., Özdemir, M.E., Dragomir, S.S.: On Hadamard-type inequalities involving several kinds of convexity. J. Inequal. Appl. (2010). https://doi.org/10.1155/2010/286845. Article Number: 286845
8. Toader, G.: Some generalizations of the convexity. In: Proceedings of the Colloquium on Approximation and Optimization, pp. 329–338. Cluj-Napoca, Romania (1984)
9. Toader, G.: On a generalization of the convexity. Mathematica **30**(53), 83–87 (1988)
10. Yang, G.-S., Tseng, K.-L., Wang, H.: A note on integral inequalities of Hadamard type for log −convex and log −concave functions. Taiwan. J. Math. **16**(2), 479–496 (2012)
11. Hadamard, J.: Étude sur les propietés des functions enti éres et en particulies d'une function considerée par Riemann. J. Math. Pure. Appl. **58**, 175–215 (1893)
12. Alomari, M.W., Darus, M., Kırmacı, U.S.: Refinements of Hadamard-type inequalities for quasi-convex functions with applications to trapezoidal formula and to special means. Comput. Math. Appl. **59**(1), 225–232 (2010)
13. Sarıkaya, M.Z., Aktan, N.: On the generalization some integral inequalities and their applications. Math. Comput. Model. **54**, 2175–2182 (2011)
14. Dragomir, S.S., Agarwal, R.P.: Two inequalities for differentiable mappings and applications to special means of real numbers and trapezoidal formula. Appl. Math. Lett. **11**(5), 91–95 (1998)
15. Dragomir, S.S., Mcandrew, A.: Refinements of the Hermite-Hadamard inequality for convex functions. J. Inequal. Pure Appl. Math. **6**(5) (2005). Article 140
16. Dragomir, S.S., Pečarić, J., Persson, L.E.: Some inequalities of Hadamard type. Soochow J. Math. **21**(3), 335–341 (1995)

17. Dragomir, S.S., Pearce, C.: Selected topics on Hermite-Hadamard inequalities and applications. RGMIA Monographs, Victoria University (17) (2000). http://ajmaa.org/RGMIA/monographs/hermitehadamard.html
18. Dragomir, S.S.: Some Jensen's type inequalities for log − convex functions of self-adjoint operators in Hilbert spaces. Bull. Malays. Math. Sci. Soc. **34**(3), 445–454 (2011)
19. Dragomir, S.S., Cerone, P., Roumeliotis, J.: A new generalization of ostrowski's integral inequality for mappings whose derivatives are bounded and applications in numerical integration and for special means. Appl. Math. Lett. **13**, 19–25 (2000)
20. Ujević, N.: A generalization of the Pre-Grüss inequality and applications to some quadrature formulae. J. Inequal. Pure Appl. Math. **3**(1) (2002). Article 13
21. Ujević, N.: Inequalities of ostrowski type and applications in numerical integration. Appl. Math. E-Notes **3**, 71–79 (2003)
22. Ujević, N.: An integral inequality for convex functions and applications in numerical integration. Appl. Math. E-Notes **5**, 253–260 (2005)

Properties of Binary Operations of n-Polygonal Fuzzy Numbers

Marwa Tuffaha and Mahmoud Alrefaei(✉)

Jordan University of Science and Technology, Irbid 22110, Jordan
mztuffaha15@sci.just.edu.jo, alrefaei@just.edu.jo

Abstract. The interest in fuzzy numbers and their applications is increasing rapidly. Many types of fuzzy numbers were studied in the literature and applied to many mathematical fields. The n-Polygonal Fuzzy Number (n-PFN) is starting to get a special attention since it generalizes the known triangular and trapezoidal fuzzy numbers and has the linearity advantages with flexibility of changing the value of n as desired. Tuffaha and Alrefaei [13] have presented a ranking function and convenient binary operations on the n-PFN with equidistant knots that satisfy the most important properties. In this paper, an equivalence relation of the set of all n-PFN's is introduced, where two n-PFN's are equivalent if and only if their ranking values are equal. Then, the set of the equivalence classes with two operations is shown to be a field that is isomorphic to the set of real numbers. Finally, matrices of n-PFN's are studied.

Keywords: n-polygonal fuzzy numbers · Multiplication of fuzzy numbers · The fuzzy ranking equivalence relation

1 Introduction

Fuzzy sets are important tools to represent fuzzy data we face in real life, as Zadeh [17] showed in the year 1965 in his introduction of Fuzzy Set Theory. Since then, the theory has been widely developed and applied to many mathematical fields [10]. In particular, the fuzzy numbers [4], which are normalized real fuzzy sets, are getting a special interest since they give a kind of generalization of the real numbers. Many types of fuzzy numbers were studied in the literature and applied in many mathematical fields including fuzzy differential equations [2,6,7,15] and fuzzy linear programming [1,3,5,8,11]. Recently, the n-polygonal fuzzy number, which generalizes the known triangular and trapezoidal fuzzy numbers, is getting a big interest and applications, especially in neural networks [9,14]. Piecewise linear fuzzy numbers were first introduced by Steyaert et al. [12] to be fuzzy numbers with piecewise linear membership functions. This linearity was shown in [12] to reduce the addition computational time in comparison with the time needed to add arbitrary fuzzy numbers. However, the authors failed to define a convenient closed multiplication operation. In 2001, Liu introduced the equidistant n-polygonal fuzzy number with arithmetic operations depending on

H. Dutta et al. (Eds.): CMES 2019, AISC 1111, pp. 256–265, 2020.
https://doi.org/10.1007/978-3-030-39112-6_20

Zadeh's extension principle [18]. Although many application of this fuzzy number with these arithmetic operations were studied [14], the multiplication based on Zadeh's extension principle was criticized to miss linearity and closeness properties [9,16], rather than not preserving the ranking values of the multiplied fuzzy numbers, which restricts the application capability. Later, in a previous work of Tuffaha and Alrefaei [13], a ranking function and convenient binary operations have been defined on the Piecewise Linear Fuzzy Number of order n (PLFN-n), which is the same as the equidistant n-polygonal fuzzy number. The binary operations were shown to overcome some disadvantages of existing arithmetic operations and to satisfy the most important properties, such as preserving the ranking values and having identities and inverses. In this paper, the aim is to show the strength of the algebraic structure by introducing a ranking equivalence relation on the set of all piecewise linear fuzzy numbers of order n, where two PLFN-n's are equivalent if and only if their ranking values are equal. Then, the set of the equivalence classes with two operations is shown to be a field that is isomorphic to the set of real numbers. On the other hand, applying the PLFN-n to other fields needs some tools that help us to deal with this new type of numbers. In this paper, matrices of PLFN-n with binary operations are also introduced and shown to be well defined.

2 Preliminaries

First, we present some definitions about the fuzzy sets and numbers:

Definition 1. *[17] A **fuzzy set** $\tilde{A}$ is a pair (X, μ), where $X \subseteq \mathbb{R}$ is a set and $\mu_{\tilde{A}} : X \to [0,1]$ is called the **membership function**, where the value of $\mu_{\tilde{A}}(x)$ represents the "**degree of membership**" of x in $\tilde{A}$.*

The definitions of piecewise linear fuzzy number is as follows.

Definition 2. *[13] If the membership function of a real fuzzy set $\tilde{A}$ is given by:*

$$f_{\tilde{A}}(x) = \begin{cases} \frac{1}{n}\left[\frac{x-a_i}{a_{i+1}-a_i}\right] + \frac{i}{n} & ; a_i \le x \le a_{i+1}, \quad i = 0, .., n-1 \\ 1 & ; a_n \le x \le b_0 \\ \frac{-1}{n}\left[\frac{x-b_i}{b_{i+1}-b_i}\right] + \frac{n-i}{n} & ; b_i \le x \le b_{i+1}, \quad i = 0, .., n-1 \\ 0 & otherwise \end{cases} \tag{1}$$

*Then, $\tilde{A}$ is called a **Piecewise Linear Fuzzy Number of Order** n (PLFN-n), which is represented by its knots: $(a_0, a_1, .., a_n; b_0, b_1, .., b_n)$. The family of all PLFN-n's is denoted by $\mathcal{PL}_n$.*

Definition 3. *Let $\tilde{P} = (a_0, a_1, .., a_n; b_0, b_1, .., b_n)$ be a PLFN-n. Then its ranking value is given by:*

$$\mathcal{R}(\tilde{P}) = \frac{1}{4n}[a_0 + 2a_1 + 2a_2 + ... + 2a_{n-1} + a_n + b_0 + 2b_1 + 2b_2 + ... + 2b_{n-1} + b_n] \tag{2}$$

The following definition gives a comparison between two PLFN-n's through their ranking values.

Definition 4. *Let $\tilde{A}$ and $\tilde{B}$ be two PLFN-n's. Then:*

- *$\tilde{A}$ and $\tilde{B}$ are said to be* **equivalent**, *denoted $\tilde{A} \approx \tilde{B}$, if $\mathcal{R}(\tilde{A}) = \mathcal{R}(\tilde{B})$.*
- *$\tilde{A}$ is said to be* **greater than or equivalent to** *$\tilde{B}$, denoted $\tilde{A} \succeq \tilde{B}$, if $\mathcal{R}(\tilde{A}) \geq \mathcal{R}(\tilde{B})$.*
- *$\tilde{A}$ is said to be* **smaller than or equivalent to** *$\tilde{B}$, denoted $\tilde{A} \preceq \tilde{B}$, if $\mathcal{R}(\tilde{A}) \leq \mathcal{R}(\tilde{B})$.*

Definition 5. *Let $\tilde{A} = (a_0, a_1, .., a_n; b_0, b_1, .., b_n), \tilde{B} = (c_0, c_1, .., c_n; d_0, d_1, .., d_n) \in \mathcal{PL}_n$. The* **addition** *of $\tilde{A}$ and $\tilde{B}$ is defined as follows:*

$$\tilde{A} \oplus \tilde{B} = (a_0 + c_0, a_1 + c_1, .., a_n + c_n; b_0 + d_0, b_1 + d_1, .., b_n + d_1)$$

The **multiplication** *of $\tilde{A}$ and $\tilde{B}$ is $\tilde{A} \otimes \tilde{B} = (e_0, e_1, .., e_n; f_0, f_1, .., f_n)$ where:*

$$f_n = \frac{1}{4n}[I + \sum_{i=1}^{n}(2i-1)X_i + 2nX_{n+1} + \sum_{i=1}^{n}(2(n+i)-1)X_{n+1+i}]$$

$$f_{i-1} = f_i - X_{n+1+i}, \ \ for\ i = n, n-1, .., 1$$

$$e_n = f_0 - X_{n+1}$$

$$e_{i-1} = e_i - X_i, \ \ for\ i = n, n-1, .., 1$$

$$and, I = \frac{1}{4n}[(a_0 + 2a_1 + .. + 2a_{n-1} + a_n + b_0 + 2b_1 + .. + 2b_{n-1} + b_n) * (c_0 + 2c_1 + .. + 2c_{n-1} + c_n + d_0 + 2d_1 + .. + 2d_{n-1} + d_n)]$$

$$X_i = (a_i - a_{i-1}) + (c_i - c_{i-1}), \ \ for\ i = n, n-1, .., 1$$

$$X_{n+1} = (b_0 - a_n) + (d_0 - c_n)$$

$$X_{n+1+i} = (b_i - b_{i-1}) + (d_i - d_{i-1}), \ \ for\ i = n, n-1, .., 1$$

Definition 6. *Let $\tilde{A} = (a_0, a_1, .., a_n; b_0, b_1, .., b_n)$ and let $\tilde{B} = (c_0, c_1, .., c_n; d_0, d_1, .., d_n) \in \mathcal{PL}_n$. We define the* **additive inverse** *of $\tilde{A}$ as follows:*

$$-\tilde{A} = (-b_n, -b_{n-1}, .., -b_1, -b_0; -a_n, -a_{n-1}, .., -a_1, -a_0)$$

The subtraction of $\tilde{B}$ from $\tilde{A}$ is defined as follows:

$$\tilde{A} \ominus \tilde{B} = \tilde{A} \oplus (-\tilde{B}) = (a_0 - d_n, a_1 - d_{n-1}, .., a_n - d_0; b_0 - c_n, b_1 - c_{n-1}, .., b_n - c_0)$$

In addition, if $\mathcal{R}(\tilde{A}) \neq 0$, the ***multiplicative inverse*** *of $\tilde{A}$, in the sense that $\tilde{A} \otimes \tilde{A}^{-1} \approx 1$, can be defined to be $\tilde{A}^{-1} = (c_0, c_1, .., c_n; d_0, d_1, .., d_n)$, where*

$$c_0 = A + \frac{1}{4n}(a_0 + 2a_1 + .. + 2a_{n-1} + a_n + b_0 + 2b_1 + .. + 2b_{n-1} - (4n-1)b_n)$$

$$A = \frac{1}{\mathcal{R}(\tilde{A})}$$

$$c_i = c_{i-1} + (b_{n-i+1} - b_{n-i}) \; for \; all \; i = 1, .., n$$

$$d_0 = c_n + (b_0 - a_n)$$

$$d_i = d_{i-1} + (a_{n-i+1} - a_{n-i}) \; for \; all \; i = 1, .., n$$

The following theorem gives some important properties of the arithmetic operations.

Theorem 1. *Let $\tilde{A} = (a_0, a_1, .., a_n; b_0, b_1, .., b_n)$ and $\tilde{B} = (c_0, c_1, .., c_n; d_0, d_1, .., d_n)$ be two PLFN-n's. Then*

1. $\mathcal{R}(\tilde{A} \oplus \tilde{B}) = \mathcal{R}(\tilde{A}) + \mathcal{R}(\tilde{B})$
2. $\mathcal{R}(\tilde{A} \ominus \tilde{B}) = \mathcal{R}(\tilde{A}) - \mathcal{R}(\tilde{B})$
3. $\mathcal{R}(-\tilde{A}) = -\mathcal{R}(\tilde{A})$
4. $\mathcal{R}(\tilde{A} \otimes \tilde{B}) = \mathcal{R}(\tilde{A}) \cdot \mathcal{R}(\tilde{B})$

Proof: The proof of properties 1, 2 and 3 is straight forward, while the multiplication operation was built in [13] to satisfy property 4.

The proof of the following theorem is straightforward based on Theorem 1.

Theorem 2. *For $\tilde{a}, \tilde{b}, \tilde{c} \in \mathcal{PL}_n$, we have:*

1. *If $\tilde{a} \oplus \tilde{b} \approx \tilde{c}$, then $\tilde{a} \approx \tilde{c} \ominus \tilde{b}$.*
2. *If $\tilde{a} \oplus \tilde{b} \preceq \tilde{c}$ and $\tilde{b} \succeq 0$, then $\tilde{a} \preceq \tilde{c}$.*
3. *If $\tilde{a} \preceq \tilde{c}$, then $\tilde{c} \ominus \tilde{a} \succeq 0$.*
4. *If $\tilde{a} \preceq \tilde{c}$ and $\tilde{b} \succeq 0$, then $\tilde{a} \otimes \tilde{b} \preceq \tilde{c} \otimes \tilde{b}$.*
5. *If $\tilde{a} \preceq 0$, then $\tilde{a} \oplus \tilde{b} \preceq \tilde{b}$.*
6. *$\tilde{a} \preceq \tilde{b}$ if and only if $-\tilde{a} \succeq -\tilde{b}$.*

3 The Ranking Equivalence Relation

In this section, showing the strength of the algebraic structure $(\mathcal{PL}_n, \oplus, \otimes)$ is intended. This is done by introducing an equivalence relation on it and proposing two binary operations on the set of equivalence classes. Then, some important results are shown.

Definition 7. *Define the ranking relation $\sim$ on $\mathcal{PL}_n$ as follows:*
For all $\tilde{N_1}, \tilde{N_2} \in \mathcal{PL}_n$: $\tilde{N_1} \sim \tilde{N_2} \Leftrightarrow \mathcal{R}(\tilde{N_1}) = \mathcal{R}(\tilde{N_2})$.

Theorem 3. *The ranking relation defined above is an equivalence relation.*

Proof:

- For all $\tilde{N} \in \mathcal{PL}_n$: $\mathcal{R}(\tilde{N}) = \mathcal{R}(\tilde{N})$. Therefore, $\tilde{N} \sim \tilde{N}$ and the relation is reflexive.
- Let $\tilde{N_1}, \tilde{N_2} \in \mathcal{PL}_n$ such that $\tilde{N_1} \sim \tilde{N_2}$, i.e. $\mathcal{R}(\tilde{N_1}) = \mathcal{R}(\tilde{N_2})$, then $\tilde{N_2} \sim \tilde{N_1}$. Therefore, the relation is symmetric.
- Let $\tilde{N_1}, \tilde{N_2}, \tilde{N_3} \in \mathcal{PL}_n$ such that $\tilde{N_1} \sim \tilde{N_2}$ and $\tilde{N_2} \sim \tilde{N_3}$, i.e. $\mathcal{R}(\tilde{N_1}) = \mathcal{R}(\tilde{N_2})$ and $\mathcal{R}(\tilde{N_2}) = \mathcal{R}(\tilde{N_3})$. Then $\mathcal{R}(\tilde{N_1}) = \mathcal{R}(\tilde{N_3})$, so $\tilde{N_1} \sim \tilde{N_3}$. Thus, the relation is transitive.

From the above points, the ranking relation is an equivalence relation.

Definition 8. *The equivalence classes on $\mathcal{PL}_n$ with respect to the equivalence relation defined above are defined by:*

$$[\tilde{N}] = \{\tilde{A} \in \mathcal{PL}_n : \mathcal{R}(\tilde{A}) = \mathcal{R}(\tilde{N})\} \quad \textit{for all } \tilde{N} \in \mathcal{PL}_n$$

We denote the set of all equivalence classes in $\mathcal{PL}_n$ by $\mathcal{F}_n = \{[\tilde{N}] : \tilde{N} \in \mathcal{PL}_n\}$.

Definition 9. *For $[\tilde{N_1}], [\tilde{N_2}] \in \mathcal{F}_n$, we define the* **addition** *on $\mathcal{F}_n$ by:*

$$[\tilde{N_1}] + [\tilde{N_2}] = \{\tilde{A} \oplus \tilde{B} : \ \mathcal{R}(\tilde{A}) = \mathcal{R}(\tilde{N_1}) \textit{ and } \mathcal{R}(\tilde{B}) = \mathcal{R}(\tilde{N_2})\}$$

Theorem 4. *$[\tilde{N_1}] + [\tilde{N_2}] = [\tilde{N_1} \oplus \tilde{N_2}]$ for all $[\tilde{N_1}], [\tilde{N_2}] \in \mathcal{F}_n$.*

Proof: Suppose $\mathcal{R}(\tilde{N_1}) = x$ and $\mathcal{R}(\tilde{N_2}) = y$. Then, by Theorem 1, $\mathcal{R}(\tilde{N_1} \oplus \tilde{N_2}) = x + y$.

1. Let $\tilde{C} \in [\tilde{N_1}] + [\tilde{N_2}]$. Then $\tilde{C} = \tilde{A} \oplus \tilde{B}$, where $\tilde{A} \in [\tilde{N_1}]$ and $\tilde{B} \in [\tilde{N_2}]$. From Theorem 1, we have:

$$\mathcal{R}(\tilde{A} \oplus \tilde{B}) = \mathcal{R}(\tilde{A}) + \mathcal{R}(\tilde{B})$$
$$\mathcal{R}(\tilde{C}) = x + y$$

Therefore, $\tilde{C} \in [\tilde{N_1} \oplus \tilde{N_2}]$.

2. Let $\tilde{C} = (e_0, e_1, .., e_n; f_0, f_1, .., f_n) \in [\tilde{N_1} \oplus \tilde{N_2}]$. Then

$$\mathcal{R}(\tilde{C}) = \frac{e_0 + 2e_1 + .. + 2e_{n-1} + e_n + f_0 + 2f_1 + .. + 2f_{n-1} + f_n}{4n} = x + y$$

We want to show that $\tilde{C} \in [\tilde{N_1}] + [\tilde{N_2}]$, i.e. we need to find $\tilde{A} \in [\tilde{N_1}]$ and $\tilde{B} \in [\tilde{N_2}]$ such that $\tilde{A} \oplus \tilde{B} = \tilde{C}$.
Let $\tilde{A} = (x, x, .., x; x, x, .., x)$. Then $\mathcal{R}(\tilde{A}) = x$ and $\tilde{A} \in [\tilde{N_1}]$. And let $\tilde{B} = (e_0 - x, e_1 - x, .., e_n - x; f_0 - x, f_1 - x, .., f_n - x)$. Then

$$\begin{aligned}
\mathcal{R}(\tilde{B}) &= \frac{1}{4n}[e_0 - x + (e_1 - x) + .. + 2(e_{n-1} - x) + e_n - x \\
&\qquad + f_0 - x + 2(f_1 - x) + .. + 2(f_{n-1} - x) + f_n - x] \\
&= \frac{1}{4n}[e_0 + 2e_1 + .. + 2e_{n-1} + e_n + f_0 + 2f_1 + .. + 2f_{n-1} + f_n] + \frac{1}{4n}[-4n.x] \\
&= x + y - x = y
\end{aligned}$$

Therefore, $\tilde{B} \in [\tilde{N}_2]$.
Moreover, $\tilde{A} \oplus \tilde{B} = (e_0, e_1, .., e_n; f_0, f_1, .., f_n) = \tilde{C}$. Hence $\tilde{C} \in [\tilde{N}_1] + [\tilde{N}_2]$.

From the above we conclude that $[\tilde{N}_1] + [\tilde{N}_2] = [\tilde{N}_1 \oplus \tilde{N}_2]$.

Theorem 5. *The set $\mathcal{F}_n$ with the addition operation given in Definition 9 is an abelian group.*

<u>**Proof:**</u>

1. $\mathcal{F}_n$ has an additive identity, that is $[\tilde{0}]$, where $\tilde{0}$ is any PLFN-n with ranking value equal to 0, since for all $[\tilde{a}] \in \mathcal{F}_n$:

$$[\tilde{a}] + [\tilde{0}] = [\tilde{0}] + [\tilde{a}] = [\tilde{a} \oplus \tilde{0}] = [\tilde{a}]$$

2. For all $[\tilde{a}] \in \mathcal{F}_n$, $[\tilde{a}]$ has an additive inverse in $\mathcal{F}_n$, that is $-[\tilde{a}] = [-\tilde{a}]$, since:

$$[\tilde{a}] + [-\tilde{a}] = [-\tilde{a}] + [\tilde{a}] = [\tilde{a} \ominus \tilde{a}] = [\tilde{0}]$$

3. The addition is associative since for all $[\tilde{a}], [\tilde{b}], [\tilde{c}] \in \mathcal{F}_n$, we have:

$$\begin{aligned}[\tilde{a}] + ([\tilde{b}] \oplus [\tilde{c}]) &= [\tilde{a}] + [\tilde{b} \oplus \tilde{c}] = [\tilde{a} \oplus (\tilde{b} \oplus \tilde{c})] \\ &= [(\tilde{a} \oplus \tilde{b}) \oplus \tilde{c}] = [\tilde{a} \oplus \tilde{b}] + [\tilde{c}] = ([\tilde{a}] + [\tilde{b}]) + [\tilde{c}]\end{aligned}$$

4. The addition is commutative since for all $[\tilde{a}], [\tilde{b}] \in \mathcal{F}_n$, we have:

$$[\tilde{a}] + [\tilde{b}] = [\tilde{a} \oplus \tilde{b}] = [\tilde{b} \oplus \tilde{a}] = [\tilde{b}] + [\tilde{a}]$$

From the above we conclude that $(\mathcal{F}_n, +)$ is an abelian group.

Definition 10. *Let $[\tilde{N}_1], [\tilde{N}_2] \in \mathcal{F}_n$. Define the* ***multiplication*** *on $\mathcal{F}_n$ by:*

$$[\tilde{N}_1] \cdot [\tilde{N}_2] = \{\tilde{A} \otimes \tilde{B} : \ \mathcal{R}(\tilde{A}) = \mathcal{R}(\tilde{N}_1) \text{ and } \mathcal{R}(\tilde{B}) = \mathcal{R}(\tilde{N}_2)\}$$

Theorem 6. $[\tilde{N}_1] \cdot [\tilde{N}_2] = [\tilde{N}_1 \otimes \tilde{N}_2]$ *for all* $[\tilde{N}_1], [\tilde{N}_2] \in \mathcal{F}_n$.

<u>**Proof:**</u> Suppose $\mathcal{R}(\tilde{N}_1) = x$ and $\mathcal{R}(\tilde{N}_2) = y$. Then $\mathcal{R}(\tilde{N}_1 \otimes \tilde{N}_2) = x \cdot y$.

1. Let $\tilde{C} \in [\tilde{N}_1] \cdot [\tilde{N}_2]$. Then $\tilde{C} = \tilde{A} \otimes \tilde{B}$, where $\tilde{A} \in [\tilde{N}_1]$ and $\tilde{B} \in [\tilde{N}_2]$. From Theorem 1, we have:

$$\begin{aligned}\mathcal{R}(\tilde{A} \otimes \tilde{B}) &= \mathcal{R}(\tilde{A}) \cdot \mathcal{R}(\tilde{B}) \\ \mathcal{R}(\tilde{C}) &= x \cdot y\end{aligned}$$

Therefore, $\tilde{C} \in [\tilde{N}_1 \otimes \tilde{N}_2]$.

2. Let $\tilde{C} = (e_0, e_1, .., e_n; f_0, f_1, .., f_n) \in [\tilde{N}_1 \otimes \tilde{N}_2]$. We want to show that $\tilde{C} \in [\tilde{N}_1] \cdot [\tilde{N}_2]$, i.e. we need to find $\tilde{A} \in [\tilde{N}_1]$ and $\tilde{B} \in [\tilde{N}_2]$ such that $\tilde{A} \otimes \tilde{B} = \tilde{C}$. Let $\tilde{A} = (x, x, .., x; x, x, .., x)$. Then $\tilde{A} \in [\tilde{N}_1]$. And let $\tilde{B} = (y[1-x]+e_0, y[1-x]+e_1, .., y[1-x]+e_n; y[1-x]+f_0, y[1-x]+f_1, .., y[1-x]+f_n)$. Then

$$\mathcal{R}(\tilde{B}) = \frac{1}{4n}[y(1-x) + e_0 + 2y(1-x) + 2e_1 + .. + 2y(1-x) + 2e_{n-1} + y(1-x) + e_n + y(1-x) + f_0 + 2y(1-x) + 2f_1 + .. + 2y(1-x) + 2f_{n-1} + y(1-x) + f_n]$$
$$= \frac{1}{4n}[4ny(1-x)] + \mathcal{R}(\tilde{C}) = y(1-x) + xy = y$$

Therefore, $\tilde{B} \in [\tilde{N}_2]$. Now, we show that $\tilde{A} \otimes \tilde{B} = \tilde{C}$.
$I = \frac{1}{4n}[4nx][4ny(1-x) + 4nxy] = 4nx[y(1-x) + xy] = 4nxy$
$X_i = (x-x) + [y(1-x) + e_i - y(1-x) - e_{i-1}] = e_i - e_{i-1}, \quad for\ all\ i = 1, .., n$
$X_{n+1} = (x-x) + [y(1-x) + f_0 - y(1-x) - e_n] = f_0 - e_n$
$X_{n+1+i} = (x-x) + [y(1-x) + f_i - y(1-x) - f_{i-1}] = f_i - f_{i-1}, \quad for\ all\ i = 1, .., n$
It is clear that $\tilde{A} \otimes \tilde{B} = \tilde{C}$. Hence $\tilde{C} \in [\tilde{N}_1] \cdot [\tilde{N}_2]$.

From the above we conclude that $[\tilde{N}_1] \cdot [\tilde{N}_2] = [\tilde{N}_1 \otimes \tilde{N}_2]$.

Theorem 7. *The structure $(\mathcal{F}_n, +, \cdot)$ is a field.*

Proof:

1. Theorem 5 shows that $(\mathcal{F}_n, +)$ is an abelian group.
2. The operation $(\cdot)$ is associative since for all $[\tilde{a}], [\tilde{b}], [\tilde{c}] \in \mathcal{F}_n$, by Theorem 6, we have:

$$[\tilde{a}] \cdot ([\tilde{b}] \cdot [\tilde{c}]) = [\tilde{a}] \cdot [\tilde{b} \otimes \tilde{c}] = [\tilde{a} \otimes (\tilde{b} \otimes \tilde{c})] = [(\tilde{a} \otimes \tilde{b}) \otimes \tilde{c}] = [\tilde{a} \otimes \tilde{b}] \cdot [\tilde{c}] = ([\tilde{a}] \cdot [\tilde{b}]) \cdot [\tilde{c}]$$

3. The operation $(\cdot)$ is commutative since for all $[\tilde{a}], [\tilde{b}] \in \mathcal{F}_n$, by Theorem 6, we have:
$$[\tilde{a}] \cdot [\tilde{b}] = [\tilde{a} \otimes \tilde{b}] = [\tilde{b} \otimes \tilde{a}] = [\tilde{b}] \cdot [\tilde{a}]$$
4. The distribution laws hold since for all $[\tilde{a}], [\tilde{b}], [\tilde{c}] \in \mathcal{F}_n$, by Theorems 4 and 6, we have $[\tilde{a}] \cdot ([\tilde{b}] + [\tilde{c}]) = [\tilde{a}] \cdot [\tilde{b} \oplus \tilde{c}] = [\tilde{a} \otimes (\tilde{b} \oplus \tilde{c})]$.
However, since both the addition and the multiplication preserve the ranking value, it can easily be seen that $\mathcal{R}[\tilde{a} \otimes (\tilde{b} \oplus \tilde{c})] = \mathcal{R}[(\tilde{a} \otimes \tilde{b}) \oplus (\tilde{a} \otimes \tilde{c})]$. Thus,

$$[\tilde{a}] \cdot ([\tilde{b}] + [\tilde{c}]) = [(\tilde{a} \otimes \tilde{b}) \oplus (\tilde{a} \otimes \tilde{c})] = [\tilde{a} \otimes \tilde{b}] + [\tilde{a} \otimes \tilde{c}] = [\tilde{a}] \cdot [\tilde{b}] + [\tilde{a}] \cdot [\tilde{c}]$$

Similarly, $([\tilde{b}] + [\tilde{c}]) \cdot [\tilde{a}] = [\tilde{b}] \cdot [\tilde{a}] + [\tilde{c}] \cdot [\tilde{a}]$.
5. $\mathcal{F}_n$ has a multiplicative identity, that is $[\tilde{1}]$, where $\tilde{1}$ is any PLFN-n with ranking value equal to 1. For all $[\tilde{a}] \in \mathcal{F}_n$, by Theorem 6, we have $[\tilde{a}] \cdot [\tilde{1}] = [\tilde{1}] \cdot [\tilde{a}] = [\tilde{a}]$.
6. For all $[\tilde{a}] \in \mathcal{F}_n$ with $[\tilde{a}] \neq [\tilde{0}]$, $[\tilde{a}]$ has a multiplicative inverse in $\mathcal{F}_n$, that is $[\tilde{a}]^{-1} = [\tilde{a}^{-1}]$, since by Theorem 6 above, we have $[\tilde{a}] \cdot [\tilde{a}]^{-1} = [\tilde{a}]^{-1} \cdot [\tilde{a}] = [\tilde{a} \otimes \tilde{a}^{-1}] = [\tilde{1}]$.

Therefore, $(\mathcal{F}_n, +, \cdot)$ is a field.

Theorem 8. *The field* $(\mathcal{F}_n, +, \cdot)$ *is isomorphic to* $\mathbb{R}$ *as rings.*

Proof: Let the map $\phi : \mathcal{F}_n \rightarrow \mathbb{R}$ with $\phi([\tilde{N}]) = \mathcal{R}(\tilde{N})$. Then ϕ is a ring homomorphism since for all $[\tilde{x}], [\tilde{y}] \in \mathcal{F}_n$, by Theorems 4 and 6, we have:

$$\phi([\tilde{x}] + [\tilde{y}]) = \phi([\tilde{x} \oplus \tilde{y}]) = \mathcal{R}(\tilde{x} \oplus \tilde{y}) = \mathcal{R}(\tilde{x}) + \mathcal{R}(\tilde{y}) = \phi([\tilde{x}]) + \phi([\tilde{y}])$$

$$\phi([\tilde{x}] \cdot [\tilde{y}]) = \phi([\tilde{x} \otimes \tilde{y}]) = \mathcal{R}(\tilde{x} \otimes \tilde{y}) = \mathcal{R}(\tilde{x}) \cdot \mathcal{R}(\tilde{y}) = \phi([\tilde{x}]) \cdot \phi([\tilde{y}])$$

Moreover, ϕ is clearly one-to-one and onto. Thus ϕ is an isomorphism and $\mathcal{F}_n \simeq \mathbb{R}$ as rings.

4 Matrices with PLFN-n

Studying fuzziness in many fields requires working with matrices of fuzzy numbers. Therefore, in this section, some concepts about the matrices with PLFN-n's are proposed here.

Definition 11

- *A* piecewise linear fuzzy matrix $\tilde{\mathbf{M}}$ *is a matrix whose entries are PLFN-n's.*
- *The set of all piecewise linear fuzzy matrices is denoted* $\mathcal{M}(\mathcal{PL}_n)$.
- *The addition and multiplication of piecewise linear fuzzy matrices are similar to those on real matrices, but using the binary operations on* $\mathcal{PL}_n$ *given in Definition 5.*

Definition 12. *Let* $\tilde{\mathbf{M}} = [\tilde{m}_{ij}]_{p \times q}, \tilde{\mathbf{N}} = [\tilde{n}_{ij}]_{p \times q} \in \mathcal{M}(\mathcal{PL}_n)$. *Then:*

1. $\tilde{\mathbf{M}}$ *and* $\tilde{\mathbf{N}}$ *are said to be* equal, *written* $\tilde{\mathbf{M}} = \tilde{\mathbf{N}}$, *if* $\tilde{m}_{ij} = \tilde{n}_{ij}$ *for all* $i = 1, .., p$ *and* $j = 1, .., q$. *Moreover, we call them* equivalent, *written* $\tilde{\mathbf{M}} \approx \tilde{\mathbf{N}}$, *if* $\tilde{m}_{ij} \approx \tilde{n}_{ij}$ *(have the same rank) for all* $i = 1, .., p$ *and* $j = 1, .., q$.
2. *A set of rows of* $\tilde{\mathbf{M}}$, $\{\tilde{\mathbf{m}}_{i_1}, \tilde{\mathbf{m}}_{i_2}, ..., \tilde{\mathbf{m}}_{i_k}\}$, *is said to be* linearly independent *if the equation:* $a_1 * \tilde{\mathbf{m}}_{i_1} \oplus a_2 * \tilde{\mathbf{m}}_{i_2} \oplus ... \oplus a_k * \tilde{\mathbf{m}}_{i_k} \approx \tilde{\mathbf{0}}$ *with* $a_1, a_2, ..., a_k \in \mathbb{R}$ *can only be satisfied by:* $a_i = 0$ *for all* $i = 1, .., k$.
3. *The* rank *of* $\tilde{\mathbf{M}}$ *is the maximal number of linearly independent rows of* $\tilde{\mathbf{M}}$.
4. *If* $p = q$, *then* $\tilde{\mathbf{M}}$ *is a* square fuzzy matrix, *and we define the* determinant *of* $\tilde{\mathbf{M}}$, *denoted* $det(\tilde{\mathbf{M}})$, *to be a PLFN-n computed in a similar way to how we compute the determinant of a real square matrix, but using the binary operations defined in Sect. 2 on* $\mathcal{PL}_n$. *Furthermore, it can easily be shown that if* $det(\tilde{\mathbf{M}}) \not\approx 0$, *then* $\tilde{\mathbf{M}}$ *has an* inverse matrix $\tilde{\mathbf{M}}^{-1}$ *such that* $\tilde{\mathbf{M}} * \tilde{\mathbf{M}}^{-1} \approx \tilde{\mathbf{I}}$, *where the square matrix* $\tilde{\mathbf{I}}$ *is a* fuzzy identity matrix *in* $\mathcal{M}(\mathcal{PL}_n)$ *which has PLFN-n's equivalent to* 1 *on the main diagonal, and all its other elements are equivalent to* 0.

Example 1. Take $\tilde{\mathbf{M}} = \begin{bmatrix} \tilde{a} = (0,2,4;5,6,7) & \tilde{b} = (5,6,7;7,8,9) \\ \tilde{c} = (-2,0,2;3,4,5) & \tilde{d} = (1,4,6;6,8,11) \end{bmatrix}$.
Then $det(\tilde{\mathbf{M}}) = (\tilde{a} \otimes \tilde{d}) \ominus (\tilde{c} \otimes \tilde{b}) = (-4,3,9;11,17,24) \not\approx 0$.

This means that $\tilde{\mathbf{M}}$ has an inverse:

$$\tilde{\mathbf{M}}^{-1} = (det(\tilde{\mathbf{M}}))^{-1} * \begin{bmatrix} \tilde{d} & -\tilde{b} \\ -\tilde{c} & \tilde{a} \end{bmatrix}$$
$$= \begin{bmatrix} \tilde{a}' = (-18.4, -8.4, -0.4; 1.6, 9.6, 19.6) & \tilde{b}' = (-16.7, -8.7, -1.7; 0.3, 7.3, 15.3) \\ \tilde{c}' = (-17.2, -9.2, -2.2; 0.8, 8.8, 17.8) & \tilde{d}' = (-17.6, -8.6, -0.6; 2.4, 9.4, 17.4) \end{bmatrix}$$

where

$$\tilde{\mathbf{M}} * \tilde{\mathbf{M}}^{-1} = \begin{bmatrix} (\tilde{a} \otimes \tilde{a}') \oplus (\tilde{b} \otimes \tilde{c}') & (\tilde{a} \otimes \tilde{b}') \oplus (\tilde{b} \otimes \tilde{d}') \\ (\tilde{c} \otimes \tilde{a}') \oplus (\tilde{d} \otimes \tilde{c}') & (\tilde{c} \otimes \tilde{b}') \oplus (\tilde{d} \otimes \tilde{d}') \end{bmatrix}$$
$$= \begin{bmatrix} (-41, -20, -2; 4, 22, 43) & (-40, -20, -2; 4, 20, 38) \\ (-45, -22, -3; 3, 22, 45) & (-42, -20, -1; 5, 22, 42) \end{bmatrix} \approx \tilde{I}$$

Note that if we find the real matrix equivalent to $\tilde{\mathbf{M}}$ by ranking all its fuzzy numbers, we get: $\mathbf{N} = \begin{bmatrix} 4 & 7 \\ 2 & 6 \end{bmatrix}$, where we notice that the value of the determinant of $\mathbf{N}$ equals the ranking value of the determinant of the fuzzy matrix $\tilde{\mathbf{M}}$, i.e. $det(\mathbf{N}) = \mathcal{R}[det(\tilde{\mathbf{M}})] = 10$ and $\mathbf{N}^{-1} = \begin{bmatrix} 0.6 & -0.7 \\ -0.2 & 0.4 \end{bmatrix} \approx \tilde{\mathbf{M}}^{-1}$.

Remark 1. Let $\tilde{\mathbf{A}}$ be a square piecewise linear fuzzy matrix with nonzero determinant, and let $\tilde{\mathbf{A}}^{-1}$ be its inverse. The inverse is not unique since any matrix $\tilde{\mathbf{B}}$ that is equivalent to $\tilde{\mathbf{A}}^{-1}$ satisfies $\tilde{\mathbf{A}} * \tilde{\mathbf{B}} \approx \tilde{\mathbf{I}}$. This means that there is a class of equivalent matrices that represents the inverse of $\tilde{\mathbf{A}}$.

5 Conclusion

In conclusion, the algebraic structure of the set of all piecewise linear fuzzy numbers of order n with its arithmetic operations introduced by Tuffaha and Alrefaei is shown to be strong. This was shown in this paper by proposing a ranking equivalence relation on it and showing that the set of the resulting equivalence classes is a field which is isomorphic to the real numbers. This means that we have a kind of generalization of the real numbers, where every real number corresponds to a class of PLFN-n's that have a ranking value equal to this real number.

Moreover, some mathematical tools to deal with PLFN-n's are developed in this paper. These are, matrices of PLFN-n, equalities and inequalities with PLFN-n's. These concepts are only an extension to those on real numbers.

Further work in the future can be done by applying this practical type of fuzzy numbers to many fields. For instance, it can be used to give a more realistic representation for the fuzziness that can be found in fuzzy linear programming (FLP) problems. The proposed approach of binary operations on PLFN-n's can be used to generalize the simplex method for solving FLP problems. Since this binary operations preserve the ranking function, therefore a more preferable solution will be obtained (i.e., the optimal solution will be consistent with the solution of the ranking linear programming which results from converting each fuzzy number to its ranking value). Moreover, this type of fuzzy numbers is also expected to get a more realistic solutions when applied to solve fuzzy differential equations, neural networks or other fields.

Acknowledgments. This work is supported by Jordan University of Science and Technology under Research Grant Number: 20170376.

References

1. Alrefaei, M., Alawneh, A., Hassan, M.: Fuzzy linear programming for supply chain management in steel industry. Appl. Math. Sci. **8**(103), 5105–5114 (2014)
2. Cabral, V., Barros, L.: On differential equations with interactive fuzzy parameter via t-norms. Fuzzy Sets Syst. **358**, 97–107 (2019)
3. Das, S.K., Mandal, T., Edalatpanah, S.: A mathematical model for solving fully fuzzy linear programming problem with trapezoidal fuzzy numbers. Appl. Intell. **46**(3), 509–519 (2017)
4. Dubois, D., Prade, H.: Operations on fuzzy numbers. Int. J. Syst. Sci. **9**(6), 613–626 (1978)
5. Fan, Y., Huang, G., Yang, A.: Generalized fuzzy linear programming for decision making under uncertainty: feasibility of fuzzy solutions and solving approach. Inf. Sci. **241**, 12–27 (2013)
6. Gomes, L.T., de Barros, L.C., Bede, B.: Fuzzy Differential Equations in Various Approaches. Springer, Cham (2015)
7. Kaleva, O.: Fuzzy differential equations. Fuzzy Sets Syst. **24**(3), 301–317 (1987)
8. Kumar, A., Kaur, J.: Fuzzy optimal solution of fully fuzzy linear programming problems using ranking function. J. Intell. Fuzzy Syst. **26**(1), 337–344 (2014)
9. Li, X., Li, D.: The structure and realization of a polygonal fuzzy neural network. Int. J. Mach. Learn. Cybern. **7**(3), 375–389 (2016)
10. Nanda, S., Das, N.: Fuzzy Mathematical Concepts. Alpha Science International, Oxford (2010)
11. Ozkok, B.A., Albayrak, I., Kocken, H.G., Ahlatcioglu, M.: An approach for finding fuzzy optimal and approximate fuzzy optimal solution of fully fuzzy linear programming problems with mixed constraints. J. Intell. Fuzzy Syst. **31**(1), 623–632 (2016)
12. Steyaert, H., Van Parys, F., Baekeland, R., Kerre, E.E.: Implementation of piecewise linear fuzzy quantities. Int. J. Intell. Syst. **10**(12), 1049–1059 (1995)
13. Tuffaha, M.Z., Alrefaei, M.H.: Arithmetic operations on piecewise linear fuzzy number. In: AIP Conference Proceedings, vol. 1991, no. 1 (2018). 020024
14. Wang, G., Li, X.: Universal approximation of polygonal fuzzy neural networks in sense of k-integral norms. Sci. China Inf. Sci. **54**(11), 2307 (2011)
15. Xu, J., Liao, Z., Hu, Z.: A class of linear differential dynamical systems with fuzzy initial condition. Fuzzy Sets Syst. **158**(21), 2339–2358 (2007)
16. Yang, Y., Wang, G., Yang, Y.: Parameters optimization of polygonal fuzzy neural networks based on GA-BP hybrid algorithm. Int. J. Mach. Learn. Cybern. **5**(5), 815–822 (2014)
17. Zadeh, L.A.: Fuzzy sets. Inf. Control **8**(3), 338–353 (1965)
18. Zadeh, L.A.: The concept of a linguistic variable and its application to approximate reasoning. Inf. Sci. **8**(3), 199–249 (1975)

Some Extension of the Mier-Keeler Fixed Point Theorems in Fuzzy Metric Space

S. Melliani(✉), M. Elomari, I. Bakhadach, and L. S. Chadli

Sultan Moulay Slimane University, 523 Beni Mellal, BP, Morocco
{s.melliani,m.elomari}@usms.ma, idris.bakhadach@gmail.com,
sa.chadli@yahoo.fr

Abstract. Mier and Keeler formulated their fixed point theorem for contractive mapping with purely metric condition. This idea was extended by numerous mathematicians. In this paper we present a simple method of proving such theorems in the fuzzy metric space and give new results.

1 Introduction

Theory of fuzzy sets was intruduced by Zadeh [23]. Kramosil and Michalek [17] introduced fuzzy metric spaces based on the notion of continuous triangular norms that were the first time applied in [24] to modify the definition of probabilistic metric spaces introduced by K. Menger [12]. By a slight modification of the Kramosil-Michalek definition, George and Veeramani [4,5] introduced and studied fuzzy metric spaces and topological spaces induced by fuzzy metric [6,7]. Many of the most important nonlinear problems of applied mathematics reduce to finding solutions of nonlinear functional equations which can be formulated in terms of finding the fixed points of a given nonlinear operator of an infinite dimensional function space X into itself. The autor in [14] establish a simple and powerful lemma that provides a criterion for sequences in metric spaces to be Cauchy. Using the lemma, it is then easily verified that the Picard iterates $\{T^n x\}$, where T is a contraction or asymptotic contraction of Meir-Keeler type, are Cauchy sequences. As an important application in this paper, M. Abtahi presentes a new and simple proofs for several known results on the existence of a fixed point for continuous and asymptotically regular self-maps of complete metric spaces satisfying a contractive condition of Meir-Keeler type. We based on the last paper and we use the notions of fuzzy metric space of George and Veeramani [4]. he concept of fuzzy sets as a generalization of the "crisp" sets [23] and later there has been much progress in the study of fuzzy sets, as many authors have proved fixed point theorems for contractions in fuzzy metric spaces. In the present paper we will try to give the version of fixed point of Meir-Keeler type in the fuzzy context.

The present paper is organized as follows: After this introduction, we presented some basic concept in Sect. 2. Somme results as lemmas take place in Sect. 3, and we finished by several results as theorems in the last section.

H. Dutta et al. (Eds.): CMES 2019, AISC 1111, pp. 266–277, 2020.
https://doi.org/10.1007/978-3-030-39112-6_21

2 Preliminaries

Definition 2.1. [4] *A benary operation* $*[0,1] \times [0,1] \to [0,1]$ *is a continuous t-norm if it satisfies the following conditions:*

1. $*$ *is commutative and associative*
2. *is continuous*
3. $a * 1 = a$, $\forall a \in [0,1]$
4. $a * b \leq c * d$ *whenever* $a \leq c$ *and* $b \leq d$ *for all* $a, b, c, d \in [0,1]$.

Definition 2.2. [4] *The 3-tuple* $(X, M, *)$ *is said to be a fuzzy metric space if* X *is an arbitrary set,* $*$ *is a continuous tnorm and* M *is a fuzzy set on* $X^2 \times [0, \infty[$ *satisfying the following conditions for all* $x, y, z \in X$ *and* $s, t > 0$:

1. $M(x, y, 0) = 0$
2. $M(x, y, t) = 1$ *for all* $t > 0$ *if and only if* $x = y$
3. $M(x, y, t) = M(y, x, t)$
4. $M(x, y, t) * M(y, z, s) \leq M(x, z, t + s)$
5. $M(x, y, .)[0, \infty[\to [0, 1]$ *is left continuous.*

The function $M(x, y, t)$ *denotes the degree of nearness between* x *and* y *with respect to* t. *So, we identify* $M(x, y, t) = 1$ *with* $x = y$ *and* $M(x, y, t) = 0$ *with* ∞.

In theorems cited further, we apply notations that are better suited to the results of the next sections of our paper. The result of Meir and Keeler was devolopped in the fuzzy case by Sidite Duraj and Elida Hoxha.

Theorem 2.3. [3] *Let* $(X, M, *)$ *be a fuzzy metric space, and* f *a mapping of* X *into itself. If given* $\epsilon > 0$, *there exists* $\alpha > 0$ *such that*

$$\epsilon - \alpha < M(x, y, t) < \epsilon \qquad \textit{implies} \quad M(fx, fy, t) \geq \epsilon$$

Then f *has a unique fixed point* x.*and* $\lim_{n \longrightarrow \infty} f^n x_0 = x$ *for* $x_0 \in X$

Theorem 2.4. [10] *Let* f *be a contractive selfmapping on a complete metric space* (X, d) *that satisfies the following condition*
for any $\alpha > 0$, *there exists* $\epsilon > 0$ *such that*

$$\alpha < d(x, y) < \epsilon + \alpha \qquad \textit{implies} \quad d(fx, fy) \leq \alpha \qquad x, y \in X$$

Then f *has a unique fixed point* x.*and* $\lim_{n \longrightarrow \infty} f^n x_0 = x$ *for each* $x_0 \in X$

both theorems are more general than the next theorem of Boyd and Wong.

Theorem 2.5. [1] *Let* (X, d) *be a complete metric space. And let* $f : X \longrightarrow X$ *satisfy*

$$d(fx, fy, t) \leq \varphi(d(fx, fy)) \qquad x, y \in X$$

where $\varphi : [0, \infty[\longrightarrow [0, \infty[$ *is upper semicontinuous from the right and such that* $\varphi(t) < t$ *for all* $t \in (0, \infty)$. *Then* f *has a unique fixed point* x.*and* $\lim_{n \longrightarrow \infty} f^n x_0 = x$ *for each* $x_0 \in X$

Clearly, our theorem in the fuzzy case is more general than the Boyd-Wong one even for metric spaces. Jachymski [16] obtained the following more general result for metric spaces.

Theorem 2.6. *([16], Corollary) Let f be a selfmapping of a complete metric space (X, d) such that $d(fy, fx) < d(y, x)$ for $x \neq y$ and $d(fy, fx) \leq \varphi(d(y, x))$ for all $x, y \in X$, where $\varphi : [0, +\infty) \longrightarrow [0, +\infty)$ satisfies condition.*
for each $\alpha > 0$, there exists $\epsilon > 0$ such that

$$\varphi(x, y) \geq \alpha \qquad on \quad (\alpha, \alpha + \epsilon)$$

Then f has a unique fixed point x.and $\lim_{n \longrightarrow \infty} f^n x_0 = x$ for each $x_0 \in X$

It appears that the simple reasoning presented in [18] applies to conditions of the MeirKeeler type. Consequently, we easily obtain extensions of the well-known theorems to the case of fuzzy metric spaces or partial fuzzy metric spaces. In addition, new results for cyclic mappings are proved. Also, the next theorem of Proinov is strongly extended in one of the Section below.

Theorem 2.7. [21] *Let (X, d) be a complete metric space, and let f be a continuous selfmapping such that $\lim_{n \longrightarrow \infty} d(f^{n+1}x_0, f^n x_0) = x$, $x_0 \in X$, and for*

$$D(x, y) = d(x, y) + \gamma[d(fx, x) + d(fy, y)] \qquad \gamma > 0$$

the following conditions are satisfied: $d(fy, fx) < D(y, x)$, $x, y \in X$,
for each $\alpha > 0$, there exists $\epsilon > 0$ such that
$\alpha < D(y, x) < \alpha + \epsilon$ implies $d(fy, fx) \leq \alpha$, $x, y \in X$.
Then f has a unique fixed point x.and $\lim_{n \longrightarrow \infty} f^n x_0 = x$ for each $x_0 \in X$

3 Lemmas

Lemma 3.1. [20] *Let $(a_n)_{n \in \mathbb{N}}$ be a nonnegative sequence such that*

$$a_{n+1} > 0 \quad yields \quad a_{n+1} < a_n \qquad n \in \mathbb{N} \tag{1}$$

Then $\lim_{n \longrightarrow \infty} a_n = 0$ iff the following condition is satisfied

for each $\epsilon > 0$, there exists $\alpha > 0$ such that

$$\alpha < a_n < \alpha + \epsilon \qquad implies \quad a_{n+1} < a_n \quad n \in \mathbb{N} \tag{2}$$

Lemma 3.2. *Let $(X, M, *)$ be a fuzzy metric space and let f be a selfmapping satisfying*

$$M(x_{n+2}, x_{n+1}, t) > 0 \text{ implies } M(x_{n+2}, x_{n+1}, t) < M(x_{n+1}, x_n, t), n \in \mathbb{N}. \tag{3}$$

Then $\lim_{n \longrightarrow \infty} M(x_{n+1}, x_n, t) = 1$ iff the following condition holds
for each $\epsilon > 0$, there exists $\alpha > 0$ such that

$$\epsilon - \alpha < M(x_{n+1}, x_n, t) < \epsilon \qquad implies \quad M(x_{n+2}, x_{n+1}, t) \geq \epsilon \quad n \in \mathbb{N} \tag{4}$$

Proof 3.3. *We apply Lemma 3.1 to* $a_n = M(x_{n+1}, x_n, t)$ *in the fuzzy case*

Definition 3.4. *Let* $(X, M, *)$ *be a fuzzy metric space, and* $0 < \epsilon < 1$. *Then a selfmapping* f *is* $\epsilon -$ *contractive if the following condition is satisfied*

$$1 - \epsilon < M(x, y, t) < 1 \Longrightarrow M(fx, fy, t) > M(x, y, t) \tag{5}$$

If $f : X \longrightarrow X$ is a $\epsilon -$ *contractive* mapping, then (3) holds for each $x \in X$. Now, from Lemma 3.2 we obtain the following:

Corollary 3.5. *Let* $(X, M, *)$ *be a fuzzy metric space, and let* f *be a* $\epsilon -$ *contractive selfmapping on* X. *Then* $\lim_{n \longrightarrow \infty} M(x_{n+1}, x_n, t) = 1$ *iff (4) holds.*

Let us consider

$$c_f(y, x, t) = min\{M(y, x, t), M(fy, y, t), M(fx, x, t)\}$$

and $M(fx, fy, t) < 1$ implies $M(fy, fx, t) > c_f(y, x, t)$
Then we obtain
$M(x_{n+2}, x_{n+1}, t) < 1$ implies

$$M\big(x_{n+2}, x_{n+1}, t\big) > c_f\big(x_{n+1}, x_n, t\big) = min\big\{M(x_{n+1}, x_n, t), M(x_{n+2}, x_{n+1}, t)\big\} \tag{6}$$

$$= M\big(x_{n+1}, x_n, t\big), n \in \mathbb{N} \tag{7}$$

(otherwise, a contradiction). Consecontly (6) yield (3) and we have the following

Corollary 3.6. *Let* $(X, M, *)$ *be a fuzzy metric space. and let* f *be a selfmapping satisfying (6) or (5) or (3) Then* $\lim n \longrightarrow \infty M(x_{n+1}, x_n, t) = 1$ *iff (4) holds* $(\epsilon - \alpha < M(x_{n+1}, x_n, t) < \epsilon)$ *can also replaced by* $\epsilon - \alpha < c_f(x_{n+1}, x_n, t) < \epsilon$ *in (4)*

Let us recall the notion of partial fuzzy metric

Definition 3.7. [15] *A partial fuzzy metric on a nonempty set* X *is a function*

$$P_M : X \times X \times [0, \infty[\rightarrow [0, 1]$$

such that for all $x, y, z \in X$ *and* $t, s > 0$

1. $M(x, y, 0) = 0$
2. $M(x, y, t) = 1$ *for all* $t > 0$ *if and only if* $x = y$
3. $M(x, y, t) = M(y, x, t)$
4. $M(x, y, t) * M(y, z, s) \leq M(x, z, t + s)$
5. $M(x, y, .)[0, \infty[\rightarrow [0, 1]$ *is left continuous.*

where $M(x, y, t) = \dfrac{t}{t + d(x; y)}$. *and* $d(x, y)$ *is partial metric If* $(X, M, *)$ *is a fuzzy partial metric space, we will say that Mp is a fuzzy partial metric on* X.

All results of the present paper for fuzzy metric spaces remain valid also for partial fuzzy metric spaces
Let us consider

$$m_f(x,y,t) = min\Big\{M(y,x,t), M(fy,y,t), M(fx,x,t), M(x,fy,2t), M(y,fx,t), \frac{[M(x,fx,t)M(y,fy,t)]}{M(x,y,t)}\Big\}$$

and $M(fy,fx,t) < 1$ implies

$$M(fy,fx,t) > m_f(y,x,t) \quad x,y \in X \text{ and } t > 0 \tag{8}$$

we have $M(x_{n+1}, x_n + 1, 2t) = 1$
and

$$\begin{aligned} M(x_{n+2},x_n,t) &\leq M(x_{n+2},x_{n+1},t/2) * M(x_{n+1},x_n,t/2) \\ &\leq \min\{M(x_{n+2},x_{n+1},t/2), M(x_{n+1},x_n,t/2)\} \\ &= c_f(x_{n+1},x_n,t). \end{aligned}$$

and

$$\begin{aligned} \frac{[M(x_{n+1},x_{n+2},t)M(x_n,x_{n+1},t)]}{M(x_{n+1},x_n,t)} &= M(x_{n+2},x_{n+1},t/2) \\ &\leq M(x_{n+2},x_n,t/2) * M(x_{n+1},x_n,t/2) \\ &\leq \min\Big\{M(x_{n+2},x_{n+1},t/2), M(x_{n+1},x_n,t/2)\Big\} \\ &= c_f(x_{n+1},x_n,t). \end{aligned}$$

Consequently, $m_f(x_{n+1},x_n,t) = c_f(x_{n+1},x_n,t)$, and the reasoning for cf applies. Each partial fuzzy metric is a fuzzy metric, and the previous reasoning, together with Corollary 2, yields the following:

Corollary 3.8. *Let $(X,M,*)$ be a partial fuzzy metric space, and let f be a selfmapping on X satisfying (8) (or (6) or (5) or (3)).*
Then $\lim_{n\longrightarrow\infty} M(x_{n+1},x_n,t) = 1$ iff (3.4) holds ($\epsilon - \alpha < M(x_{n+1},x_n,t) < \epsilon$) can also replaced by
$\epsilon - \alpha < m_f(x_{n+1},x_n,t) < \epsilon$ or by $\epsilon - \alpha < m_f(x_{n+1},x_n,t) < \epsilon$ in (4)).

Lemma 3.9. *Let $(X,M,*)$ be a fuzzy metric space and let f be a selfmapping on X satisfying the following conditions:*

$$M(x_{n+k+1},x_{k+1},t) < 1 \qquad \textit{implies} \tag{9}$$

$M(x_{n+k+1},x_{k+1},t) > c_f(x_{n+k},x_k,t), k,n \in \mathbb{N}$,
for each $\epsilon > 0$, there exists $\alpha > 0$ such that

$$\epsilon - \alpha < c_f(x_{n+k},x_k,t) < \epsilon \qquad \textit{implies} \qquad M(x_{n+k+1},x_{k+1}) \geq \epsilon, k,n \in \mathbb{N}. \tag{10}$$

Then $\lim_{n,m\longrightarrow\infty} M(x_n, x_m, t) = 1$ *In addition,* c_f *can be replaced by* M *in (9) or (10) (so also in both of them). Similarly, if* M *is a partial fuzzy metric, then cf can be replaced by* m_f *in (9) or (10).*

Proof 3.10. *Let* $x_0 \in X$ *be an arbitrary point. Consider the sequence* $\{x_n\} = \{f^n x_0\}$. *We will prove that* $\{x_n\}$ *is a Cauchy sequence in* X. *Let* $M_n = M(x_n, x_{n+1}, t), t > 0$
If for some n, $M_n = M(x_n, x_{n+1}, t) = 1$ *Then we have the result immidiatly. Now suppose that* $M_n \neq 1$ *for all* n.
if $c_f(x, y, t) = 1$ *then for some* $x, y, \in X$, *we have that* $c_f(x, fx, t) = 1$.
Now if $c_f(x, y, t) < 1$ *for all* $x, y, \in X$, *than by (9) we have* $c_f(x, x, t) < \epsilon < M(fx, fy, t)$. *For any* $n \in \mathbb{N}$

$$\begin{aligned} M_n &= M(x_n, x_{n+1}, t) = M(fx_{n-1}, fx_n, t) > c_f(x_{n-1}, x_n, t) \\ &= \min\{M(x_{n-1}, x_n, t), M(x_{n-1}, fx_{n-1}, t), M(x_n, fx_n, t)\} \\ &= \min\{M(x_{n-1}, x_n, t), M(x_{n-1}, x_n, t), M(x_n, x_{n+1}, t)\} \\ &= \min\{M(x_{n-1}, x_n, t), M(x_n, x_{n+1}, t)\} \\ &= \min\{M_{n-1}, M_n\}. \end{aligned}$$

Thus the sequence $\{M_n\}$ *is strictly increasing. Since* $\{M_n\} \subset [0, 1]$ *then* $\{M_n\}$ *converges to some* $s \in [0, 1]$, *where* $\sup\{M_n\}$.
If $s < 1$ *than exists* $\delta > 0$ *and* $m \in N$ *such that for* $n < m$, $s - \delta < M_{n-1} = M(x_{n-1}, x_n, t) \leq s$. *By the condition (10) and* $c_f(x, x, t) < \epsilon < M(fx, fy, t)$, *we have*
$s - \delta < M_{n-1} = c_f(x_{n-1}, x_n, t) \leq s \Longrightarrow M(fx_{n-1}, fx_n, t) \geq s$
But $M(fx_{n-1}, fx_n, t) = M(x_n, x_{n+1}, t) = M_n \geq s$ *that is a contradiction.*
So $s = 1 = \lim_{n\longrightarrow\infty} M(x_n, x_{n+1}, t)$
For $p \in \mathbb{N}$, $M(x_n, x_{n+p}, t) > M(x_n, x_{n+1}, t/p) * M(x_{n+1}, x_{n+2}, t/p) * ... * M(x_{n+p-1}, x_{n+p}, t/p)$ *then* $\lim_{n\longrightarrow\infty} M(x_n, x_{n+p}, t) = 1$. *So the sequence* $\{x_n\}$ *is Cauchy sequence in* $(X, M, *)$ *hence the result.*

Lemma 3.11. *Let* $(X, M, *)$ *be a fuzzy metric space, and let* f *be a selfmapping on* X *satisfying the following conditions for a fixed* $t \in \mathbb{N}$

$$M(x_{nt+k+2}, x_{k+1}, s) < 1 \qquad \textit{implies} \tag{11}$$

$M(x_{nt+k+2}, , x_{k+1}, s) < c_f(x_{nt+k+1}, x_k, s)$, $k \in \mathbb{N}$, $n \in \mathbb{N} \cup \{0\}$
for each $\epsilon > 0$ *there exists* $\alpha > 0$ *such that*

$$\epsilon - \alpha < c_f(x_{nt+k+1}, x_k, s) < \epsilon \qquad \textit{implies} \qquad \epsilon < M(x_{nt+k+2}, x_{k+1}, s) \tag{12}$$

$$, k \in \mathbb{N}, n \in \mathbb{N} \cup \{0\} \tag{13}$$

Then $\lim_{m,n\longrightarrow\infty} M(x_n, x_m, s) = 1$. *In addition,* c_f *can be replaced by* M *in (11) or (12)(so also in both of them). Similarly, if* M *is a partial fuzzy metric, then* c_f *can be replaced by* m_f *in (11) or (12).*

Proof 3.12. *Clearly, if (11) and (12) hold, then, in particular, (15) and (4) are satisfied. Therefore, we have* $\lim_{m,n\longrightarrow\infty} M(x_{n+1}, x_n, s) = 1$ *(see Corollaries 3.5 or 3.6). Suppose that* $\lim_{k,n\longrightarrow\infty} M(x_{nt+k+2}, x_{k+1}, s) = 1$ *is false. Then, for an infinite subset* $\mathbb{K}$ *of* $\mathbb{N}$ *and each* $k \in \mathbb{K}$*, there exists* $n \in \mathbb{N}$ *such that* $M(x_{(n+1)t+k+2}, x_{k+1}, s) < \epsilon < 1$*. Let* $n = n(k)$ *be the smallest such number. For large* k*, we obtain (see (11)).*

$$\begin{aligned}
\epsilon > M(x_{(n+1)t+k+2}, x_{k+1}, s) &> c_f(x_{(n+1)t+k+1}, x_k, s) = M(x_{(n+1)t+k+1}, x_k, s) \\
&\geq M(x_{(n+1)t+k+1}, x_{(n+1)t+k}, s/(n+1)t+1) * ... * M(x_{nt+k+2}, x_{k+1}, s/(n+1)t+1) \\
&* M(x_{k+1}, x_k, s/(n+1)t+1) \\
&\geq M(x_{(n+1)t+k+1}, x_{(n+1)t+k}, s/(n+1)t+1) * ... * \epsilon * M(x_{k+1}, x_k, s/(n+1)t+1)
\end{aligned}$$

Therefore (see Corollary 3.5), we have
$\epsilon > M(x_{(n+1)t+k+2}, x_{k+1}, s) > c_f(x_{(n+1)t+k+1}, x_k, s) > \epsilon - \alpha$

for large k*. Now, condition (12) yields*

$$\epsilon > M(x_{(n+1)t+k+2}, x_{k+1}, s) \geq \epsilon$$

a contradiction. Therefore $\lim_{n,k\longrightarrow\infty} M(x_{(n+1)t+k+2}, x_{k+1}, s) = 0$*.*
For any $p \in \{3, ..., t\}$*, we have.*

$$\begin{aligned}
M(x_{nt+k+p}, x_{k+1}, s) \geq\ & M(x_{nt+k+p}, x_{nt+k+p-1}, s/nt+p-1) \\
& * ... * M(x_{nt+k+2}, x_{k+1}, s/nt+p-1)
\end{aligned}$$

and Lemma 3.2 yields $\lim_{m,n\longrightarrow\infty} M(x_{nt+k+p}, x_{k+1}, s) = 1$*, that is, the proof of our lemma is completed.*

It can be seen that Lemma 3.9 is a consequence of Lemma 3.11 for $t = 1$.
For a mapping $\beta : [0,1] \times [0,1] \longrightarrow [0,1]$ continuous at $(1,1)$ and such that $\beta(1,1) = 1$, let us consider

$$D_f(x, y, t) = M(x, y, t) + \beta(M(fx, x, t), M(fy, y, t)) \tag{14}$$

and

$$M(fx, fy, t) < 1 \Rightarrow M(fx, fy, t) > D_f(x, y, t) \qquad x, y \in X \tag{15}$$

Lemma 3.13. *Let* $(X, M, *)$ *be a fuzzy metric space, and let* f *be be such a selfmapping on* X *that* $\lim_{n\longrightarrow\infty} M(x_{n+1}, x_n, t) = 1$ *and the following conditions hold for a fixed* $t \in \mathbb{N}$

$$M(x_{nt+k+2}, x_{k+1}, s) < 1 \qquad \textit{implies} \tag{16}$$

$M(x_{nt+k+2},, x_{k+1}, s) < D_f(x_{nt+k+1}, x_k, s)$, $k \in \mathbb{N}$, $n \in \mathbb{N} \cup \{0\}$
for each $\epsilon > 0$ *there exists* $\alpha > 0$ *such that*

$$\epsilon-\alpha < D_f(x_{nt+k+1}, x_k, s) < \epsilon \quad \textit{implies} \quad \epsilon < M(x_{nt+k+2}, x_{k+1}, s), k \in \mathbb{N}, n \in \mathbb{N}\cup\{0\} \quad (17)$$

Then $\lim_{m,n\longrightarrow\infty} M(x_n, x_m, s) = 1$. *In addition,* M *can be replaced by* c_f *(or by* $m_f(x, y)$ *if* M *is a partial fuzzy metric) in (14) for (16) or (17).*

Proof 3.14. *Suppose that* $\lim_{k,n\longrightarrow\infty} M(x_{nt+k+2}, x_{k+1}, s) = 1$ *is false. Then, for an infinite subset* $\mathbb{K}$ *of* $\mathbb{N}$ *and each* $k \in \mathbb{K}$*, there exists* $n \in \mathbb{N}$ *such that* $M(x_{(n+1)t+k+2}, x_{k+1}, s) < \epsilon < 1$. *Let* $n = n(k)$ *be the smallest such number. For large* k*, we obtain (see (16)).*

$$\begin{aligned}
\epsilon &> M(x_{(n+1)t+k+2}, x_{k+1}, s) > D_f(x_{(n+1)t+k+1}, x_k, s) \\
&= M(x_{(n+1)t+k+1}, x_k, s) * \beta(M(x_{(n+1)t+k+2}, x_{(n+1)t+k+1}, s), M(x_{k+1}, x_k, s)) \\
&\geq M(x_{(n+1)t+k+1}, x_{(n+1)t+k}, s/(n+1)t+1) * ... * M(x_{nt+k+2}, x_{k+1}, s/(n+1)t+1) \\
&* \; M(x_{k+1}, x_k, s/(n+1)t+1) * \beta(M(x_{(n+1)t+k+2}, x_{(n+1)t+k+1}, s), M(x_{k+1}, x_k, s)) \\
&\geq M(x_{(n+1)t+k+1}, x_{(n+1)t+k}, s/(n+1)t+1) * ... * \epsilon * M(x_{k+1}, x_k, s/(n+1)t+1) * \beta(..,..)
\end{aligned}$$

for large k. *Therefore (*β *is continuous at* $(1, 1)$ *and such that* $\beta(1, 1) = 1$*), we have*

$$\epsilon > M(x_{(n+1)t+k+2}, x_{k+1}, s) > D_f(x_{(n+1)t+k+1}, x_k, s) > \epsilon - \alpha$$

for large k. *Our condition (17) yields*

$$\epsilon > M(x_{(n+1)t+k+2}, x_{k+1}, s) \geq \epsilon$$

a contradiction. Therefore, $\lim_{k,n\longrightarrow\infty} M(x_{nt+k+2}, x_{k+1}, s) = 1$. *Now, we follow the final part of the proof of Lemma 3.11.*

Definition 3.15. *A selfmapping* f *on a fuzzy metric space* $(X, M, *)$ *is* $1-$*continuous at* x *if* $\lim_{n\longrightarrow\infty} M(x, x_n, t) = 1$ *implies* $\lim_{n\longrightarrow\infty} M(fx, fx_n, t) = 1$ *for each sequence* $(x_n)_{n\in\mathbb{N}}$ *in* X*;* f *is* $1-$*continuous if it is* $1-$*continuous at each point* $x \in X$.

Lemma 3.16. *Let* $(X, M, *)$ *be fuzzy metric space and let* f *be a selfmapping on* X. *If* f *is contractive, then* f *has at most one fixed point; the same holds if* f *satisfies (6) or (8) and* M *satisfies* $M(x, x, t) \geq M(x, y, t)$, $x, y \in X$ *or if* M *is a fuzzy metric and (15) holds. If* f *is 1-continuous at* x *(e.g., if* f *is contractive) and* $\lim_{n\longrightarrow\infty} M(x, f^n x_0, t) = 1$ *then* $x = fx$ *and* $M(x, x, t) = 1$

Proof 3.17. *If x, y are fixed points of f and $M(x,x,t) \geq M(x,y,t), \quad x,y \in X$ holds, then we obtain*

$$m_f(x,y,t) = \min\Big\{M(y,x,t), M(fy,y,t), M(fx,x,t), M(x,fy,2t), M(y,fx,t), \frac{[M(x,fx,t)M(y,fy,t)]}{M(x,y,t)}\Big\}$$
$$= c_f(x,y,t) = M(x,y,t)$$

In addition, if $x \neq y$, then each of conditions (5), (6), and (8) yields

$$1 > M(x,y,t) = M(fx,fy,t) > M(x,y,t)$$

a contradiction. If M is a fuzzy metric and (15) holds, then we have

$$1 > M(x,y,t) = M(fx,fy,t) > D_f(x,y,t) = M(x,y,t) * \beta(1,1) = M(x,y,t)$$

also a contradiction. Let us consider $x \in X$ with $\lim_{n \longrightarrow \infty} M(x,x_n,t) = 1$. Then we have

$$M(fx,x,t) \geq M(fx,x_{n+1},t/2) * M(x_{n+1},x,t/2)$$

and in view of 1-continuity of f at x we also obtain $\lim_{n \longrightarrow \infty} M(fx,x_{n+1},t) = 1$, that is $\lim_{n \longrightarrow \infty} M(fx,x,t) = 1$.

4 Theorems

Let us recall ([18], Definition 2.3) that a d-metric space (X,p) is 0-complete if for each sequence $(x_n)_{n \in \mathbb{N}}$ in X with $\lim_{n,m \longrightarrow \infty} p(x_n,x_m) = 0$, there exists $x \in X$ such that $\lim_{n \longrightarrow \infty} p(x,x_n) = 0$. Now, we are ready to extend the Ciric theorem [2] and the Matkowski Theorem 1.5.1 in [10] to the case of fuzzy metric spaces (and c_f in place of p).

Theorem 4.1. *Let f be a $1-$continuous selfmapping on a $1-$complete fuzzy metric space $(X,M,*)$. Assume that (6) or (5) holds and the following condition is satisfied:*
for each $\epsilon > 0$, there exists $\alpha > 0$ such that

$$\epsilon - \alpha < c_f(x,y,t) < \epsilon \quad \text{implies} \quad M(fx,fy,t) \geq \epsilon \quad x,y \in X \tag{1}$$

Then f has a unique fixed point, say x, and $\lim_{n \longrightarrow \infty} M(x, f^n x_0, t) = M(x,x,t) = 1$, $x_0 \in X$

Proof 4.2. *Our space is $1-$complete, and, therefore, the sequence $(f^n x_0)_{n \in \mathbb{N}}$ converges (Lemma 3.9) to a unique fixed point of f (Lemma 3.16).*

Lemma 29 from [19] and the previous theorem yield the following result.

Theorem 4.3. *Let h be a selfmapping on a $1-$complete fuzzy metric space $(X, M, *)$ such that $f = h^s$ (for some $s \in \mathbb{N}$) satisfies the assumptions of Theorem 1. Then h has a unique fixed point, say x, and $\lim_{n \longrightarrow \infty} M(x, h^n x_0, t) = M(x, x, t) = 1$, $x_0 \in X$.*

Lemma 3.11 enables us to extend the previous theorems to the case of fuzzy cyclic mappings.
The idea was introduced by Kirk et al. [11], and we apply Definition 2.5 from [18]. fixed $t \in \mathbb{N}$, we put $t++ = 1$ and $j++ = j+1$ for $j \in \{1, ..., t-1\}$. Then $f : X \longrightarrow X$ is cyclic if $X = X_1 \cup ... \cup X_t$ and $f(X_j) \subset X_j++$, $j = 1, ..., t$.

Theorem 4.4. *Let f be a $1-$continuous fuzzy cyclic selfmapping on a $1-$complete fuzzy metric space $(X, M, *)$, and let the following conditions be satisfied:*

$$M(fx, fy, t) < 1 \qquad \textit{implies} \tag{2}$$

$M(fx, fy, t) > c_f(x, y, t), \quad x \in X_j,\ y \in X_{j++},\ j = 1, ..., t$
for each $\epsilon > 0$, there exists $\alpha > 0$ such that

$$\epsilon - \alpha < c_f(x, y, t) < \epsilon \quad \textit{implies} \quad M(fx, fy, t) \geq \epsilon, \tag{3}$$

$$x \in X_j,\ y \in X_{j++},\ j = 1, ..., t. \tag{4}$$

Then f has a unique fixed point, say x, and $\lim_{n \longrightarrow \infty} M(x, f^n x_0, t) = M(x, x, t) = 1$, $x_0 \in X$

Proof 4.5. *Our space is $1-$complete, and, therfore, the sequence $(f^n x_0)_{n \in \mathbb{N}}$ converges (Lemma 3.11) to a unique fixed point of f (Lemma 3.16).*

An analogue of theorem 2 for fuzzy cyclic mappings is the following consequence of Theorem 3.3 and Lemma 29 of [19].

Theorem 4.6. *Let f be a $1-$continuous fuzzy cyclic selfmapping on a $1-$complete fuzzy metric space $(X, M, *)$, such that $f = h^s$ (for some $s \in \mathbb{N}$) satisfies the assumptions of Theorem 3. Then h has a unique fixed point, say x, and*

$$\lim_{n \longrightarrow \infty} M(x, h^n x_0, t) = M(x, x, t) = 1, \ x_0 \in X.$$

Let us note that a fuzzy partial metric space $(X, M, *)$ is $1-$complete iff $(X, M, *)$ treated as a fuzzy metric space is $1-$complete (see [19], Corollary 4, Proposition 5).

Remark 4.7. *In view of Lemma 3.9, c_f can be replaced by M in any of conditions of Theorems 1, 2, 3, and 4; if M is a fuzzy partial metric, then in view of Lemma 3.11, c_f can be replaced by m_f in any condition of those theorems. Theorem 1 for m_f becomes an extension of a theorem of Jachymski ([16], Theorem 2) in fuzzy case to the case of fuzzy partial metric spaces.*

Theorem 4.8. *Let $(X, M, *)$ be a $1-$complete fuzzy metric space, and let f be a $1-$continuous fuzzy cyclic selfmapping on X such that $\lim_{n\longrightarrow\infty} M(f^{n+1}x_0, f^n x_0, t) = 1$, $x_0 \in X$, assume that the following conditions hold:*

$$M(fx, fy, r) < 1 \quad \textit{implies} \tag{5}$$

$$M(fx, fy, r) > D_f(x, y, r), \qquad x \in X_j,\ y \in X_{j++},\ j = 1, ..., t$$

for each $\epsilon > 0$, there exists $\alpha > 0$ such that

$$\epsilon - \alpha < D_f(x, y, r) < \epsilon \quad \textit{implies} \quad M(fx, fy, r) \geq \epsilon, \tag{6}$$

$$x \in X_j,\ y \in X_{j++},\ j = 1, ..., t. \tag{7}$$

Then f has a fixed point, say x, such that $\lim_{n\longrightarrow\infty} M(x, f^n x_0, r) = 1$, $x_0 \in X$ and x is unique if M is a fuzzy metric. In addition, $M(x, y, r)$ can be replaced by $c_f(x, y, t)$ (or by $m_f(x, y, r)$ if M is a partial fuzzy metric) in (14) for (5) or (6) (so also for both of them).

Proof 4.9. *We apply Lemmas 3.13 and 3.16.*

theorem 5 with $t = 1$ yield the followin one.

Theorem 4.10. *Let $(X, M, *)$ be a $1-$complete fuzzy metric space, and let f be a $1-$continuous fuzzy cyclic selfmapping on X such that $\lim_{n\longrightarrow\infty} M(f^{n+1}x_0, f^n x_0, r) = 1$, $x_0 \in X$, assume that the following conditions hold:*

$$M(fx, fy, r) < 1 \quad \textit{implies} \quad M(fx, fy, r) > D_f(x, y, r) \qquad x, y \in X \tag{8}$$

for each $\epsilon > 0$, there exists $\alpha > 0$ such that

$$\epsilon - \alpha < D_f(x, y, r) < \epsilon \quad \textit{implies} \quad M(fx, fy, r) \geq \epsilon, \qquad x, y \in X. \tag{9}$$

Then f has a fixed point, say x, such that $\lim_{n\longrightarrow\infty} M(x, f^n x_0, r) = 1$, $x_0 \in X$ and x is unique if M is a fuzzy metric. In addition, $M(x, y, r)$ can be replaced by $c_f(x, y, t)$ (or by $m_f(x, y, r)$ if M is a partial fuzzy metric) in (14) for (8) or (9) (so also for both of them).

We can easily present extensions of the previous theorems for $f = h^s$ (see Theorems 2 and 4).

Theorem 6 is a further extension of Theorem 4.2 in [21].

References

1. Boyd, D.W., Wong, J.S.W.: On nonlinear contractions. Proc. Am. Math. Soc. **20**, 458–464 (1969)

2. Ciric, L.: A new fixed-point theorem for contractive mappings. Publ. Inst. Math. (Belgr.) **30**(44), 25–27 (1981)
3. Duraj, S., Hoxha, E.: A Meir-Keeler type fixed point theorem in fuzzy metric spaces. J. Adv. Math. **12**(11)
4. George, A., Veeramani, P.: On some results in fuzzy metric spaces. Fuzzy Sets Syst. **64**, 395–399 (1994)
5. George, A., Veeramani, P.: On some results of analysis for fuzzy metric spaces. Fuzzy Sets Syst. **90**, 365–368 (1970)
6. Gregori, V., Romaguera, S.: Some properties of fuzzy metric spaces. Fuzzy Sets Syst. **115**, 485–489 (2000)
7. Gregori, V., Romaguera, S., Veeramani, P.: A note on intuitionistic fuzzy metric spaces. Chaos, Solitons Fractals **28**, 902–905 (2006)
8. Gregori, V., Sapena, A.: On fixed-point theorems in fuzzy metric spaces. Fuzzy Sets Syst. **125**, 245–252 (2002)
9. Heilpern, S.: Fuzzy mappings and fixed point theorems. J. Math. Anal. Appl. **83**(9), 566–569 (1981)
10. Kuczma, M., Choczewski, B., Ger, R.: Iterative Functional Equations. Encyclopedia of Mathematics and Its Applications, vol. 32. Cambridge University Press, Cambridge (1990)
11. Kirk, W.A., Srinivasan, P.S., Veeramani, P.: Fixed points for mappings satisfying cyclical contractive conditions. Fixed Point Theory **4**, 79–89 (2003)
12. Menger, K.: Statistical metrics. Proc. Nat. Acad. sci. (USA) **28**, 535–537 (1942)
13. Mihet, D.: Fuzzy $\psi-$contractive mappings in non-archimedean fuzzy metric spaces. Fuzzy Sets Syst. **159**, 739–744 (2008)
14. Meir, A., Keeler, E.: A theorem on contraction mappings. J. Math. Anal. Appl. **28**, 326–329.1,3 (1969)
15. Amer, F.J.: Fuzzy partial metric spaces. Computational Analysis, vol. 155, pp. 153–161. Springer (2016)
16. Jachymski, J.: Equivalent conditions and the Meir-Keeler type theorems. J. Math. Anal. Appl. **194**, 293–303 (1995)
17. Kramosil, J., Michalek, J.: Fuzzy metric and statistical metric spaces. Kybernetika **11**, 326–334 (1975)
18. Pasicki, L.: The Boyd-Wong idea extended. Fixed Point Theory Appl. **2016**, 63 (2016)
19. Pasicki, L.: Fixed point theorems for contracting mappings in partial metric spaces. Fixed Point Theory Appl. **2014**, 185 (2014)
20. Pasicki, L.: Some extension of the Mier-Keeler thoerem. Fixed Point Theory Appl., 1 (2017)
21. Proinov, P.D.: Fixed point theorems in metric spaces. Nonlinear Anal. **64**, 546–557 (2006)
22. Razani, A.: A contraction theorem in fuzzy metric spaces. Fixed Point Theory Appl. **3**, 257–265 (2005)
23. Zadeh, L.A.: Fuzzy sets. Inform. Control **8**, 338–353 (1965)
24. Schweizer, B., Sklar, A.: Statistical metric spaces. Pacific J. Math. **10**, 313–334 (1960)

Unified Fractional Integral Formulae Involving Generalized Multiindex Bessel Function

Mehar Chand[1] and Zakia Hammouch[2(✉)]

[1] Department of Mathematics, Baba Farid College, Bathinda 151001, India
mehar.jallandhra@gmail.com
[2] Department of Mathematics, Faculty of Sciences and Techniques, Moulay Ismail University of Meknes, 52000 Errachidia, Morocco
hammouch.zakia@gmail.com

Abstract. The aim of the present paper is to establish generalized fractional integral formulae involving generalized multiindex Bessel function $J_{(\nu_j)_m,q}^{(\lambda_j)_m,\gamma}(z)$. Then their image formulae (Beta transform, Laplace transform and Whittaker transform) are also established. The results obtained here are quite general in nature and capable of yielding a very large number of known and (presumably) new results.

1 Introduction and Preliminaries

The Fractional Order Calculus (FOC) constitutes the branch of mathematics dealing with differentiation and integration under an arbitrary order of the operation, i.e. the order can be any real or even complex number, not only the integer one [1–3]. Although the FOC represents more than 300-year-old issue [4,5], its great consequences in contemporary theoretical research and real world applications have been widely discussed relatively recently (see [6–13,39,40]). The idea of non-integer derivative was mentioned for the first time probably in a letter from Leibniz to L'Hospital in 1695. Later on, the pioneering works related to FOC have been elaborated by personalities such as Euler, Fourier, Abel, Liouville or Riemann. The interested reader can find the more detailed historical background of FOC in [1].

According to [4, 14], the reason why FOC remained practically unexplored for engineering applications and why only pure mathematics were privileged to deal with it for so long can be seen in multiple definitions of FOC, such as missing simple geometrical interpretation, absence of solution methods for fractional order differential equations and seeming adequateness of the Integer Order Calculus (IOC) for majority of problems. Nowadays, the situation is going better and the FOC provides efficient tool for many issues related to fractal dimension,"infinite memory", chaotic behaviour, etc. Thus, the FOC has already come in useful in engineering areas such as bioengineering, viscoelasticity, electronics, robotics, control theory and signal processing [14]. Several control applications are available e.g. in [15–17].

A great number of additional results of fractional calculus was presented in the twentieth century, but at this point we only concentrate on one more, given by Caputo

H. Dutta et al. (Eds.): CMES 2019, AISC 1111, pp. 278–290, 2020.
https://doi.org/10.1007/978-3-030-39112-6_22

and first used extensively in [18]. Given a function f with an $n-1$ absolute continuous derivative, Caputo defined a fractional derivative by

$$D_*^{\alpha} f(x) = \frac{1}{\Gamma(n-\alpha)} \int_0^t (t-s)^{n-\alpha-1} \left(\frac{d}{ds}\right)^n f(s)ds \tag{1.1}$$

today usually named *Caputo fractional derivative*. The derivative 1.1 is strongly connected to the Riemann-Liouville fractional derivative (see [19]) and is today frequently used in applications. This is because using the Caputo derivative one can specify the initial conditions of fractional differential equations in classical form, i.e.

$$y^{(k)}(0) = b_k, \qquad k = 0, 1, ..., n-1, \tag{1.2}$$

in contrast to differential equations containing the Riemann-Liouville differential operator (see [19]). While the operator D_*^{α} is denoted today as Caputo operator, Rabotnov had already introduced this differential operator into the Russian viscoelastic literature in [20], a year before Caputo's paper was published.

By the second half of the twentieth century the field of fractional calculus had grown to such extent, that in 1974 the first conference concerned solely with the theory and applications of fractional calculus was held in New Haven [21]. In the same year the first book on fractional calculus by Oldham and Spanier [1] was published. A number of additional books have appeared since then, the most popular the ones by Miller and Ross [2], Samko et al. [22] and Podlubny [3]. In 1998 the first issue of the mathematical journal "Fractional calculus & applied analysis" was printed. This journal is solely concerned with topics on the theory of fractional calculus and its applications. Finally in 2004 the large conference "Fractional differentiation and its applications" was held in Bordeaux, where no less than 104 talks were given in the field of fractional calculus.

From its birth - a simple question from L'Hospital to Leibniz – to its today's wide use in numerous scientific fields fractional calculus has come a long way. Even though its nearly as old as classical calculus itself, it flourished mainly over the last decades because of its good applicability on models describing complex real life problems (see recent work [23–27]). And even though the term fractional calculus is a misnomer we will use it throughout this text, which will be concerned with theoretical and, more importantly, numerical aspects of problems arising in this field.

For our present study we start by recalling the previous work. The Bessel-Maitland function $J_\nu^\lambda(z)$ is given as (see Marichev [28]):

$$J_\nu^\lambda(z) = \sum_{n=0}^{\infty} \frac{(-z)^n}{\Gamma(\nu+\lambda n+1)n!}, \qquad \lambda > 0; z \in \mathbb{C} \tag{1.3}$$

The generalized form of Bessel function $J_{\nu,\mu}^\lambda$ is given by Jain and Agarwal [29] as:

$$J_{\nu,\mu}^\lambda(z) = \sum_{n=0}^{\infty} \frac{(-1)^r \left(\frac{z}{2}\right)^{\nu+2\mu+2n}}{\Gamma(\nu+\mu+\lambda n+1)\Gamma(\mu+n+1)}, \tag{1.4}$$
$$\lambda > 0, \nu, \mu \in \mathbb{C}; z \in \mathbb{C} \setminus (-\infty, 0].$$

Further, Pathak [30] gave the following more generalized form of the generalized Bessel-Maitland function $J^{\lambda,\gamma}_{\nu,\mu}(.)$ as:

$$J^{\lambda,\gamma}_{\nu,q}(z) = \sum_{n=0}^{\infty} \frac{(\gamma)_{qn}(-z)^n}{\Gamma(\nu+\lambda n+1)n!}, \tag{1.5}$$
$$\lambda,\nu,\gamma \in \mathbb{C}, \Re(\lambda) \geq 0, \Re(\nu) \geq -1, \Re(\gamma) \geq 0 \text{ and } q \in (0,1) \cup \mathbb{N}.$$

If ν is replaced by $\nu-1$ and z by $-z$ then generalized Bessel-Maitland function given in Eq. (1.5) reduces to well known Mittag-Leffler function as follows:

$$J^{\lambda,\gamma}_{\nu-1,q}(-z) = E^{\gamma,q}_{\lambda,\nu}(z), \tag{1.6}$$
$$\lambda,\nu,\gamma \in \mathbb{C}, \Re(\lambda) > 0, \Re(\nu) > 0, \Re(\gamma) > 0; q \in (0,1) \cup \mathbb{N},$$

where $E^{\gamma,q}_{\lambda,\nu}(z)$ denotes generalized Mittag-Lefller function, was introduced by Shukla and Prajapati [31].

If $q=1, \gamma=1, \nu$ is replaced by $\nu-1$ and z by $-z$ then generalized Bessel-Maitland function given in Eq. (1.5) reduces to Mittag-Leffler function, studied by Wiman [32] as follows:

$$J^{\lambda,1}_{\nu-1,1}(-z) = E_{\lambda,\nu}(z), \qquad \lambda,\nu \in \mathbb{C}, \Re(\lambda) > 0, \Re(\nu) > 0. \tag{1.7}$$

The generalized multiindex Bessel function $J^{(\lambda_j)_m,\gamma}_{(\nu_j)_m,q}(z)$ studied by [33] is defined as follows:

$$J^{(\lambda_j)_m,\gamma}_{(\nu_j)_m,q}(z) = \sum_{n=0}^{\infty} \frac{(\gamma)_{qn}}{\prod_{j=1}^{m}\Gamma(\lambda_j n+\nu_j+1)} \frac{(-z)^n}{n!}, \tag{1.8}$$

where $m \in \mathbb{N}, \lambda_j, \nu_j, \gamma, q, z \in \mathbb{C}(j=1,...,m)$ such that $\sum_{j=1}^{m} \Re(\lambda_j) > \max\{0; \Re(q)-1\}; q > 0, \Re(\nu_j) > -1, \Re(\gamma) > 0$ and $q \in (0,1) \cup \mathbb{N}$.

On setting $m=1, q=0, \lambda_1=1, \nu_1=\nu$ and replace z by $\frac{z^2}{4}$ in (1.8), we have

$$J^{1,\gamma}_{\nu,0}\left[\frac{z^2}{4}\right] = \left(\frac{2}{z}\right)^{\nu} J_\nu[z], \tag{1.9}$$

where $J_\nu[z]$ is a well known Bessel function of first kind defined by (see [34])

$$J_\nu[z] = \sum_{n=0}^{\infty} \frac{(-1)^n \left(\frac{z}{2}\right)^{2n+\nu}}{\Gamma(n+\nu+1)}, \qquad \nu \in \mathbb{C}; z \in \mathbb{C} \setminus (-\infty,0]. \tag{1.10}$$

For more details about the Bessel function one may refer to earlier work by Erdélyi et al. [35] and Watson [36].

The Fox-Wright function ${}_p\Psi_q$ defined as

$$\begin{aligned} {}_p\Psi_q[z] &= {}_p\Psi_q\left[\begin{matrix}(a_1,\alpha_1),...,(a_p,\alpha_p);\\(b_1,\beta_1),...,(b_q,\beta_q);\end{matrix} z\right] \\ &= {}_p\Psi_q\left[\begin{matrix}(a_i,\alpha_i)_{1,p};\\(b_j,\beta_j)_{1,q};\end{matrix} z\right] = \sum_{n=0}^{\infty}\frac{\prod_{i=1}^{p}\Gamma(a_i+\alpha_i n)}{\prod_{j=1}^{q}\Gamma(b_j+\beta_j n)}\frac{z^n}{n!}, \end{aligned} \tag{1.11}$$

where the coefficients $\alpha_1,...,\alpha_p$, $\beta_1,...,\beta_q \in \mathbb{R}^+$ such that

$$1+\sum_{j=1}^{q}\beta_j-\sum_{i=1}^{p}\alpha_i \geq 0. \tag{1.12}$$

2 Fractional Integration

In this section, some fractional integral formulas involving generalized multiindex Bessel function $J_{(\nu_j)_{m},q}^{(\lambda_j)_{m},\gamma}(z)$ are established. To do this, we need to recall the following pair of generalized fractional integral operators introduced by Katugampola [37], which are presented in Lemma 1.

Lemma 1. *Let $\Omega = [a,b](-\infty < a < b < \infty)$ be a finite interval on the real axis $\mathbb{R}$. The generalized fractional integral ${}^{\rho}I_{a+}^{\sigma}f$ of order $\sigma \in \mathbb{C}$ for $x > a$ and $\Re(\sigma) > 0$ is defined as:*

$$\left({}^{\rho}I_{a+}^{\sigma}f\right)(x) = \frac{(\rho)^{1-\sigma}}{\Gamma(\sigma)}\int_a^x \frac{t^{\rho}f(t)}{(x^{\rho}-t^{\rho})^{1-\sigma}}dt, \tag{1.13}$$

similarly the generalized fractional integral ${}^{\rho}I_{b-}^{\sigma}f$ of order $\sigma \in \mathbb{C}$ for $x < b$ and $\Re(\sigma) > 0$ is defined as:

$$\left({}^{\rho}I_{b-}^{\sigma}f\right)(x) = \frac{(\rho)^{1-\sigma}}{\Gamma(\sigma)}\int_x^b \frac{t^{\rho}f(t)}{(t^{\rho}-x^{\rho})^{1-\sigma}}dt. \tag{1.14}$$

In our investigation, we choose $a = b = 0$ the above Lemma 1 reduces to the following form:

Lemma 2. *The generalized fractional integral ${}^{\rho}I_{0+}^{\sigma}f$ of order $\sigma \in \mathbb{C}$ for $x > 0$ and $\Re(\sigma) > 0$ is defined as*

$$\left({}^{\rho}I_{0+}^{\sigma}f\right)(x) = \frac{(\rho)^{1-\sigma}}{\Gamma(\sigma)}\int_0^x \frac{t^{\rho}f(t)}{(x^{\rho}-t^{\rho})^{1-\sigma}}dt, \tag{1.15}$$

similarly the generalized fractional integral ${}^{\rho}I_{0-}^{\sigma}f$ *of order* $\sigma \in \mathbb{C}$ *for* $x < 0$ *and* $\Re(\sigma) > 0$ *is defined as*

$$\left({}^{\rho}I_{0-}^{\sigma}f\right)(x) = \frac{(\rho)^{1-\sigma}}{\Gamma(\sigma)} \int_x^0 \frac{t^{\rho} f(t)}{(t^{\rho} - x^{\rho})^{1-\sigma}} dt. \tag{1.16}$$

The main results are given in the following theorems.

Theorem 1. *Let* $x > 0, m \in \mathbb{N}; \sigma, \rho, \lambda_j, \nu_j, \gamma, q, z \in \mathbb{C}(j = 1,...,m)$ *such that* $\sum_{j=1}^{m} \Re(\lambda_j) > \max\{0; \Re(q) - 1\}; q > 0, \Re(\nu_j) > -1, \Re(\gamma) > 0$ *and* $q \in (0,1) \cup \mathbb{N}$, *then*

$$\left({}^{\rho}I_{0+}^{\sigma} t^{\lambda} J_{(\nu_j)_m,q}^{(\lambda_j)_m,\gamma}(t^{\nu})\right)(x) = \frac{x^{\lambda+\rho\sigma+1}}{(\rho)^{\sigma}\Gamma(\gamma)} {}_2\Psi_{m+1}\left[\begin{array}{c} (\gamma,q), \left(\frac{\lambda+1}{\rho}+1, \frac{\nu}{\rho}\right) \\ (\nu_j+1,\lambda_j)_{j=1}^{m}, \left(\frac{\lambda+1}{\rho}+\sigma+1, \frac{\nu}{\rho}\right) \end{array}\middle| -x^{\nu}\right] \tag{1.17}$$

Proof. For convenience, we denote the left-hand side of the result (1.17) by $\mathscr{I}$. Using (1.8), and changing the order of integration and summation, then

$$\mathscr{I} = \sum_{n=0}^{\infty} \frac{(\gamma)_{qn}}{\prod_{j=1}^{m} \Gamma(\lambda_j n + \nu_j + 1)} \frac{(-1)^n}{n!} \left({}^{\rho}I_{0+}^{\sigma} t^{n\nu+\lambda}\right), \tag{1.18}$$

applying the fractional derivative formula given in Eq. (1.15), the above Eq. (1.18) reduces to

$$\mathscr{I} = \sum_{n=0}^{\infty} \frac{(\gamma)_{qn}}{\prod_{j=1}^{m} \Gamma(\lambda_j n + \nu_j + 1)} \frac{(-1)^n}{n!} \frac{(\rho)^{1-\sigma}}{\Gamma(\sigma)} \int_0^x \frac{t^{\rho+n\nu+\lambda}}{(x^{\rho} - t^{\rho})^{1-\sigma}} dt. \tag{1.19}$$

Put $t^{\rho} = x^{\rho} z$ in Eq. (1.19) and by proper substitution, we have

$$\mathscr{I} = \sum_{n=0}^{\infty} \frac{(\gamma)_{qn}}{\prod_{j=1}^{m} \Gamma(\lambda_j n + \nu_j + 1)} \frac{(-1)^n}{n!} \frac{(\rho)^{-\sigma}}{\Gamma(\sigma)} x^{\rho+n\nu+\rho\sigma+\lambda+1} \int_0^1 z^{\frac{n\nu+\lambda+1}{\rho}} (1-z)^{\sigma-1} dz, \tag{1.20}$$

after simplification, the above Eq. (1.20) reduces to

$$\mathscr{I} = \frac{x^{\lambda+\rho\sigma+1}}{(\rho)^{\sigma}\Gamma(\gamma)} \sum_{n=0}^{\infty} \frac{\Gamma(\gamma+qn)\Gamma\left(\frac{\lambda+1}{\rho}+1+\frac{\nu}{\rho}n\right)}{\prod_{j=1}^{m} \Gamma(\lambda_j n + \nu_j + 1)\Gamma\left(\frac{\lambda+1}{\rho}+1+\sigma+\frac{\nu}{\rho}n\right)} \frac{(-x^{\nu})^n}{n!}, \tag{1.21}$$

interpreting the above result in the view of (1.11), we have the required result. □

Theorem 2. *Let* $x < 0, m \in \mathbb{N}; \sigma, \rho, \lambda_j, v_j, \gamma, q, z \in \mathbb{C}(j = 1,...,m)$ *such that* $\sum_{j=1}^{m} \Re(\lambda_j) > \max\{0; \Re(q)-1\}; q > 0, \Re(v_j) > -1, \Re(\gamma) > 0$ *and* $q \in (0,1) \cup \mathbb{N}$, *then*

$$\left({}^{\rho}I_{0-}^{\sigma} t^{\lambda} J_{(v_j)_{m,q}}^{(\lambda_j)_{m},\gamma}(t^{v})\right)(x) = (-1)^{\sigma} \frac{x^{\lambda+\rho\sigma+1}}{(\rho)^{\sigma}\Gamma(\gamma)}\, {}_2\Psi_{m+1}\left[\begin{array}{c} (\gamma,q),\left(\frac{\lambda+1}{\rho}+1,\frac{v}{\rho}\right) \\ (v_j+1,\lambda_j)_{j=1}^{m},\left(\frac{\lambda+1}{\rho}+\sigma+1,\frac{v}{\rho}\right) \end{array}\Bigg| -x^{v}\right] \quad (1.22)$$

Proof. The proof of the Theorem 2, would run parallel to those of Theorem 1, so we omit the proof.

2.1 Special Cases

By assigning the values to the parameters, the above results established in Eqs. (1.17) and (1.22) reduces to the following form:

Choose $m = 1$, the results in Eqs. (1.17) and (1.22) reduces to the following form:

Corollary 1. *Let* $x > 0; \sigma, \rho, \lambda, v, \gamma, q, z \in \mathbb{C}$ *such that* $\Re(\lambda) > \max\{0; \Re(q)-1\}; q > 0, \Re(v) > -1, \Re(\gamma) > 0$ *and* $q \in (0,1) \cup \mathbb{N}$, *then*

$$\left({}^{\rho}I_{0+}^{\sigma} t^{\lambda} J_{v,q}^{\lambda,\gamma}(t^{v})\right)(x) = \frac{x^{\lambda+\rho\sigma+1}}{(\rho)^{\sigma}\Gamma(\gamma)}\, {}_2\Psi_2\left[\begin{array}{c} (\gamma,q),\left(\frac{\lambda+1}{\rho}+1,\frac{v}{\rho}\right) \\ (v+1,\lambda),\left(\frac{\lambda+1}{\rho}+\sigma+1,\frac{v}{\rho}\right) \end{array}\Bigg| -x^{v}\right] \quad (1.23)$$

Corollary 2. *Let* $x < 0; \sigma, \rho, \lambda, v, \gamma, q, z \in \mathbb{C}$ *such that* $\Re(\lambda) > \max\{0; \Re(q)-1\}; q > 0, \Re(v) > -1, \Re(\gamma) > 0$ *and* $q \in (0,1) \cup \mathbb{N}$, *then*

$$\left({}^{\rho}I_{0-}^{\sigma} t^{\lambda} J_{v,q}^{\lambda,\gamma}(t^{v})\right)(x) = (-1)^{\sigma}\frac{x^{\lambda+\rho\sigma+1}}{(\rho)^{\sigma}\Gamma(\gamma)}\, {}_2\Psi_2\left[\begin{array}{c} (\gamma,q),\left(\frac{\lambda+1}{\rho}+1,\frac{v}{\rho}\right) \\ (v+1,\lambda),\left(\frac{\lambda+1}{\rho}+\sigma+1,\frac{v}{\rho}\right) \end{array}\Bigg| -x^{v}\right] \quad (1.24)$$

If v is replaced by $v-1$ and z by $-z$ then the results given in Corollaries 1 and 2 reduces to the following form:

Corollary 3. *Let* $x > 0; \sigma, \rho, \lambda, v, \gamma, q, z \in \mathbb{C}$ *such that* $\Re(\lambda) > 0, \Re(v) > 0, \Re(\gamma) > 0$ *and* $q \in (0,1) \cup \mathbb{N}$, *then*

$$\left({}^{\rho}I_{0+}^{\sigma} t^{\lambda} E_{\lambda,v}^{\gamma,q}(t^{v})\right)(x) = \frac{x^{\lambda+\rho\sigma+1}}{(\rho)^{\sigma}\Gamma(\gamma)}\, {}_2\Psi_2\left[\begin{array}{c} (\gamma,q),\left(\frac{\lambda+1}{\rho}+1,\frac{v}{\rho}\right) \\ (v,\lambda),\left(\frac{\lambda+1}{\rho}+\sigma+1,\frac{v}{\rho}\right) \end{array}\Bigg| x^{v}\right] \quad (1.25)$$

Corollary 4. *Let* $x < 0; \sigma, \rho, \lambda, v, \gamma, q, z \in \mathbb{C}$ *such that* $\Re(\lambda) > 0, \Re(v) > 0, \Re(\gamma) > 0$ *and* $q \in (0,1) \cup \mathbb{N}$, *then*

$$\left({}^{\rho}I_{0-}^{\sigma} t^{\lambda} E_{\lambda,v}^{\gamma,q}(t^{v})\right)(x) = (-1)^{\sigma}\frac{x^{\lambda+\rho\sigma+1}}{(\rho)^{\sigma}\Gamma(\gamma)}\, {}_2\Psi_2\left[\begin{array}{c} (\gamma,q),\left(\frac{\lambda+1}{\rho}+1,\frac{v}{\rho}\right) \\ (v,\lambda),\left(\frac{\lambda+1}{\rho}+\sigma+1,\frac{v}{\rho}\right) \end{array}\Bigg| x^{v}\right] \quad (1.26)$$

If we choose $q=\gamma=1$, ν is replaced by $\nu-1$ and z by $-z$ then the results given in Corollaries 1 and 2 reduces to the following form:

Corollary 5. *Let* $x>0;\sigma,\rho,\lambda,\nu,z\in\mathbb{C}$ *such that* $\Re(\lambda)>0,\Re(\nu)>0,\Re(\gamma)>0$ *and* $q\in(0,1)\cup\mathbb{N}$, *then*

$$\left({}^{\rho}I_{0+}^{\sigma}t^{\lambda}E_{\lambda,\nu}(t^{\nu})\right)(x)=\frac{x^{\lambda+\rho\sigma+1}}{(\rho)^{\sigma}\Gamma(\gamma)}{}_2\Psi_2\left[\begin{matrix}(1,1),\left(\frac{\lambda+1}{\rho}+1,\frac{\nu}{\rho}\right)\\(\nu,\lambda),\left(\frac{\lambda+1}{\rho}+\sigma+1,\frac{\nu}{\rho}\right)\end{matrix}\middle|x^{\nu}\right]\tag{1.27}$$

Corollary 6. *Let* $x<0;\sigma,\rho,\lambda,\nu,z\in\mathbb{C}$ *such that* $\Re(\lambda)>0,\Re(\nu)>0,\Re(\gamma)>0$ *and* $q\in(0,1)\cup\mathbb{N}$, *then*

$$\left({}^{\rho}I_{0-}^{\sigma}t^{\lambda}E_{\lambda,\nu}(t^{\nu})\right)(x)=(-1)^{\sigma}\frac{x^{\lambda+\rho\sigma+1}}{(\rho)^{\sigma}\Gamma(\gamma)}{}_2\Psi_2\left[\begin{matrix}(1,1),\left(\frac{\lambda+1}{\rho}+1,\frac{\nu}{\rho}\right)\\(\nu,\lambda),\left(\frac{\lambda+1}{\rho}+\sigma+1,\frac{\nu}{\rho}\right)\end{matrix}\middle|x^{\nu}\right]\tag{1.28}$$

3 Image Formulas Associated with Integral Transform

In this section we establish certain theorems involving the results obtained in the previous section associated with integral transforms like Beta transform, Laplace transform and Whittaker transform.

3.1 Beta Transform

The Beta transform of $f(z)$ is defined as [38]

$$B\{f(z):a,b\}=\int_0^1 z^{a-1}(1-z)^{b-1}f(z)dz\tag{1.29}$$

Theorem 3. *Let* $x>0,m\in\mathbb{N};\sigma,\rho,\lambda_j,\nu_j,\gamma,q,z\in\mathbb{C}(j=1,...,m)$ *such that* $\sum_{j=1}^{m}\Re(\lambda_j)>\max\{0;\Re(q)-1\};q>0,\ \Re(\nu_j)>-1,\Re(\gamma)>0$ *and* $q\in(0,1)\cup\mathbb{N}$, $\Re(l)>0,\Re(k)>0>\rho>0$, *then*

$$\begin{aligned}B\left\{\left({}^{\rho}I_{0+}^{\sigma}t^{\lambda}J_{(\nu_j)_{m},q}^{(\lambda_j)_{m},\gamma}(tz)^{\nu}\right)(x):l,k\right\}&=\Gamma(k)\frac{x^{\lambda+\rho\sigma+1}}{(\rho)^{\sigma}\Gamma(\gamma)}\\&\times{}_3\Psi_{m+2}\left[\begin{matrix}(\gamma,q),\left(\frac{\lambda+1}{\rho}+1,\frac{\nu}{\rho}\right),(l,\nu)\\(\nu_j+1,\lambda_j)_{j=1}^{m},\left(\frac{\lambda+1}{\rho}+\sigma+1,\frac{\nu}{\rho}\right),(l+k,\nu)\end{matrix}\middle|-x^{\nu}\right]\end{aligned}\tag{1.30}$$

Proof. For convenience, we denote the left-hand side of the result (1.30) by $\mathscr{B}$. Using the definition of beta transform, the LHS of (1.30) becomes:

$$\mathscr{B}=\int_0^1 z^{l-1}(1-z)^{k-1}\left({}^{\rho}I_{0+}^{\sigma}t^{\lambda}J_{(\nu_j)_m,q}^{(\lambda_j)_m,\gamma}(tz)^{\nu}\right)(x)dz \tag{1.31}$$

further using (1.8) and then changing the order of integration and summation, which is valid under the conditions of Theorem 1, then

$$\mathscr{B}=\sum_{n=0}^{\infty}\frac{(\gamma)_{qn}}{\prod_{j=1}^{m}\Gamma(\lambda_j n+\nu_j+1)}\frac{(-1)^n}{n!}\left({}^{\rho}I_{0+}^{\sigma}t^{n\nu+\lambda}\right)(x)\int_0^1 z^{l+n\nu-1}(1-z)^{k-1}dz, \tag{1.32}$$

applying the result (1.15), after simplification Eq. (1.19) reduced to

$$\mathscr{B}=\frac{x^{\lambda+\rho\sigma+1}}{(\rho)^{\sigma}\Gamma(\gamma)}\sum_{n=0}^{\infty}\frac{\Gamma(\gamma+qn)\Gamma\left(\frac{\lambda+1}{\rho}+1+\frac{\nu}{\rho}n\right)}{\prod_{j=1}^{m}\Gamma(\lambda_j n+\nu_j+1)\Gamma\left(\frac{\lambda+1}{\rho}+1+\sigma+\frac{\nu}{\rho}n\right)}\times\frac{(x^{\nu})^n}{n!}\int_0^1 z^{l+n\nu-1}(1-z)^{k-1}dz \tag{1.33}$$

applying the definition of beta transform, Eq. (1.33) reduced to

$$\mathscr{B}=\frac{x^{\lambda+\rho\sigma+1}}{(\rho)^{\sigma}\Gamma(\gamma)}\sum_{n=0}^{\infty}\frac{\Gamma(\gamma+qn)\Gamma\left(\frac{\lambda+1}{\rho}+1+\frac{\nu}{\rho}n\right)}{\prod_{j=1}^{m}\Gamma(\lambda_j n+\nu_j+1)\Gamma\left(\frac{\lambda+1}{\rho}+1+\sigma+\frac{\nu}{\rho}n\right)}\times\frac{(x^{\nu})^n}{n!}\frac{\Gamma(l+\nu n)\Gamma(k)}{\Gamma(l+k+\nu n)} \tag{1.34}$$

$$\mathscr{B}=\Gamma(k)\frac{x^{\lambda+\rho\sigma+1}}{(\rho)^{\sigma}\Gamma(\gamma)}{}_3\Psi_{m+2}\left[\begin{matrix}(\gamma,q),\left(\frac{\lambda+1}{\rho}+1,\frac{\nu}{\rho}\right),(l,\nu)\\(\nu_j+1,\lambda_j)_{j=1}^{m},\left(\frac{\lambda+1}{\rho}+\sigma+1,\frac{\nu}{\rho}\right),(l+k,\nu)\end{matrix}\middle|-x^{\nu}\right] \tag{1.35}$$

Theorem 4. *Let* $x<0,m\in\mathbb{N};\sigma,\rho,\lambda_j,\nu_j,\gamma,q,z\in\mathbb{C}(j=1,...,m)$ *such that* $\sum_{j=1}^{m}\Re(\lambda_j)>\max\{0;\Re(q)-1\};q>0,\Re(\nu_j)>-1,\Re(\gamma)>0$ *and* $q\in(0,1)\cup\mathbb{N}$, $\Re(l)>0,\Re(k)>0>\rho>0$, *then*

$$B\left\{\left({}^{\rho}I_{0-}^{\sigma}t^{\lambda}J_{(\nu_j)_m,q}^{(\lambda_j)_m,\gamma}(zt)^{\nu}\right)(x):l,k\right\}=(-1)^{\sigma}\Gamma(k)\frac{x^{\lambda+\rho\sigma+1}}{(\rho)^{\sigma}\Gamma(\gamma)}\times{}_3\Psi_{m+2}\left[\begin{matrix}(\gamma,q),\left(\frac{\lambda+1}{\rho}+1,\frac{\nu}{\rho}\right),(l,\nu)\\(\nu_j+1,\lambda_j)_{j=1}^{m},\left(\frac{\lambda+1}{\rho}+\sigma+1,\frac{\nu}{\rho}\right),(l+k,\nu)\end{matrix}\middle|-x^{\nu}\right] \tag{1.36}$$

Proof. The proof of the Theorem 4, would run parallel to those of Theorem 3, so we omit the proof.

3.2 Laplace Transform

The Laplace transform of f(z) is defined as [38]:

$$Ł\{f(z)\} = \int_0^\infty e^{-sz} f(z)dz \tag{1.37}$$

Theorem 5. *Let* $x > 0, m \in \mathbb{N}; \sigma, \rho, \lambda_j, \nu_j, \gamma, q, z \in \mathbb{C}(j = 1,...,m)$ *such that* $\sum_{j=1}^{m} \Re(\lambda_j) > \max\{0; \Re(q) - 1\}; q > 0, \Re(\nu_j) > -1, \Re(\gamma) > 0$ *and* $q \in (0,1) \cup \mathbb{N}$, $\Re(l) > 0$ *and* $\rho > 0$*, then*

$$L\left\{z^{l-1}\left({}^{\rho}I_{0+}^{\sigma} t^{\lambda} J_{(\nu_j)_m,q}^{(\lambda_j)_m,\gamma}(tz)^{\nu}\right)(x)\right\} =$$
$$\frac{x^{\lambda+\rho\sigma+1}}{s^{l}(\rho)^{\sigma}\Gamma(\gamma)} {}_3\Psi_{m+1}\left[\begin{array}{c} (\gamma,q), \left(\frac{\lambda+1}{\rho}+1, \frac{\nu}{\rho}\right), (l,\nu) \\ (\nu_j+1,\lambda_j)_{j=1}^{m}, \left(\frac{\lambda+1}{\rho}+\sigma+1, \frac{\nu}{\rho},\right) \end{array}\middle| -\left(\frac{x}{s}\right)^{\nu}\right] \tag{1.38}$$

Proof. For convenience, we denote the left-hand side of the result (1.38) *by* $\mathscr{L}$*. Using the definition of beta transform, the LHS of* (1.38) *becomes:*

$$\mathscr{L} = \int_0^\infty e^{-sz} z^{l-1}\left({}^{\rho}I_{0+}^{\sigma} t^{\lambda} J_{(\nu_j)_m,q}^{(\lambda_j)_m,\gamma}(tz)^{\nu}\right)(x)dz \tag{1.39}$$

further using (1.8) *and then changing the order of integration and summation, which is valid under the conditions of Theorem* 1*, then applying the result* (1.15)*, after simplification Eq.* (1.19) *reduced to*

$$\mathscr{L} = \frac{x^{\lambda+\rho\sigma+1}}{(\rho)^{\sigma}\Gamma(\gamma)} \sum_{n=0}^{\infty} \frac{\Gamma(\gamma+qn)\Gamma\left(\frac{\lambda+1}{\rho}+1+\frac{\nu}{\rho}n\right)}{\prod_{j=1}^{m}\Gamma(\lambda_j n+\nu_j+1)\Gamma\left(\frac{\lambda+1}{\rho}+1+\sigma+\frac{\nu}{\rho}n\right)} \times \frac{(k^{q-\frac{\alpha}{k}}x^{\nu})^n}{n!} \int_0^\infty e^{-sz} z^{l+n\nu-1} dz \tag{1.40}$$

Equation (1.40) reduced to

$$\mathscr{L} = \frac{x^{\lambda+\rho\sigma+1}}{(\rho)^{\sigma}\Gamma(\gamma)} \sum_{n=0}^{\infty} \frac{\Gamma(\gamma+qn)\Gamma\left(\frac{\lambda+1}{\rho}+1+\frac{\nu}{\rho}n\right)}{\prod_{j=1}^{m}\Gamma(\lambda_j n+\nu_j+1)\Gamma\left(\frac{\lambda+1}{\rho}+1+\sigma+\frac{\nu}{\rho}n\right)} \times \frac{(x^{\nu})^n}{n!} \frac{\Gamma(l+\nu n)}{s^{l+\nu n}} \tag{1.41}$$

$$\mathscr{B} = \frac{x^{\lambda+\rho\sigma+1}}{s^{l}(\rho)^{\sigma}\Gamma(\gamma)} {}_3\Psi_{m+1}\left[\begin{array}{c} (\gamma,q), \left(\frac{\lambda+1}{\rho}+1, \frac{\nu}{\rho}\right), (l,\nu) \\ (\nu_j+1,\lambda_j)_{j=1}^{m}, \left(\frac{\lambda+1}{\rho}+\sigma+1, \frac{\nu}{\rho},\right) \end{array}\middle| -\left(\frac{x}{s}\right)^{\nu}\right] \tag{1.42}$$

Theorem 6. *Let* $x < 0, m \in \mathbb{N}; \sigma, \rho, \lambda_j, v_j, \gamma, q, z \in \mathbb{C}(j = 1,...,m)$ *such that* $\sum_{j=1}^{m} \Re(\lambda_j) > \max\{0; \Re(q)-1\}; q > 0, \Re(v_j) > -1, \Re(\gamma) > 0$ *and* $q \in (0,1) \cup \mathbb{N}$, $\Re(l) > 0$, $\rho > 0$, *then*

$$L\left\{z^{l-1}\left({}^{\rho}I_{0-}^{\sigma} t^{\lambda} J_{(v_j)_m,q}^{(\lambda_j)_m,\gamma}(zt)^{v}\right)(x)\right\} = (-1)^{\sigma} \frac{x^{\rho\sigma+\sigma+\lambda+1}}{s^{l}(\rho)^{\sigma}\Gamma(\gamma)} \times {}_3\Psi_{m+1}\left[\begin{array}{c}(\gamma,q),\left(\frac{\lambda+1}{\rho}+1,\frac{v}{\rho}\right),(l,v)\\ (v_j+1,\lambda_j)_{j=1}^{m},\left(\frac{\lambda+1}{\rho}+\sigma+1,\frac{v}{\rho}\right)\end{array}\middle| -\left(\frac{x}{s}\right)^{v}\right]. \tag{1.43}$$

Proof. The proof of the Theorem 6, would run parallel to those of Theorem 5, so we omit the proof.

3.3 Whittaker Transform

Theorem 7. *Let* $x > 0, m \in \mathbb{N}; \sigma, \rho, \lambda_j, v_j, \gamma, q, z \in \mathbb{C}(j = 1,...,m)$ *such that* $\sum_{j=1}^{m} \Re(\lambda_j) > \max\{0; \Re(q)-1\}; q > 0, \Re(v_j) > -1, \Re(\gamma) > 0$ *and* $q \in (0,1) \cup \mathbb{N}$, *then*

$$\int_0^{\infty} z^{\xi-1} e^{-\eta z/2} W_{\tau,\omega}(\eta z)\left\{\left({}^{\rho}I_{0+}^{\sigma} t^{\lambda} J_{(v_j)_m,q}^{(\lambda_j)_m,\gamma}((zt)^{v})\right)(x)\right\} dz = \frac{x^{\lambda+\rho\sigma+1}}{\eta^{\xi}(\rho)^{\sigma}\Gamma(\gamma)} \times {}_4\Psi_{m+2}\left[\begin{array}{c}(\gamma,q),\left(\frac{\lambda}{\rho}+1,\frac{v}{\rho}\right),(1/2+\omega+\xi,v),(1/2-\omega+\xi,v)\\ (v_j+1,\lambda_j)_{j=1}^{m},\left(\frac{\lambda}{\rho}+\sigma+1,\frac{v}{\rho}\right),(1/2-\tau+\xi,v)\end{array}\middle| -\left(\frac{x}{\eta}\right)^{v}\right] \tag{1.44}$$

Proof. For convenience, we denote the left-hand side of the result (1.44) by $\mathscr{W}$. Then using the result from (1.21) after changing the order of integration and summation, we get

$$\mathscr{W} = \frac{x^{\rho+\lambda+\rho\sigma}}{(\rho)^{\sigma}\Gamma(\gamma)} \sum_{n=0}^{\infty} \frac{\Gamma(\gamma+qn)\Gamma\left(\frac{\lambda}{\rho}+1+\frac{v}{\rho}n\right)}{\prod_{j=1}^{m}\Gamma(\lambda_j n+v_j+1)\Gamma\left(\frac{\lambda}{\rho}+\sigma+1+\frac{v}{\rho}n\right)} \times \frac{(x^{v})^{n}}{n!}\int_0^{\infty} z^{nv+\xi-1} e^{-\eta z/2} W_{\tau,\omega}(\eta z) dz \tag{1.45}$$

by substituting η z = t, (1.45) becomes:

$$\mathscr{W} = \frac{x^{\lambda+\rho\sigma+1}}{(\rho)^{\sigma}\Gamma(\gamma)} \sum_{n=0}^{\infty} \frac{\Gamma(\gamma+qn)\Gamma\left(\frac{\lambda}{\rho}+1+\frac{v}{\rho}n\right)}{\prod_{j=1}^{m}\Gamma(\lambda_j n+v_j+1)\Gamma\left(\frac{\lambda}{\rho}+\sigma+1+\frac{v}{\rho}n\right)} \times \frac{(x^{v})^{n}}{n!}\frac{1}{\eta^{nv+\xi}}\int_0^{\infty} t^{nv+\xi-1} e^{-\eta z/2} W_{\tau,\omega}(t) dt \tag{1.46}$$

Now we use the following integral formula involving Whittaker function

$$\int_0^\infty t^{\nu-1} e^{-t/2} W_{\tau,\omega}(t)dt = \frac{\Gamma(1/2+\omega+\nu)\Gamma(1/2-\omega+\nu)}{\Gamma(1/2-\tau+\nu)}, \quad \left(\Re(\nu \pm \omega) > \frac{-1}{2}\right) \tag{1.47}$$

Then we have

$$\mathscr{W} = \frac{x^{\lambda+\rho\sigma+1}}{\eta^{\xi}(\rho)^{\sigma}\Gamma(\gamma)} \sum_{n=0}^{\infty} \frac{\Gamma(\gamma+qn)\Gamma\left(\frac{\lambda}{\rho}+1+\frac{\nu}{\rho}n\right)}{\prod_{j=1}^{m}\Gamma(\lambda_j n+\nu_j+1)\Gamma\left(\frac{\lambda}{\rho}+\sigma+1+\frac{\nu}{\rho}n\right)} \frac{1}{n!} \times \frac{\Gamma(1/2+\omega+\xi+n\nu)\Gamma(1/2-\omega+\xi+n\nu)}{\Gamma(1/2-\tau+\xi+n\nu)} \left(\frac{x}{\eta}\right)^{n\nu} \tag{1.48}$$

interpreting the above Eq. (1.48) with the help of (1.11), we have the required result.

Theorem 8. *Let* $x < 0, m \in \mathbb{N}; \sigma, \rho, \lambda_j, \nu_j, \gamma, q, z \in \mathbb{C} (j = 1, ..., m)$ *such that* $\sum_{j=1}^{m} \Re(\lambda_j) > \max\{0; \Re(q)-1\}; q > 0, \Re(\nu_j) > -1, \Re(\gamma) > 0$ *and* $q \in (0,1) \cup \mathbb{N}$, *then*

$$\int_0^\infty z^{\xi-1} \exp^{-\delta z/2} W_{\tau,\omega}(\eta z) \left\{ \left({}^{\rho}I_{0-}^{\sigma} t^{\lambda} J_{(\nu_j)_{m,q}}^{(\lambda_j)_{m,\gamma}}((zt)^{\nu})\right)(x)\right\} dz = (-1)^{\sigma} \frac{x^{\lambda+\rho\sigma+1}}{(\rho)^{\sigma}\Gamma(\gamma)} \times {}_4\Psi_{m+2}\left[\begin{matrix} (\gamma,q), \left(\frac{\lambda}{\rho}+1, \frac{\nu}{\rho}\right), (1/2+\omega+\xi, \nu), (1/2-\omega+\xi, \nu) \\ (\nu_j+1, \lambda_j)_{j=1}^{m}, \left(\frac{\lambda}{\rho}+\sigma+1, \frac{\nu}{\rho}\right), (1/2-\tau+\xi, \nu) \end{matrix} \middle| -\left(\frac{x}{\eta}\right)^{\nu}\right] \tag{1.49}$$

Proof. The proof of this theorem would run parallel as those of Theorem 7.

4 Conclusion

In this paper, we established some image formulas by applying generalized fractional integral operators generalized multiindex Bessel function $J_{(\nu_j)_{m,q}}^{(\lambda_j)_{m,\gamma}}(z)$. Then, some more image formulas are derived by employing the integral transform. All the results are in the form of Fox's Wright function, hence all the results are in series form. By giving some particular value to the parameters, we obtained some special cases of our findings.

References

1. Oldham, K.B., Spanier, J.: The fractional Calculus: Theory and Applications of Differentiation and Integration of Arbitrary Order. Academic Press, New York (1974)
2. Miller, K.S., Ross, B.: An Introduction to the Fractional Calculus and Fractional Differential Equations. Wiley, New York (1993)
3. Podlubny, I.: Fractional Differential Equations. Academic Press, New York (1999)
4. Chen, Y., Petráš, I., Xue, D.: Fractional order control. In: A Tutorial Proceedings of 2009 American Control Conference, St. Louis, MO, USA (2009)

5. Petráš, I.: Stability of fractional-order systems with rational orders: a survey. Fractional Calc. Appl. Anal. **12**, 269–298 (2009)
6. Agarwal, P., Chand, M., Singh, G.: Certain fractional kinetic equations involving the product of generalized k-Bessel function. Alexandria Eng. J. **55**, 3053–3059 (2016)
7. Singh, G., Agarwal, P., Chand, M., Jain, S.: Certain fractional kinetic equations involving generalized k-Bessel function. Trans. A. Razmadze Math. Inst. (2018). https://doi.org/10.1016/j.trmi.2018.03.001
8. Agarwal, P., Ntouyas, S.K., Jain, S., Chand, M., Singh, G.: Fractional kinetic equations involving generalized k-Bessel function via Sumudu transform. Alexandria Eng. J. **57**(3), 1937–1942 (2017)
9. Al-Bassam, M.A., Luchko, Y.K.: On generalized fractional calculus and its application to the solution of integro-differential equations. J. Fract. Calc. **7**, 69–88 (1995)
10. Choi, J., Agarwal, P., Mathur, S., Purohit, S.D.: Certain new integral formulas involving the generalized Bessel function. Bull. Korean Math. Soc. **51**(4), 995–1003 (2014)
11. Choi, J., Agarwal, P.: A note on fractional integral operator associated with multiindex Mittag-Leffler functions. Filomat. **30**(7), 1931–1939 (2016)
12. Atangana, A., Gómez-Aguilar, J.F.: Decolonisation of fractional calculus rules: breaking commutativity and associativity to capture more natural phenomena. Eur. Phys. J. Plus. **133**, 1–23 (2018)
13. Atangana, A.: Non validity of index law in fractional calculus: a fractional differential operator with Markovian and non-Markovian properties. Phys. A **505**, 688–706 (2018)
14. Gutiérrez, R.E., Rosário, J.M., Machado, J.T.: Fractional order calculus: basic concepts and engineering applications. Math. Probl. Eng. **2010**, 19 (2010)
15. Axtell, M., Bise, M.E.: Fractional calculus applications in control systems. In: Proceedings of the: National Aerospace and Electronics Conference, Dayton, OH, USA, p. 1990 (1990)
16. Hamamc, S.E.: Stabilization using fractional order PI and PID controllers. Nonlinear Dyn. **51**, 329–343 (2008)
17. Hamamci, S.E., Koksa, M.: Calculation of all stabilizing fractional-order PD controllers for integrating time delay systems. Comput. Math. Appl. **59**, 1621–1629 (2010)
18. Caputo, M.: Linear models of dissipation whose q is almost frequency independent II. Geophys. J. Royal Astr. Soc. **13**, 529–539 (1967)
19. Kilbas, A.A., Srivastava, H.M., Trujillo, J.J.: Theory and applications of fractional differential equations. In: North-Holland Mathematics Studies, vol. 204. Elsevier, Amsterdam (2006)
20. Rabotnov, Y.N.: Creep problems in structural members. In: North-Holland Series in Applied Mathematics and Mechanics, vol. 7 (1969)
21. Rose, B.: Fractional calculus and its applications. In: Proceedings of the International Conference Held at the University of New Haven (1974)
22. Samko, S.G., Kilbas, A.A., Marichev, O.I.: Fractional Integrals and Derivatives: Theory and Applications. Gordon and Breach Science Publishers, New York (1993)
23. Escamilla, A.C., Gómez-Aguilar, J.F., Baleanu, D., Córdova-Fraga, T., Escobar-Jiménez, R.F., Olivares-Peregrino, V.H., Qurash, M.M.A.: Bateman-feshbach tikochinsky and caldirola-kanai oscillators with new fractional differentiation. Entropy **19**(2), 1–13 (2017)
24. Escamilla, A.C., Torres, F., Gómez-Aguilar, J.F., Escobar-Jiménez, R.F., Guerrero-Ramírez, G.V.: On the trajectory tracking control for an scara robot manipulator in a fractional model driven by induction motors with PSO tuning. Multibody Syst. Dyn. **43**(3), 257–277 (2017)
25. Escamilla, A.C., Gómez-Aguilar, J.F., Torres, L., Escobar-Jiménez, R.F.: A numerical solution for a variable-order reaction-diffusion model by using fractional derivatives with non-local and non-singular kernel. Phys. A **491**, 406–424 (2018)
26. Gómez-Aguilar, J.F.: Chaos in a nonlinear bloch system with Atangana-Baleanu fractional derivatives. Numer. Methods Partial Differ. Eq. **33**, 1–23 (2017)

27. Gómez-Aguilar, J.F., Yépez-Martínez, H., Torres-Jiménez, J., Córdova-Fraga, T., Escobar-Jiménez, R.F., Olivares-Peregrino, V.H.: Homotopy perturbation transform method for non-linear differential equations involving to fractional operator with exponential kernel. Adv. Differ. Eq. **68**, 1–18 (2017)
28. Marichev, O.I.: Handbook of Integral Transforms and Higher Transcendental Functions. Chichester: Ellis, Horwood. Wiley, New York (1983)
29. Jain, S., Agarwal, P.: A new class of integral relations involving general class of polynomials and I-functions. Walialak J. Sci. Tech. **12**(11), 1009–1018 (2015)
30. Pathak, R.S.: Certain convergence theorem and asymptotic properties of a generalization of Lommel and Bessel transformations. Proc. Nat. Acad. Sci. India. Sect. A **36**(1), 81 (1966)
31. Shukla, A.K., Prajapati, J.C.: On a generalized Mittag-Leffler function and its properties. J. Math. Anal. Appl. **336**, 797–811 (2007)
32. Wiman, A.: Uber de fundamental satz in der theorie der funktionen $E_\alpha(x)$. Acta Math. **29**, 191–201 (1905)
33. A note on fractional integral operator associated with multiindex Mittag-Leffler functions. Filomat **30**(7), 1931–1939 (2016)
34. Kilbas, A.A., Srivastava, S.M., Trujillo, J.J.: Theory and Applications of Fractional Differential Equations: North-Holland Mathematics Studies, vol. 204. Elsevier, Amsterdam (2006)
35. Erdéldyi, A., Magnus, W., Oberhettinger, F., Tricomi, F.G.: Higher Transcendental Functions. Krieger Pubisher, Melbourne (1981)
36. Watson, G.N.: A Treatise on the Theory of Bessel Functions, 2nd edn. Cambridge University Press, Cambridge (1996)
37. Katugampola, U.N.: New approach to a generalized fractional integral. Appl. Math. Comput. **218**(3), 860–865 (2011)
38. Sneddon, I.N.: The Use of Integral transforms. Tata McGraw-Hill, Delhi (1979)
39. Chand, M., Agarwal, P., Hammouch, Z.: Certain sequences involving product of k-Bessel function. Int. J. Appl. Comput. Math. **4**, 101 (2018)
40. Chand, M., Hammouch, Z., Asamoah, J.K.K., Baleanu, D.: Certain fractional integrals and solutions of fractional kinetic equations involving the product of S-function. In: Taş, K., Baleanu, D., Machado, J. (eds.) Mathematical Methods in Engineering. Nonlinear Systems and Complexity, vol. 24. Springer, Cham (2019)

(T, φ, λ) – Statistical Convergence of Order β

Ekrem Savaş(✉)

Department of Mathematics, Uşak University, Uşak, Turkey
ekremsavas@yahoo.com

Abstract. In this paper we introduce (T, φ, λ) - statistically convergent of order β. In addition, we establish some connections between (T, φ, λ) - statistically convergent of order β and the λ – strong convergence of order β.

Keywords: Modulus function · φ-function · Statistically convergent of order β · λ – strong convergence of order β · Matrix transformations

2010 Mathematics Subject Classification: Primary 40H05 · Secondary 40C05

1 Introduction

Before we present the new definition and main theorems we shall state a few known definitions.

Let s denote the set of all real and complex sequences $x = (x_k)$. By l_∞ and c, we denote the Banach spaces of bounded and convergent sequences $x = (x_n)$ normed by $||x|| = \sup_n |x_n|$, respectively. A linear functional ν on l_∞ is said to be a Banach limit if it has the following properties:

(1) $\nu(\xi) \geq 0$ if $n \geq 0$ (i.e. $\xi_n \geq 0$ for all n),
(2) $\nu(e) = 1$ where $e = (1, 1, \ldots)$,
(3) $\nu(D\xi) = \nu(\xi)$, where the shift operator D is defined by $D(\xi_n) = \{\xi_{n+1}\}$.

Let B be the set of all Banach limits on l_∞. A sequence $\xi \in \ell_\infty$ is said to be almost convergent if all of its Banach limits coincide. Let $\hat{c}$ denote the space of almost convergent sequences.

Lorentz [8] has shown that

$$\hat{c} = \left\{ x \in l_\infty : \lim_m \varphi_{m,n}(\xi) \text{ exists uniformly in } n \right\}$$

where

$$\varphi_{m,n}(\xi) = \frac{\xi_n + \xi_{n+1} + \xi_{n+2} + \cdots + \xi_{n+m}}{m+1}.$$

The space $[\hat{c}]$ of strongly almost convergent sequences was introduced by Maddox [9] and also independently by Freedman et al. [5] as follows:

$$[\hat{c}] = \left\{ x \in l_\infty : \lim_m \varphi_{m,n}(|\xi - L|) = 0, \text{ uniformly in } n, \text{ for some } L \right\}.$$

H. Dutta et al. (Eds.): CMES 2019, AISC 1111, pp. 291–298, 2020.
https://doi.org/10.1007/978-3-030-39112-6_23

If $\xi = (\xi_k)$ is a sequence and $T = (t_{nk})$ is an infinite matrix, then $T\xi$ is the sequence whose nth term is given by $T_n(\xi) = \sum_{k=0}^{\infty} t_{nk}\xi_k$. Thus we say that ξ is T-summable to L if $\lim_{n\to\infty} T_n(\xi) = L$. Let E and F be two sequence spaces and $T = (t_{nk})$ an infinite matrix. If for each $\xi \in E$ the series $T_n(\xi) = \sum_{k=0}^{\infty} t_{nk}\xi_k$ converges for each n and the sequence $T\xi = T_n(\xi) \in E$ we say that T maps E into F. By (E, F) we denote the set of all matrices which maps E into F, and in addition if the limit is preserved then we denote the class of such matrices by $(E, F)_{reg}$.

Let $\lambda = (\lambda_i)$ be a non-decreasing sequence of positive numbers tending to ∞ such that

$$\lambda_{i+1} \leq \lambda_i + 1, \lambda_1 = 1.$$

The collection of such sequence λ will be denoted by Δ.

The generalized de la Valée-Poussin mean is defined by

$$T_j(\xi) = \frac{1}{\lambda_j} \sum_{k \in I_j} \xi_k$$

where $I_j = [j - \lambda_j + 1, j]$. A sequence $\xi = (\xi_k)$ is said to be (V, λ)-summable to a number L, if $T_j(\xi) \to L$ as $j \to \infty$ (see [11]).

Quite Recently, Malkowsky and Savaş [11] defined the space $[V, \lambda]$ of λ-strongly convergent sequences as follows:

$$[V, \lambda] = \left\{ \xi = (\xi_k) : \lim_j \frac{1}{\lambda_j} \sum_{k \in I_j} |\xi_k - L| = 0, \text{ for some } L \right\}.$$

Note that in the special case where $\lambda_j = j$, the space $[V, \lambda]$ reduces the space w of strongly Cesàro summable sequences which is defined by

$$w = \left\{ \xi = (\xi_k) : \lim_j \frac{1}{j} \sum_{k=1}^{j} |\xi_k - L|) = 0, \text{ for some } L \right\}.$$

More results on λ - strong convergence can be seen from [12,18–23].

By a φ-function we understand a continuous non-decreasing function $\varphi(u)$ defined for $u \geq 0$ and such that $\varphi(0) = 0, \varphi(u) > 0$, for $u > 0$ and $\varphi(u) \to \infty$ as $u \to \infty$, (see, [25]).

A φ-function φ is called non weaker than a φ-function ψ if there are constants $c, b, k, l > 0$ such that $c\psi(lu) \leq b\varphi(ku)$, (for all large u) and we write $\psi \prec \varphi$. (for all large u), (see, [25]).

A φ-function φ is said to satisfy (Δ_2)-condition, (for all large u) if there exists constant $K > 1$ such that $\varphi(2u) \leq K\varphi(u)$.

The following definition was giving in [23].

Definition 1.1. *Let $\Delta = (\mu_j)$ be same as above, φ be given φ-function and f be given modulus function, respectively. Moreover, let $\boldsymbol{T} = (t_{nk}(i))$ be the generalized three parametric real matrix and $0 < \beta \leq 1$ be given. Then we write,*

$$V_\lambda^\beta(T, \varphi, f, p)_0 = \left\{ \xi = (\xi_k) : \lim_j \frac{1}{\lambda_j^\beta} \sum_{n \in I_j} f\left(\left| \sum_{k=1}^{\infty} t_{nk}(i) \varphi(|\xi_k|) \right| \right)^{p_n} = 0, \ \textit{uniformly in } i \right\}.$$

If $\xi \in V_\lambda^\beta(T, \varphi, f, p)_0$, the sequence ξ is said to be λ - strong (T, φ) - convergent of order β to zero with respect to a modulus f.

2 Main Result

The idea of statistically convergence of sequence of real numbers was introduced by Fast [4] in 1951. Schoenberg [24] studied statistically convergence as a summability method and listed some of the elementary properties of statistical convergence. Both of these authors noted that if a bounded sequence is statistically convergent to L, then it is Cesàro summable to L. Recently in [3], Connor extended the definition of statistical convergence to T-statistical convergence by using nonnegative regular matrices.

Definition 2.1 ([4,6]). *A complex number sequence ξ is said to be statistically convergent to the number L if for every $\varepsilon > 0$*

$$\lim_n \frac{1}{n} |k \leq n : |\xi_k - L| \geq \varepsilon\}| = 0,$$

where by $k \leq n$ we mean that $k = 0, 1, 2, ..., n$ and the vertical bars indicate the number of elements in the enclosed set. In this case we write $st_1 - \lim \xi = L$ or $\xi_k \to L(st_1)$.

Statistical convergence turned out to be one of the most active areas of research in summability theory after the work of Fridy [6] and Šalát [15].

In another direction, a new type of convergence called λ - statistical convergence was introduced in [13] as follows.

Definition 2.2. *A sequence (ξ_k) of real numbers is said to be λ - statistically convergent to L (or, S_λ-convergent to L) if for any $\varepsilon > 0$,*

$$\lim_{j \to \infty} \frac{1}{\lambda_j} |\{k \in I_j : |\xi_k - L| \geq \varepsilon\}| = 0$$

where $|A|$ denotes the cardinality of $A \subset \mathbb{N}$.

In [13] the relation between λ - statistical convergence and statistical convergence was established among other things.

Recently Savas [17] defined almost λ-statistical convergence by using the notion of (V, λ)-summability to generalize the concept of statistical convergence.

Assume that T is a non-negative regular summability matrix. Then the sequence $\xi = (\xi_k)$ is called T-statistically convergent to L provided that, for every $\varepsilon > 0$, (see, [7])

$$lim_j \sum_{n:|\xi_k - L| \geq \varepsilon} t_{jn} = 0$$

We denote this by $st_T - lim_n x_k = L$.

Let $\mathbf{T} = (t_{nk}(i))$ be the generalized three parametric real matrix and the sequence $\xi = (\xi_k)$, the φ - function $\varphi(u)$ and a positive number $\varepsilon > 0$ be given. We write, for all i

$$K((T, \varphi), \varepsilon, i) = \{n \in I_j : \sum_{k=1}^{\infty} t_{nk}(i)\varphi(|\xi_k|) \geq \varepsilon\}.$$

The sequence ξ is said to be (T, φ, λ) - statistically convergent of order β to a number zero if for every $\varepsilon > 0$

$$lim_j \frac{1}{\lambda_j^\beta} \mu(K((T, \varphi), \varepsilon, i)) = 0, \text{uniformly in}\, i$$

where $\mu(K((T, \varphi, \lambda), \varepsilon, i))$ denotes the number of elements belonging to $K((T, \varphi, \lambda), \varepsilon, i)$. We denote by $S_\lambda^\beta((T, \varphi))_0$, the set of sequences $\xi = (\xi_k)$ which are (T, φ, λ) – statistical convergent of order β to zero.

If we take $T = I$ and $\varphi(x) = x$ respectively, then $S_\lambda^\beta(T, \varphi)_0$ reduce to $S_{\lambda_0}^\beta$ which was defined as follows, (see, Colak and Bektas [2]).

$$S_{\lambda_0}^\beta = \left\{ \xi = (\xi_k) : lim_j \frac{1}{\lambda_j^\beta} |\{k \in I_j : |\xi_k| \geq \varepsilon\}| = 0 \right\}.$$

If we take $T = I$, $\varphi(x) = x$ and $\beta = 1$ respectively, then we get the following (see, Mursaleen [2]):

$$S_{\lambda_0} = \left\{ x = (x_k) : lim_j \frac{1}{\lambda_j} |\{k \in I_j : |\xi_k| \geq \varepsilon\}| = 0 \right\}.$$

Remark 1. *(i) If for all i,*

$$t_{nk}(i) := \{ \begin{matrix} \frac{1}{n}, & if \quad n \geq k, \\ 0, & otherwise. \end{matrix}$$

then $S_\lambda^\beta((T, \varphi))_0$ reduce to $S_\lambda^\beta((C, \varphi))_0$, i.e., uniform (C, φ) – statistical convergence of order β to zero.
(ii) If for all i,

$$t_{nk}(i) := \{ \begin{matrix} \frac{p_k}{P_n}, & if \quad n \geq k, \\ 0, & otherwise. \end{matrix}$$

then $S_\lambda^\beta((T,\varphi))_0$ *reduce to* $S_\lambda^\beta((N,p),\varphi))_0$, *i.e., uniform* $((N,p),\varphi)$ – *statistical convergence of order* β *to zero, where* $p = p_k$ *is a sequence of nonnegative numbers such that* $p_0 > 0$ *and*

$$P_i = \sum_{k=0}^{n} p_k \to \infty (n \to \infty).$$

We now have

Theorem 2.1. *If* $\psi \prec \varphi$ *then* $S_\lambda^\beta(T,\psi,i)_0 \subset S_\lambda^\beta(T,\varphi,i)_0$.

Proof. By our assumptions we have $\psi(|\xi_k|) \le b\varphi(c|x_k|)$ and we have for all i,

$$\sum_{k=1}^{\infty} t_{nk}(i)\psi(|\xi_k|) \le b\sum_{k=1}^{\infty} t_{nk}(i)\varphi(c|\xi_k|) \le K\sum_{k=1}^{\infty} t_{nk}(i)\varphi(|\xi_k|)$$

for $b, c > 0$, where the constant K is connected with properties of φ. Thus, the condition $\sum_{k=1}^{\infty} t_{nk}(i)\psi(|\xi_k|) \ge \varepsilon$ implies the condition $\sum_{k=1}^{\infty} t_{nk}(i)\varphi(|\xi_k|) \ge \varepsilon$ and in consequence we get

$$\mu(K((T,\varphi),\varepsilon,i)) \subset \mu(K((T,\psi),\varepsilon,i))$$

and

$$lim_j \frac{1}{\lambda_j^\beta}\mu\Big(K((T,\varphi),\varepsilon)) \le lim_j \frac{1}{\lambda_j^\beta}\mu(K((T,\psi),\varepsilon,i)\Big).$$

This completes the proof. □

Theorem 2.2. *If* $0 < \beta \le \gamma \le 1$ *then* $S_\lambda^\beta(T,\varphi,i)_0 \subset S_\lambda^\gamma(T,\varphi,i)_0$.

Proof. Let $0 < \beta \le \gamma \le 1$. Then

$$\frac{1}{\lambda_j^\beta}\mu(K(T,\varphi),\varepsilon,i) \le \frac{1}{\lambda_j^\gamma}\mu(K(T,\varphi),\varepsilon,i)$$

for every $\epsilon > 0$ and finally we have that $S_\lambda^\beta(T,\varphi)_0 \subset S_\lambda^\gamma(T,\varphi,i)_0$. This proves the theorem. □

Theorem 2.3. *(a) If the matrix* T, *functions* f *and* φ *be given, then*

$$V_\lambda^\beta(T,\varphi,f)_0 \subset S_\lambda^\beta(T,\varphi,i)_0.$$

(b) If the φ *- function* $\varphi(u)$ *and the matrix* T *are given, and if the modulus function* f *is bounded, then*

$$S_\lambda^\beta(T,\varphi,i)_0 \subset V_\lambda^\beta(T,\varphi,f)_0.$$

(c) If the φ *- function* $\varphi(u)$ *and the matrix* T *are given, and if the modulus function* f *is bounded, then*

$$S_\lambda^\beta(T,\varphi,i)_0 = V_\lambda^\beta(T,\varphi,f)_0.$$

Proof. (a) Let f be a modulus function and let ε be a positive numbers. We write the following inequalities, for all i:

$$\frac{1}{\lambda_j^\beta}\sum_{n\in I_r} f\left(\left|\sum_{k=1}^{\infty} t_{nk}(i)\varphi(|\xi_k|)\right|\right)^{p_n}$$

$$\begin{aligned}
&= \frac{1}{\lambda_j^\beta}\sum_{n\in I_r^1} f\left(\left|\sum_{k=1}^{\infty} t_{nk}(i)\varphi(|\xi_k|)\right|\right)^{p_n} \\
&\geq \frac{1}{\lambda_j^\beta}\sum_{n\in I_j^1} [f(\varepsilon)]^{p_n} \\
&\geq \frac{1}{\lambda_j^\beta}\sum_{n\in I_j^1} \min\left([f(\varepsilon)]^{\inf p_n}, [f(\varepsilon)]^H\right) \\
&\geq \frac{1}{\lambda_j^\beta}\mu(K_\lambda^j(T,\varphi),\varepsilon)\min\left([f(\varepsilon)]^{\inf p_n}, [f(\varepsilon)]^H\right),
\end{aligned}$$

where

$$I_j^1 = \left\{n \in I_j : \sum_{k=1}^{\infty} t_{nk}(i)\varphi(|\xi_k|) \geq \varepsilon\right\}.$$

Finally, if $x \in V_\lambda^\beta((T,\varphi), f, p)_0$ then $x \in S_\lambda^\beta(T,\varphi,i)_0$.

(b) Let us suppose that $x \in S_\lambda^\beta(T,\varphi,i)_0$. If the modulus function f is a bounded function, then there exists an integer M such that $f(x) < M$ for $x \geq 0$. Let us take

$$I_j^2 = \left\{n \in I_j : \sum_{k=1}^{\infty} t_{nk}(i)\varphi(|\xi_k|) < \varepsilon\right\}.$$

Thus we have

$$\frac{1}{\lambda_j^\beta}\sum_{n\in I_j} f\left(\left|\sum_{k=1}^{\infty} t_{nk}(i)\varphi(|\xi_k|)\right|\right)^{p_n}$$

$$\begin{aligned}
&\leq \frac{1}{\lambda_j^\beta}\sum_{n\in I_j^1} f\left(\left|\sum_{k=1}^{\infty} t_{nk}(i)\varphi(|\xi_k|)\right|\right)^{p_n} \\
&+ \frac{1}{\lambda_j^\beta}\sum_{n\in I_j^2} f\left(\left|\sum_{k=1}^{\infty} t_{nk}(i)\varphi(|x_k|)\right|\right)^{p_n} \\
&\leq \frac{1}{\lambda_j^\beta}\sum_{n\in I_j^1} max(M^h, M^H) + \frac{1}{\lambda_j^\beta}\sum_{n\in I_j^2} [f(\varepsilon)]^{p_n} \\
&\leq max(M^h, M^H)\frac{1}{\lambda_j^\beta}\mu(K((A,\varphi),\varepsilon)) + max([f(\varepsilon)]^h, [f(\varepsilon)]^H).
\end{aligned}$$

Taking the limit as $\varepsilon \to 0$, we obtain that $x \in V_\lambda^\beta(T,\varphi,f,p)_0$.

The proof of (c) follows from (a) and (b).
This completes the proof. □

Theorem 2.4. *If a sequence $x = (\xi_k)$ is $S^\beta(T, \varphi, i)$ - convergent to L and*

$$\lim inf_j \left(\frac{\lambda_j^\beta}{j}\right) > 0$$

then it is $S_\lambda^\beta(T, \varphi, i)$ convergent to L, where

$$S^\beta(T, \varphi, i) = \{\xi = (\xi_k) : lim_j \frac{1}{j^\beta} \mu(K((T, \varphi), \varepsilon, i)) = 0\}.$$

Proof. For a given $\varepsilon > 0$, we have, for all i

$$\{n \in I_j : \sum_{k=0}^{\infty} t_{nk}(i)\varphi(|\xi_k - L|) \geq \varepsilon\} \subseteq \{n \leq j : \sum_{k=0}^{\infty} t_{nk}(i)\varphi(|\xi_k - L|) \geq \varepsilon\}.$$

Hence we have,

$$K(T, \varphi, \varepsilon) \subseteq K(T, \varphi, \varepsilon).$$

Finally the proof follows from the following inequality:

$$\frac{1}{j}\mu(K(T, \varphi, \varepsilon)) \geq \frac{1}{j}\mu(K(T, \varphi, \varepsilon)) = \frac{\lambda_j^\beta}{j}\frac{1}{\lambda_j^\beta}\mu(K((T, \varphi), \varepsilon, i).$$

This completes the proof. □

Theorem 2.5. *If $\lambda \in \triangle$ be such that for a particular $\beta, 0 < \beta < 1$, $\lim_j \frac{j - \lambda_j}{j^\beta} =$ 0 then $S_\lambda^\beta(T, \varphi, i) \subset S^\beta(T, \varphi, i)$.*

Proof. Let $\varepsilon > 0$ be given. Since $lim_j \frac{j - \lambda_j}{j^\beta} = 0$, we can choose $m \in N$ such that $\frac{j-\lambda_j}{j^\beta} < \frac{\delta}{2}$, for all $j \geq m$. Now observe that for all i,

$$\frac{1}{j^\beta}\left|\left\{n \leq j : \sum_{k=0}^{\infty} t_{nk}(i)\varphi(|\xi_k - L|) \geq \epsilon\right\}\right| = \frac{1}{j^\beta}\left|\left\{n \leq j - \lambda_j : \sum_{k=0}^{\infty} t_{nk}(i)\varphi(|\xi_k - L|) \geq \epsilon\right\}\right|$$
$$+ \frac{1}{j^\beta}\left|\left\{n \in I_j : \sum_{k=0}^{\infty} t_{nk}(i)\varphi(|\xi_k - L|) \geq \epsilon\right\}\right|$$
$$\leq \frac{j - \lambda_j}{j^\beta} + \frac{1}{j^\beta}\left|\left\{n \in I_j : \sum_{k=0}^{\infty} t_{nk}(i)\varphi(|\xi_k - L|) \geq \epsilon\right\}\right|$$
$$\leq \frac{\delta}{2} + \frac{1}{\lambda_j{}^\beta}\left|\left\{n \in I_j : \sum_{k=0}^{\infty} t_{nk}(i)\varphi(|\xi_k - L|) \geq \epsilon\right\}\right|,$$

for all $j \geq m$. □

References

1. Banach, S.: Theorie des Operations Lineaires (Warszawa) (1932)
2. Colak, R., Bektas, C.A.: μ-statistical convergence of order α. Acta Math. Scientia **31B**(3), 953–959 (2011)
3. Connor, J.: On strong matrix summability with respect to a modulus and statistical convergent. Canad. Math. Bull. **32**(2), 194–198 (1989)
4. Fast, H.: Sur la convergence statistique. Colloq. Math. **2**, 241–244 (1951)
5. Freedman, A.R., Sember, J.J., Raphel, M.: Some Cesaro-type summability spaces. Proc. London Math. Soc. **37**, 508–520 (1978)
6. Fridy, J.A.: On statistical convergence. Analysis **5**, 301–313 (1985)
7. Kolk, E.: Matrix summability of statistically convergent sequences. Analysis **18**, 77–83 (1993)
8. Lorentz, G.G.: A contribution to the theory of divergent sequences. Acta. Math. **80**, 167–190 (1948)
9. Maddox, I.J.: Spaces of strongly summable sequences. Quart. J. Math. **18**, 345–355 (1967)
10. Maddox, I.J.: Sequence spaces defined by a modulus. Math. Proc. Camb. Philos. Soc. **100**, 161–166 (1986)
11. Malkowsksy, E., Savaş, E.: Some λ- sequence spaces defined by a modulus. Archivum Math. **36**, 219–228 (2000)
12. Mursaleen, M., Çakan, C., Mohiuddine, S.A., Savaş, E.: Generalized statistical convergence and statistical core of double sequences. Acta Math. Sin. (Engl. Ser.) **26**(11), 2131–2144 (2010)
13. Mursaleen, M.: λ-statistical convergence. Math. Slovaca **50**, 111–115 (2000)
14. Ruckle, W.H.: FK spaces in which the sequence of coordinate vectors in bounded. Cand. J. Math. **25**, 973–978 (1973)
15. Šalát, T.: On statistically convergent sequences of real numbers. Math. Slovaca **30**, 139–150 (1980)
16. Savaş, E.: On some generalized sequence spaces defined by a modulus. Indian J. Pur. Appl. Math. **30**(5), 459–464 (1999)
17. Savaş, E.: Strong almost convergence and almost λ-statistical convergence. Hokkaido Math. J. **24**(3), 531–536 (2000)
18. Savaş, E.: Some sequence spaces and statistical convergence. Inter. J. Math. Math. Sci. **29**, 303–306 (2002)
19. Savaş, E.: On some sequence spaces and A-statistical convergence. Inter. J. Math. Math. Sci. **29**, 303–306 (2002)
20. Savaş, E., Kiliçman, A.: A note on some strongly sequence spaces. Abstr. Appl. Anal. **2011**, 8 (2011). Art. ID 598393
21. Savaş, E.: On some sequence spaces and A - statistical convergence. In: 2nd Strathmore International Mathematics Conference 1216 August 2013, Nairobi, Kenya (2013)
22. Savaş, E.: On asymptotically λ -statistical equivalent sequences of fuzzy numbers. New Math. Nat. Comput. **3**(3), 301–306 (2007)
23. Savaş, E.: A new sequence space of order α defined by modulus function. In: The Conference Fourth International Conference on Computational Mathematics and Engineering Sciences CMES-2019, Antalya/Turkey, 20–22 April 2019, pp. 301–306 (2019)
24. Schoenberg, I.J.: The integrability of certain functions and related summability methods. Am. Math. Monthly **66**, 361–375 (1959)
25. Waszak, A.: On the strong convergence in sequence spaces. Fasciculi Math. **33**, 125–137 (2002)

Unitary Operators on the Bergman Space

Chinmayee Padhy[1], Pabitra Kumar Jena[2], Susanta Kumar Paikray[1(✉)], and Hemen Dutta[3]

[1] Department of Mathematics, Veer Surendra Sai University of Technology, Burla 768018, Odisha, India
chinmayee.padhy83@gmail.com, skpaikray_math@vssut.ac.in

[2] P. G. Department of Mathematics, Berhampur University, Bhanjabihar, Ganjam, Berhampur 760007, Odisha, India
pabitramath@gmail.com

[3] Department of Mathematics, Gauhati University, Guwahati 781014, Odisha, India
hemen_dutta08@rediffmail.com

Abstract. The main objective of this paper is to explore some sufficient conditions for elementary operators on the Bergman space to be the average of unitaries. Moreover, some sufficient conditions for positive operators on the Bergman space are also proved to be unitarily equivalent.

Keywords: Toeplitz operators · Unitary operators · Unitarily equivalent operators · Bergman space and elementary operators

2010 Mathematics Subject Classification: 47B38 · 47B35 · 47B47

1 Introduction

The Bergman space is the Hilbert space of all holomorphic functions f on the open unit disk $\mathbb{D} = \{z \in \mathbb{C} : |z| < 1\}$, denoted as $A^2(\mathbb{D})$ for which

$$\|f\|_{A^2(\mathbb{D})} = \Big(\int_{\mathbb{D}} |f(z)|^2 dA(z)\Big)^{\frac{1}{2}} < \infty,$$

where $dA(z)$ is the normalized Lebesgue area measure on the open unit disk $\mathbb{D}$. If $h(z) = \sum_{n=0}^{\infty} a_n z^n$ and $k(z) = \sum_{n=0}^{\infty} b_n z^n$ are two functions in $A^2(\mathbb{D})$, then the inner product of h and k is given by

$$\langle h, k\rangle = \int_{\mathbb{D}} h(z)\overline{k(z)}dA(z) = \sum_{n=0}^{\infty} \frac{a_n \overline{b_n}}{n+1} .$$

The Bergman reproducing kernel is the function $K_z \in A^2(\mathbb{D})$ for $z \in \mathbb{D}$ such that $f(z) = \langle f, K_z\rangle$ for all $f \in A^2(\mathbb{D})$ and normalized reproducing kernel k_z is the function $\frac{K_z}{\|K_z\|_2}$. Here the norm $\|.\|_2$ and the inner product $\langle .,. \rangle$ are taken

H. Dutta et al. (Eds.): CMES 2019, AISC 1111, pp. 299–306, 2020.
https://doi.org/10.1007/978-3-030-39112-6_24

in the space $L^2(\mathbb{D}, dA)$. For any integer, $n \geq 0$, let $e_n(z) = \sqrt{n+1}z^n$. Then, $\{e_n\}_{n=0}^{\infty}$ forms an orthonormal basis for $A^2(\mathbb{D})$. The Toeplitz operator T_ϕ with symbol $\phi \in L^\infty(\mathbb{D})$ on $A^2(\mathbb{D})$ is defined by $T_\phi f = P(\phi f)$; here P is an orthogonal projection from $L^2(\mathbb{D}, dA)$ onto $A^2(\mathbb{D})$.

Let $Aut(\mathbb{D})$ be the Lie group of all automorphisms (biholomorphic mappings) of $\mathbb{D}$. We can define for each $a \in \mathbb{D}$, an automorphism ϕ_a in $Aut(\mathbb{D})$ such that

(i) $(\phi_a \, o \, \phi_a)(z) \equiv z$;
(ii) $\phi_a(0) = a, \phi_a(a) = 0$;
(iii) ϕ_a has a unique fixed point in $\mathbb{D}$.

In fact, $\phi_a(z) = \frac{a-z}{1-\bar{a}z}$ for all a and z in $\mathbb{D}$. It is easy to verify that the derivative of ϕ_a at z is equal to $-k_a(z)$. It implies the real Jacobian determinant of ϕ_a at z is $J_{\phi_a(z)} = |k_a(z)|^2 = \frac{(1-|a|^2)^2}{|1-\bar{a}z|^4}$. Given $a \in \mathbb{D}$ and f (any measurable function on $\mathbb{D}$), let us define a function $U_a f$ on $\mathbb{D}$ by $U_a f(w) = k_a(w) f(\phi_a(w))$. Notice that U_a is a bounded linear operator on $L^2(\mathbb{D}, dA)$ and $A^2(\mathbb{D})$ for all $a \in \mathbb{D}$. Further, it can be verified that $U_a^2 = I$, the identity operator, $U_a^* = U_a, U_a(A^2(\mathbb{D})) \subset A^2(\mathbb{D})$ and $U_a((A^2(\mathbb{D}))^\perp) \subset (A^2(\mathbb{D}))^\perp$ for all $a \in \mathbb{D}$. Thus, $U_a P = P U_a$ for all $a \in \mathbb{D}$ (see [8]). Let $H^\infty(\mathbb{D})$ denote the space of bounded analytic functions on $\mathbb{D}$. Let $\mathcal{L}(H)$ denote the algebra of bounded linear operators from a Hilbert space H into itself.

Let H and K be nonzero complex Hilbert spaces. The tensor product of $x \in H$ and $y \in K$ is a conjugate bilinear functional $x \otimes y : H \times K \longrightarrow \mathbb{C}$ defined by $(x \otimes y)(u, v) = \langle x, u \rangle \langle y, v \rangle$ for every $(u, v) \in H \times K$. The collection of all (finite) sums of tensors $x_i \otimes y_i$ with $x_i \in H$ and $y_i \in K$, denoted by $H \otimes K$, is a complex linear space equipped with an inner product $\langle .,. \rangle : (H \otimes K) \times (H \otimes K) \longrightarrow \mathbb{C}$ defined, for arbitrary $\sum_{i=1}^{N} x_i \otimes y_i$ and $\sum_{j=1}^{M} w_j \otimes z_j$ in $H \otimes K$, by $\left\langle \sum_{i=1}^{N} x_i \otimes y_i, \sum_{j=1}^{M} w_j \otimes z_j \right\rangle = \sum_{i=1}^{N} \sum_{j=1}^{M} \langle x_i, w_j \rangle \langle y_i, z_j \rangle$ (the same notation for the inner products on H, K and $H \otimes K$). The tensor product on $H \otimes K$ of two operators T in $\mathcal{L}(H)$ and S in $\mathcal{L}(K)$ is the operator $T \otimes S : H \otimes K \longrightarrow H \otimes K$ defined by $(T \otimes S) \sum_{i=1}^{N} x_i \otimes y_i = \sum_{i=1}^{N} T x_i \otimes S y_i$ for every $\sum_{i=1}^{N} x_i \otimes y_i \in H \otimes K$, which lies in $\mathcal{L}(H \otimes K)$. The complete inner product space $H \otimes K$ is denoted by $H \hat{\otimes} K$, which is the tensor product space of H and K. The extension of $T \otimes S$ over the Hilbert space $H \hat{\otimes} K$ denoted by $T \hat{\otimes} S$, is the tensor product of T and S on the tensor product space, which lies in $\mathcal{L}(H \hat{\otimes} K)$ (see [3,5,7]).

Let $T \in \mathcal{L}(H)$ and $U|T|$ be the polar decomposition of T. Some interesting concepts regarding elementary operators have been found in [2].

In this article, we explain some sufficient conditions for elementary operators on the Bergman space to be the average of unitaries and further discuss some conditions for positive operators to be unitarily equivalent on the Bergman space. Moreover, unitariness of elementary operators on the Bergman space as well as on the tensor product of Bergman spaces are focused in Sect. 2. Finally, in Sect. 3, we present some sufficient conditions for positive operators on the Bergman space to be unitarily equivalent.

2 Elementary Operators on $A^2(\mathbb{D})$

In this section, we obtain some sufficient conditions for elementary operators on $A^2(\mathbb{D})$ to be the average of unitary operators.

For $U_{a_i} \in \mathcal{L}(A^2(\mathbb{D})), a_i \in \mathbb{D}$ and $i = 1, 2, \cdots, n$, the n – tuple $\overline{U}$ defined by $\overline{U} = (U_{a_1}, U_{a_2}, \cdots U_{a_n})$.

Theorem 1. *Let $\|\phi\|_\infty \leq 1$ for a positive function $\phi \in L^\infty(\mathbb{D})$ and $\|T_{1+\phi}\| < 1$. Then the elementary operator $\delta_{\overline{U},\overline{U}}(T_\phi)$ can be represented as n times the average of $2n$ unitary operators.*

Proof. Since $\phi \geq 0$, we have $T_\phi \geq 0$ on $A^2(\mathbb{D})$. Then by ([1], Theorem 3.1), for every unitary operator W on $A^2(\mathbb{D})$, we obtain $\|W - T_\phi\| \leq \|I + T_\phi\| = \|T_{1+\phi}\| < 1$.

So T_ϕ is invertible. Let T_ϕ has the polar decomposition as $T_\phi = VP$ with P as positive operator and V as partial isometry on $A^2(\mathbb{D})$. Since T_ϕ is invertible, V becomes unitary and P is a positive invertible operator on the Bergman space $A^2(\mathbb{D})$.

Since $\|T_\phi\| \leq 1$, it implies $\|P\| \leq 1$. Therefore, $I - P^2 \geq 0$ and $\|I - P^2\| \leq 1$. Let us define two operators as $W_1 = P + i(I - P^2)^{\frac{1}{2}}$ and $W_2 = P - i(I - P^2)^{\frac{1}{2}}$. One can easily observe that, $W_1^* = W_2$, $W_1 W_1^* = P^2 + I - P^2 = I$, and $W_1^* W_1 = I$. Hence, $W_1 W_1^* = W_1^* W_1 = I$ and also $W_2 W_2^* = W_2^* W_2 = I$. That implies, W_1 and W_2 are unitary operators on $A^2(\mathbb{D})$. So, $T_\phi = VP = V(\frac{W_1+W_2}{2}) = \frac{1}{2}(VW_1 + VW_2) = \frac{V_1+V_2}{2}$, where $V_1 = VW_1$ and $V_2 = VW_2$ are two unitary operators on $A^2(\mathbb{D})$. Further, $T_{\phi \circ \phi_{a_i}} = U_{a_i}(\frac{V_1+V_2}{2})U_{a_i} = \frac{U_{a_i}V_1U_{a_i} + U_{a_i}V_2U_{a_i}}{2} = \frac{W_1^i + W_2^i}{2}$, where $W_1^i = U_{a_i}V_1U_{a_i}$ and $W_2^i = U_{a_i}V_2U_{a_i}$ are unitary operators for $i = 1, 2, 3, \cdots n$. Thus,

$$\begin{aligned}
\delta_{\overline{U},\overline{U}}(T_\phi) &= \sum_{i=1}^{n} U_{a_i} T_\phi U_{a_i} \\
&= U_{a_1} T_\phi U_{a_1} + U_{a_2} T_\phi U_{a_2} + \cdots U_{a_n} T_\phi U_{a_n} \\
&= T_{\phi \circ \phi_{a_1}} + T_{\phi \circ \phi_{a_2}} + \cdots + T_{\phi \circ \phi_{a_n}} \\
&= \sum_{i=1}^{n} T_{\phi \circ \phi_{a_i}} \\
&= \sum_{i=1}^{n} \frac{W_1^i + W_2^i}{2}.
\end{aligned}$$

Hence, the theorem is proved. □

Corollary 1. *Let $\phi_j \in L^\infty(\mathbb{D})$ for $j = 1, 2, \cdots, n$ having $\phi_j \geq 0$ with $\|\phi_j\|_\infty \leq \frac{1}{n}$ and $\left\| T_{1+\sum_{j=1}^{n} \phi_j} \right\| < 1$. Then the elementary operator $\delta_{\overline{U},\overline{U}}\left(T_{\sum_{j=1}^{n} \phi_j}\right)$ is n times as an average of $2n$ unitary operators.*

Proof. Since $\sum_{j=1}^n T_{\phi_j} = T_{\sum_{j=1}^n \phi_j}$ for $\phi_j \in L^\infty(\mathbb{D})$ and $\|\sum_{j=1}^n \phi_j\|_\infty \leq 1$ as $\|\phi_j\|_\infty \leq \frac{1}{n}$, then

$$\begin{aligned}\delta_{\overline{U},\overline{U}}(T_{\sum_{j=1}^n \phi_j}) &= \sum_{i=1}^{n} U_{a_i}(T_{\sum_{j=1}^n \phi_j})U_{a_i} \\ &= \sum_{i=1}^{n} U_{a_i}T_\psi U_{a_i}, \quad \text{where } \psi = \sum_{j=1}^{n} \phi_j.\end{aligned}$$

By our assumption $\psi \geq 0$, $\|\psi\|_\infty \leq 1$ and $\|T_{1+\psi}\| < 1$, it follows from Theorem 1 that $\delta_{\overline{U},\overline{U}}\left(T_{\sum_{j=1}^{n} \phi_j}\right)$ is n times the average of $2n$ unitary operators. □

Corollary 2. *Let $\phi_j \in H^\infty(\mathbb{D})$ for $j = 1, 2, \cdots, n$. Suppose that $\phi_j \geq 0$, $\|\phi_j\|_\infty \leq 1$ and $\|T_{1+\prod_{j=1}^n \phi_j}\| < 1$. Then the elementary operator $\delta_{\overline{U},\overline{U}}(T_{\prod_{j=1}^n \phi_j})$ is n times the average of $2n$ unitary operators.*

Proof. Since $\phi_j \in H^\infty(\mathbb{D})$ and $\|\phi_j\|_\infty \leq 1$ for $j = 1, 2, \cdots, n$ then, $\prod_{j=1}^n T_{\phi_j} = T_{\prod_{j=1}^n \phi_j}$ and $\|\prod_{j=1}^n \phi_j\|_\infty \leq 1$ respectively. Now

$$\begin{aligned}\delta_{\overline{U},\overline{U}}(T_{\prod_{j=1}^n \phi_j}) &= \sum_{i=1}^{\infty} U_{a_i}T_{\prod_{j=1}^n \phi_j}U_{a_i} \\ &= \sum_{i=1}^{\infty} U_{a_i}T_\Xi U_{a_i}, \quad \text{where } \Xi = \prod_{j=1}^{n} \phi_j \\ &= \sum_{i=1}^{\infty} T_{\Xi \circ \phi_{a_i}}.\end{aligned}$$

Therefore, the corollary is evident from Theorem 1. □

For unitary operators $U_i, V_i, W_i, X_i \in \mathcal{L}(A^2(\mathbb{D})), i = 1, 2, \cdots, n$, the $n-$tuples $\overline{U\hat{\otimes}V}$ and $\overline{W\hat{\otimes}X}$ are defined on $A^2(\mathbb{D})\hat{\otimes}A^2(\mathbb{D})$ respectively, by

$\overline{U\hat{\otimes}V} = (U_1\hat{\otimes}V_1, U_2\hat{\otimes}V_2, \cdots, U_n\hat{\otimes}V_n)$ *and* $\overline{W\hat{\otimes}X} = (W_1\hat{\otimes}X_1, W_2\hat{\otimes}X_2, \cdots, W_n\hat{\otimes}X_n)$.
We introduce here the elementary operator on tensor product Bergman space induced by $\overline{U\hat{\otimes}V}$ and $\overline{W\hat{\otimes}X}$.

Corollary 3. *Let ϕ and $\psi \in L^\infty(\mathbb{D})$ with $\|\phi\|_\infty \leq 1, \|\psi\|_\infty \leq 1$. Suppose $T_\phi = Q|T_\phi|, T_\psi = R|T_\psi|$ are polar decompositions of T_ϕ and T_ψ respectively. If T_ϕ and T_ψ are of finite rank, then the elementary operator $\delta_{\overline{U\hat{\otimes}V},\overline{W\hat{\otimes}X}}(T_\phi\hat{\otimes}T_\psi)$ on $A^2(\mathbb{D})\hat{\otimes}A^2(\mathbb{D})$ can be expressed as n times the average of $4n$ unitary operators.*

Proof. Assume $T_\phi = Q|T_\phi|, T_\psi = R|T_\psi|$ are polar decompositions of T_ϕ and T_ψ respectively. Given T_ϕ and T_ψ are finite rank operators, this implies $\dim\ker(T_\phi) = \dim\ker(T_{\overline{\phi}})$ and $\dim\ker(T_\psi) = \dim\ker(T_{\overline{\psi}})$. Since T_ϕ and T_ψ are contractions, so Q, R are unitary. Now

$$\begin{aligned}\delta_{\overline{U\hat{\otimes}V},\overline{W\hat{\otimes}X}}(T_\phi\hat{\otimes}T_\psi) &= \sum_{i=1}^{n}(U_i\hat{\otimes}V_i)(T_\phi\hat{\otimes}T_\psi)(W_i\hat{\otimes}X_i)\\ &= \sum_{i=1}^{n}(U_iT_\phi W_i)\hat{\otimes}(V_iT_\psi X_i).\end{aligned}$$

Therefore, T_ϕ is the average of two unitary operators and T_ψ is also the average of another two unitary operators. Hence, the result follows from Theorem 1. □

3 Unitarily Equivalent Operators

In this section, we present some sufficient conditions for positive operators on the Bergman space $A^2(\mathbb{D})$ to be unitarily equivalent.

Theorem 2. *Let A, B be two positive operators and U, V be two unitary operators on $A^2(\mathbb{D})$. Suppose $\|B^{\frac{1}{2}}Uk_z\| \leq \|A^{\frac{1}{2}}k_z\| \leq \frac{1}{\|B^{\frac{1}{2}}k_z\|}\langle Re(V^*B)k_z, k_z\rangle$ for all $k_z \in A^2(\mathbb{D})$ and $U(B+\alpha I)^{\frac{1}{2}}V$ has a single limit point in its spectrum for $\alpha > 0$. Then A and B are unitarily equivalent.*

Proof. Since A, B are two positive operators on $A^2(\mathbb{D})$ and $\|B^{\frac{1}{2}}Uk_z\| \leq \|A^{\frac{1}{2}}k_z\| \leq \frac{1}{\|B^{\frac{1}{2}}k_z\|}\langle Re(V^*B)k_z, k_z\rangle$ for all $k_z \in A^2(\mathbb{D})$, then

$$\langle U^*BUk_z, k_z\rangle^{\frac{1}{2}} \leq \langle Ak_z, k_z\rangle^{\frac{1}{2}} \leq \frac{1}{\|B^{\frac{1}{2}}k_z\|}\langle Re(V^*B)k_z, k_z\rangle$$

for all $k_z \in A^2(\mathbb{D})$. This implies $U^*BU \leq A$. Moreover, from Heinz Inequality [4], we obtain

$$\begin{aligned}\langle Ak_z, k_z\rangle^{\frac{1}{2}} &\leq \frac{1}{\|B^{\frac{1}{2}}k_z\|}\langle Re(V^*B)k_z, k_z\rangle\\ &= \frac{1}{\|B^{\frac{1}{2}}k_z\|}Re\langle(V^*B)k_z, k_z\rangle\\ &\leq \frac{1}{\|B^{\frac{1}{2}}k_z\|}|\langle(V^*B)k_z, k_z\rangle|\\ &\leq \frac{1}{\langle Bk_z, k_z\rangle^{\frac{1}{2}}}\langle Bk_z, k_z\rangle^{\frac{1}{2}}\langle BVk_z, Vk_z\rangle^{\frac{1}{2}}\\ &= \langle V^*BVk_z, k_z\rangle^{\frac{1}{2}}.\end{aligned}$$

This implies, $A \leq V^*BV$. Therefore, $U^*BU \leq A \leq V^*BV$. Since $B \geq 0$, choose $\alpha > 0$ such that $(B + \alpha I)^{\frac{1}{2}}$ turns out to be positive and invertible. Let's define $S = U^*(B + \alpha I)^{\frac{1}{2}}V$. Then $SS^* \leq A + \alpha I \leq S^*S$. Hence, S is hyponormal. Further, $|S| = V^*(B + \alpha I)^{\frac{1}{2}}V$. Let $S = W|S|$ be the polar decomposition of S with $|S|$ as positive and W as a partial isometry operator on $A^2(\mathbb{D})$. Since S is invertible, the partial isometry W becomes a unitary operator (say V). Now $S = V|S| = VV^*(B + \alpha I)^{\frac{1}{2}}V = (B + \alpha I)^{\frac{1}{2}}V = UU^*(B + \alpha I)^{\frac{1}{2}}V = US$. This implies, $U = I$. Therefore, by our assumption, we obtain S is a hyponormal operator on $A^2(\mathbb{D})$ with a single limit point in its spectrum. Hence it follows from [6], that S is normal. The normality of S implies that A and B are unitarily equivalent. □

Corollary 4. *Let A, B be two positive operators and U, V be two unitary operators on $A^2(\mathbb{D})$. Suppose $\|B^{\frac{1}{2}}Uk_z\| \leq \|A^{\frac{1}{2}}k_z\| \leq \frac{1}{\|B^{\frac{1}{2}}k_z\|}\langle Re(V^*B)k_z, k_z\rangle$ for all $k_z \in A^2(\mathbb{D})$. If $(B + \alpha I)^{\frac{1}{2}}$ is compact for $\alpha > 0$, then A and B are unitarily equivalent.*

Proof. Since $\|B^{\frac{1}{2}}Uk_z\| \leq \|A^{\frac{1}{2}}k_z\| \leq \frac{1}{\|B^{\frac{1}{2}}k_z\|}\langle Re(V^*B)k_z, k_z\rangle$ for all $k_z \in A^2(\mathbb{D})$, then by Theorem 2, $U^*BU \leq A \leq V^*BV$. Put $S = U^*(B + \alpha I)^{\frac{1}{2}}V$. Then $SS^* \leq A + \alpha I \leq S^*S$. Thus, S is hyponormal. By our assumption $(B + \alpha I)^{\frac{1}{2}}$ is compact and this implies S is a compact operator on $A^2(\mathbb{D})$. Since S is hyponormal and also S is normal. Hence, the result follows. □

Corollary 5. *Let A_i, B_i be positive operators on $A^2(\mathbb{D})$ for $i = 1, 2, 3, \cdots, n$ and U, V be two unitary operators on $A^2(\mathbb{D})$. Suppose that $\left\|\left(\sum_{i=1}^n B_i\right)^{\frac{1}{2}} Uk_z\right\| \leq \left\|\left(\sum_{i=1}^n A_i\right)^{\frac{1}{2}} k_z\right\| \leq \frac{1}{\left\|\left(\sum_{i=1}^n B_i\right)^{\frac{1}{2}} k_z\right\|}\left\langle Re\left(V^*\sum_{i=1}^n B_i\right)k_z, k_z\right\rangle$ for all $k_z \in A^2(\mathbb{D})$ and the operator $\left(\sum_{i=1}^n (B_i + \alpha_i I)\right)^{\frac{1}{2}}$ is compact on $A^2(\mathbb{D})$ for all $\alpha_i > 0$. Then $\sum_{i=1}^n A_i$ and $\sum_{i=1}^n B_i$ are unitarily equivalent.*

Proof. Since A_i, B_i are positive operators on $A^2(\mathbb{D})$ for $i = 1, 2, 3, \cdots, n$ and $\left\|\left(\sum_{i=1}^n B_i\right)^{\frac{1}{2}} Uk_z\right\| \leq \left\|\left(\sum_{i=1}^n A_i\right)^{\frac{1}{2}} k_z\right\| \leq \frac{1}{\left\|\left(\sum_{i=1}^n B_i\right)^{\frac{1}{2}} k_z\right\|}\left\langle Re\left(V^*\sum_{i=1}^n B_i\right)k_z, k_z\right\rangle$ for all $k_z \in A^2(\mathbb{D})$, then $\langle U^*\sum_{i=1}^n B_iUk_z, k_z\rangle^{\frac{1}{2}} \leq \langle\sum_{i=1}^n A_ik_z, k_z\rangle^{\frac{1}{2}}$

$\leq \frac{1}{\|(\sum_{i=1}^{n} B_i)^{\frac{1}{2}} k_z\|} \langle Re(V^* \sum_{i=1}^{n} B_i) k_z, k_z \rangle$ for all $k_z \in A^2(\mathbb{D})$. From the left hand inequality, we get $U^* \sum_{i=1}^{n} B_i U \leq \sum_{i=1}^{n} A_i$ and from the Heinz inequality [4], we obtain

$$
\begin{aligned}
\langle (\sum_{i=1}^{n} A_i) k_z, k_z \rangle^{\frac{1}{2}} &\leq \frac{1}{\|(\sum_{i=1}^{n} B_i)^{\frac{1}{2}} k_z\|} \langle Re(V^* (\sum_{i=1}^{n} B_i)) k_z, k_z \rangle \\
&= \frac{1}{\|(\sum_{i=1}^{n} B_i)^{\frac{1}{2}} k_z\|} Re\langle (V^* \sum_{i=1}^{n} B_i) k_z, k_z \rangle \\
&\leq \frac{1}{\|(\sum_{i=1}^{n} B_i)^{\frac{1}{2}} k_z\|} |\langle (V^* \sum_{i=1}^{n} B_i) k_z, k_z \rangle| \\
&\leq \frac{1}{\langle (\sum_{i=1}^{n} B_i) k_z, k_z \rangle^{\frac{1}{2}}} \langle (\sum_{i=1}^{n} B_i) k_z, k_z \rangle^{\frac{1}{2}} \langle (\sum_{i=1}^{n} B_i) V k_z, V k_z \rangle^{\frac{1}{2}} \\
&= \langle V^* (\sum_{i=1}^{n} B_i) V k_z, k_z \rangle^{\frac{1}{2}}.
\end{aligned}
$$

This implies, $\sum_{i=1}^{n} A_i \leq V^*(\sum_{i=1}^{n} B_i) V$. Therefore, $U^*(\sum_{i=1}^{n} B_i) U \leq \sum_{i=1}^{n} A_i \leq V^*(\sum_{i=1}^{n} B_i) V$. Since $B_i \geq 0$, choose $\alpha_i > 0$ such that $(\sum_{i=1}^{n}(B_i + \alpha_i I))^{\frac{1}{2}}$ turns out to be positive. Put $S = U^*(\sum_{i=1}^{n}(B_i + \alpha_i I))^{\frac{1}{2}} V$. Then $SS^* \leq \sum_{i=1}^{n}(A_i + \alpha_i I) \leq S^* S$. Hence, S is hyponormal. By our assumption $\left(\sum_{i=1}^{n}(B_i + \alpha_i I)\right)^{\frac{1}{2}}$ is compact on $A^2(\mathbb{D})$. Hence it follows from Corollary 4 that S is normal. The normality of S implies that $\sum_{i=1}^{n} A_i$ and $\sum_{i=1}^{n} B_i$ are unitarily equivalent. □

Corollary 6. *Let A_i, B_i be positive operators on $A^2(\mathbb{D})$ for $i = 1, 2, 3, \cdots, n$ and V be a unitary operators on $A^2(\mathbb{D})$. Suppose that* $\left\| \left(\sum_{i=1}^{n} B_i \right)^{\frac{1}{2}} k_z \right\|$

$\leq \left\| \left(\sum_{i=1}^{n} A_i \right)^{\frac{1}{2}} k_z \right\| \leq \frac{1}{\left\| \left(\sum_{i=1}^{n} B_i \right)^{\frac{1}{2}} k_z \right\|} \left\langle Re \left(V^* \sum_{i=1}^{n} B_i \right) k_z, k_z \right\rangle$ *for all* $k_z \in$ $A^2(\mathbb{D})$ *and the operator* $\left(\sum_{i=1}^{n} (B_i + \alpha_i I) \right)^{\frac{1}{2}}$ *is compact on* $A^2(\mathbb{D})$ *for all* $\alpha_i > 0$. *Then* $\sum_{i=1}^{n} A_i$ *and* $\sum_{i=1}^{n} B_i$ *are unitarily equivalent.*

Proof. By taking $U = I$ in Corollary 5, we get $\sum_{i=1}^{n} B_i \leq \sum_{i=1}^{n} A_i \leq V^* \sum_{i=1}^{n} B_i V$. Put $T = (\sum_{i=1}^{n} (B_i + \alpha_i I))^{\frac{1}{2}} V$. Now $TT^* \leq \sum_{i=1}^{n} (A_i + \alpha_i I) \leq T^*T$. Thus, T is hyponormal. Hence, the result follows. □

References

1. Aiken, J.G., Erdos, J.A., Goldstein, J.A.: Unitary approximation of positive operators. Ill. J. Math. **24**, 61–72 (1980)
2. Botelho, F., Jamison, J.: Elementary operators and the Aluthge transform. Linear Algebra Appl. **432**, 275–282 (2010)
3. Das, N., Jena, P.K.: On the range and kernel of Toeplitz and little Hankel operators. Methods Funct. Anal. Topology **19**, 55–67 (2013)
4. Heinz, E.: On an inequality for linear operators in Hilbert space. Report on Operator Theory and Group Representations, National Academy of Sciences-National Research Council, Washington, DC, vol. 387, pp. 27–29 (1995)
5. Kubrusly, C.S.: A concise introduction to tensor product. Far East J. Math. Sci. **22**, 137–174 (2006)
6. Stampfli, J.G.: Hyponormal operators. Pac. J. Math. **12**, 1453–1458 (1962)
7. Tanahashi, K., Cho, M.: Tensor products of log-hyponormal and of class $A(s,t)$ operators. Glasg. Math. J. **46**, 91–95 (2004)
8. Zhu, K.: Operator Theory in Function Spaces. Marcel Dekker, New York (1990)

Some Relations Between the Sets of *f*-Statistically Convergent Sequences

Rifat Çolak(✉)

Department of Mathematics, Firat University, Elazig, Turkey
rftcolak@gmail.com

Abstract. In this study we first established the relations between *f*-density and *g*-density of a subset of the set of positive integers for any modulus functions *f* and *g*. Using the obtained facts we establish the relationship between the sets S_f and S_g of statistically convergent and BS_f and BS_g of statistically bounded sequences which defined by modulus functions *f* and *g*.

Keywords: Density · Modulus function · Statistical convergence · Statistical boundedness

1 Introduction

The idea of statistical convergence first appeared in a study by Zygmund [17] in 1935. This concept was introduced for the first time by Steinhaus [16] and Fast [5] independently. After then the concept studied by Schoenberg [15]. In the last decades and under different names the subject was discussed in many different theories such as in Fourier analysis theory, number theory, ergodic theory, measure theory, trigonometric series and Banach spaces. It was further investigated from the sequence spaces and summability theory point of view and via summability theory by Fridy [6], Connor [4], Salát [14] and many others.

The idea of a modulus function was structured by Nakano [12] in 1953. Ruckle [13] and Maddox [10] have introduced some sequence spaces by using a modulus function. Other than them, Ghosh and Srivastava [8], Bhardwaj and Singh [2] and some others used a modulus function in order to establish a number of sequence spaces. Statistical boundedness and some generalizations have been studied by some mathematicians (see [9]).

Throughout the study, l_∞ and c will denote the spaces of bounded and convergent sequences of real numbers, respectively.

Now, we will remember some concepts and definitions which are needful in the study.

Let $\mathbb{N}$ be the set of positive integers. The natural density of a set $H \subseteq \mathbb{N}$ is defined by

$$\delta(H) = \lim_{n\to\infty} \frac{1}{n} |\{k \leq n \ : \ k \in H\}|$$

H. Dutta et al. (Eds.): CMES 2019, AISC 1111, pp. 307–314, 2020.
https://doi.org/10.1007/978-3-030-39112-6_25

where $|\{k \leq n \ : \ k \in H\}|$ indicates the number of elements of H not exceeding n. One easily may see that $\delta(\mathbb{N}) = 1$ and $\delta(K) = 0$ if $K \subset \mathbb{N}$ is a finite set and $\delta(K^c) = \delta(\mathbb{N}) - \delta(K) = 1 - \delta(K)$, where $K^c = \mathbb{N} - K$.

A number sequence (x_k) is said to be statistically convergent to the number l if for each $\varepsilon > 0$, the set $\{k \in \mathbb{N} \ : |x_k - l| \geq \varepsilon\}$ has natural density zero. We write $S - \lim x_k = l$ for a sequence (x_k) which is statistically convergent to l.

A number sequence (x_k) is said to be statistically bounded, if $\delta(\{k \in \mathbb{N} \ : \ |x_k| > M\}) = 0$ for some real number $M > 0$.

Throughout the paper notations S and BS symbolize the classes of statistically convergent and statistically bounded sequences of real numbers, respectively (see [7, 15]).

A function $f \ : [0, \infty) \to [0, \infty)$ is called a modulus or modulus function if

(i) $f(x) = 0$ if and only if $x = 0$,
(ii) $f(x + y) \leq f(x) + f(y)$, for all $x \geq 0$, $y \geq 0$,
(iii) f is increasing,
(iv) f is continuous from the right at 0.

From these properties it is clear that a modulus function must be continuous everywhere on $[0, \infty)$. Any modulus may be bounded or unbounded. For instance, $f(x) = \frac{x}{1+x}$ is bounded but $f(x) = x^p$, where $0 < p \leq 1$, is unbounded.

Aizpuru et al. presented the following definition of f - density of a set A in [1]. The $f - density$ of a set $A \subseteq \mathbb{N}$ is defined by

$$\delta_f(A) = \lim_{n\to\infty} \frac{f(|\{k \leq n \ : \ k \in A\}|)}{f(n)}$$

if the limit exists, where f is an unbounded modulus function.

The $f-$ density of A reduces to the natural density $\delta(A)$ of A in case $f(x) = x$.

It is well known that $\delta(A) + \delta(\mathbb{N} - A) = 1$ for the natural density. But this result is not true for f-density, i.e. $\delta_f(A) + \delta_f(\mathbb{N} - A) = 1$ does not hold in general. Indeed, if we take unbounded modulus $f(x) = \log(x + 1)$ and $A = \{2n \ : \ n \in \mathbb{N}\}$ then $\delta_f(A) = \delta_f(\mathbb{N} - A) = 1$.

However, we have the fact that if $\delta_f(A) = 0$ then $\delta_f(\mathbb{N} - A) = 1$ in case of $f-$ density.

For any finite set A, $f-$ density has similar properties with naturel density of the set, that is $\delta_f(A) = 0$ and $\delta_f(A) + \delta_f(\mathbb{N} - A) = 1$.

Given any unbounded modulus function f and given a set $A \subseteq \mathbb{N}$, $\delta_f(A) = 0$ implies that $\delta(A) = 0$ (see [1]). But the converse need not be true, in general. Indeed for the unbounded modulus $f(x) = \log(x + 1)$ and set $A = \left\{i^2 \ : \ i = 1, 2, 3, \ldots\right\}$ we have $\delta(A) = 0$ but $\delta_f(A) = \frac{1}{2}$. However $\delta(A) = 0$ implies $\delta_f(A) = 0$ is always true if $A \subseteq \mathbb{N}$ is finite, regardless of selection of unbounded modulus f.

In this study we establish the relations between S_f and S_g, BS_f and BS_g, S_f and BS_g for different modulus functions f and g under some conditions on the considered modulus functions. However the relations between the sets S, S_f, BS and BS_f are known already for a modulus f.

2 Definitions and Basic Results

We need the following facts in the sequel.

Lemma 2.1 ([11]). The limit $\lim_{t\to\infty} \frac{f(t)}{t} = \beta$ exists for any modulus f.

The following Theorem gives the relation between $f-$ densities of a set of positive integers for different modulus functions. This helps us to construct the relations between statistically convergent and statistically bounded sequence sets defined by modulus functions.

Theorem 2.2. Let f and g be two unbounded modulus functions. Then for a set $A \subseteq \mathbb{N}$

(*i*) if

$$\lim_{t\to\infty} \frac{f(t)}{g(t)} > 0 \tag{1}$$

then $\delta_g(A) = 0$ implies $\delta_f(A) = 0$ in case the limit in (1) exists;

(*ii*) if

$$0 < \lim_{t\to\infty} \frac{f(t)}{g(t)} = \alpha < \infty \tag{2}$$

then $\delta_g(A) = 0 \Leftrightarrow \delta_f(A) = 0$ in case the limit exists.

Proof. Let $A \subseteq \mathbb{N}$, $A(n) = \{k \leq n : k \in A\}$ and f, g be two unbounded modulus functions.

(*i*) Suppose that $\lim_{t\to\infty} \frac{f(t)}{g(t)} = \alpha > 0$. Then given any $\varepsilon > 0$ there exists a real number t_0 such that $(\alpha - \varepsilon)\, g(t) < f(t) < (\alpha + \varepsilon)\, g(t)$ if $t > t_0$ (We may choose $\varepsilon > 0$ so small that $\alpha - \varepsilon > 0$). Therefore we have the inequality $f(t) < 2\alpha g(t)$ if $t > t_0$. Now we may write the inequality

$$\frac{g(|A(n)|)}{g(n)} \geq \frac{1}{2\alpha} \frac{f(|A(n)|)}{f(n)} \frac{f(n)}{g(n)}$$

if $|A(n)| > t_0$ (so $n > t_0$). Since $\lim_{t\to\infty} \frac{f(t)}{g(t)} = \alpha > 0$ from the above inequality we obtain $\lim_{n\to\infty} \frac{f(|A(n)|)}{f(n)} = 0$ if $\lim_{n\to\infty} \frac{g(|A(n)|)}{g(n)} = 0$. This means that $\delta_f(A) = 0$ if $\delta_g(A) = 0$.

(*ii*) We may write the following equality

$$\frac{g(|A(n)|)}{g(n)} = \frac{g(|A(n)|)}{f(|A(n)|)} \cdot \frac{f(|A(n)|)}{f(n)} \cdot \frac{f(n)}{g(n)}.$$

Suppose $\lim_{t\to\infty} \frac{f(t)}{g(t)} = \alpha$ $(0 < \alpha < \infty)$ and so that $\lim_{t\to\infty} \frac{g(t)}{f(t)} = 1/\alpha$. Using this fact from the above equality we obtain $\lim_{n\to\infty} \frac{f(|A(n)|)}{f(n)} = 0$ if and only if $\lim_{n\to\infty} \frac{g(|A(n)|)}{g(n)} = 0$. This means that $\delta_f(A) = 0$ if and only if $\delta_g(A) = 0$.

Corollary 2.3. For any $A \subseteq \mathbb{N}$ and any unbounded modulus f providing

$$\lim_{t\to\infty} \frac{f(t)}{t} > 0 \tag{3}$$

we have $\delta_f(A) = 0 \Leftrightarrow \delta(A) = 0$.

Proof. It is known that $\delta_f(A) = 0$ implies $\delta(A) = 0$ for any unbounded modulus f (see [1]). Taking $g(t) = t$ in Theorem 2.2 (i) we obtain $\delta(A) = 0$ implies $\delta_f(A) = 0$. (Note that the limit given in (3) exists by Lemma 2.1).

3 Main Results

In this section we will give the main results of this study in which the relations between the sets S_f and S_g, BS_f and BS_g, S_f and BS_g will be obtained for different unbounded modulus functions under some conditions.

Using $f-$ density we recall the following definition which is given by Aizpuru et al. in [1].

Definition 3.1. Let f be an unbounded modulus function. Then it is said that the sequence (x_k) is *f-statistically convergent* to l or *$S_f -$ convergent to l*, if

$$\lim_{n\to\infty} \frac{1}{f(n)} f(|\{k \leq n \ : \ |x_k - l| \geq \varepsilon\}|) = 0,$$

that is $\delta_f(\{k \in \mathbb{N} \ : \ |x_k - l| \geq \varepsilon\}) = 0$ for every $\varepsilon > 0$. We write $S_f - \lim x_k = l$ if (x_k) is f-statistically convergent to l.

Throughout, S_f will denote the class of sequences which are f-statistically convergent. We write S instead of S_f in case $f(t) = t$.

In the light of above knowledge, we may give the following two results (see [1]).

Lemma 3.2. If a sequence (x_k) is f -statistically convergent, then its $S_f -$lim is unique.

Lemma 3.3. Let f, g be two unbounded modulus functions. If $S_f - \lim x_k = l$ and $S_g - \lim x_k = l'$ then $l = l'$.

It easily can be checked that every convergent sequence is f-statistically convergent for any unbounded modulus f. But the converse is not true. For example, the sequence (x_k) defined by

$$x_k = \begin{cases} 0, \ k = n^2 \\ & n \in \mathbb{N} \\ 2, \ k \neq n^2 \end{cases}$$

is not convergent, but it is f-statistically convergent to 2 for modulus $f(x) = x^p$, $0 < p \leq 1$.

Theorem 3.4. Let f and g be two unbounded modulus functions. Then

(*i*) if (1) holds then a sequence (x_k) is f -statistically convergent (with same limit) if it is $g-$ statistically convergent, that is $S_g \subseteq S_f$ and the inclusion may be strict.

(*ii*) if (2) holds then a sequence (x_k) is f-statistically convergent if and only if it is $g-$ statistically convergent, that is $S_g = S_f$.

Proof. (*i*) Suppose (x_k) is $g-$ statistically convergent to l, that is $S_g - \lim x_k = l$. Define $A = \{k \in \mathbb{N} : |x_k - l| \geq \varepsilon\}$. Then

$$\delta_g(A) = \lim_{n\to\infty} \frac{g(|\{k \leq n : |x_k - l| \geq \varepsilon\}|)}{g(n)} = 0$$

and this implies

$$\delta_f(A) = \lim_{n\to\infty} \frac{f(|\{k \leq n : |x_k - l| \geq \varepsilon\}|)}{f(n)} = 0$$

if (1) holds by Theorem 2.2 (*i*). This means that (x_k) is $f-$ statistically convergent to l.

Proof of (*ii*) follows from Theorem 2.2 (*ii*)

The following example shows that the inclusion $S_g \subseteq S_f$ may be strict at least for some special modulus functions f and g. It can easily be shown that the sequence (x_k) defined by

$$x_k = \begin{cases} k, & k = n^2 \\ & \qquad n \in \mathbb{N} \\ 1 + \frac{1}{k}, & k \neq n^2 \end{cases} \tag{4}$$

is $f-$ statistically convergent to 1, but it is not $g-$ statistically convergent, where $g(t) = \log(t + 1)$ and $f(t) = t^{1/2}$. Therefore $(x_k) \in S_f - S_g$ and so that the inclusion $S_g \subset S_f$ is strict.

Note that in Theorem 3.4 also $S_f - \lim x_k = S_g - \lim x_k$ holds if $S_g - \lim x_k$ exists. Unfortunately the existence of $S_f - \lim x_k$ does not require the presence of $S_g - \lim x_k$.

Corollary 3.5. Let f be an unbounded modulus function. If (3) holds then $S_f = S$.

Proof. The inclusion $S_f \subseteq S$ is given in [1] for any unbounded modulus f. To show that $S \subseteq S_f$, let (3) be hold and the sequence (x_k) be statistically convergent to l. Define $A = \{k \in \mathbb{N} : |x_k - l| \geq \varepsilon\}$. Using Corollary 2.3 the inclusion $S \subseteq S_f$ follows from the fact "$\delta(A) = 0$ implies $\delta_f(A) = 0$".

Note that from Corollary 3.5 we obtain that $S - \lim x_k = S_f - \lim x_k$ in case $S_f - \lim x_k$ exists.

The following definition may be given by using $f-$ density.

Definition 3.6. Let f be an unbounded modulus function. Then a sequence (x_k) is said to be *f-statistically Cauchy sequence*, if there exists a positive integer N such that $\delta_f(\{k \in \mathbb{N} : |x_k - x_N| \geq \varepsilon\}) = 0$ for every $\varepsilon > 0$.

Theorem 3.7. Let f and g be two unbounded modulus functions. Then

(i) if (1) holds then a $g-$ statistically Cauchy sequence is f-statistically Cauchy sequence;
(ii) if (2) holds then a sequence (x_k) is $g-$ statistically Cauchy sequence if and only if it is f -statistically Cauchy sequence.

Taking $A = \{k \in \mathbb{N} : |x_k - x_N| \geq \varepsilon\}$ the proof follows from Theorem 2.2 (i) and (ii).

One may to refer [1] for the next result.

Corollary 3.8. Let f be an unbounded modulus function. Then a sequence of real numbers is f-statistically convergent if and only if it is an f-statistically Cauchy sequence.

Bhardwaj et al. presented the following definition in [3].

Definition 3.9. Let f be an unbounded modulus function. Then a number sequence (x_k) is said to be $f-$ *statistically bounded*, if $\delta_f(\{k \in \mathbb{N} : |x_k| > M\}) = 0$ for some real number $M > 0$.

The set of all f-statistically bounded sequences of real numbers will be denoted by BS_f.

Lemma 3.10. [3] Any bounded sequence is f-statistically bounded for any unbounded modulus f, that is $l_\infty \subset BS_f$ but the converse does not hold in general and the inclusion may be strict.

Theorem 3.11. Let f and g be two unbounded modulus functions. Then

(i) if (1) holds then a $g-$ statistically bounded sequence is f-statistically bounded, that is $BS_g \subseteq BS_f$ and the inclusion may be strict;
(ii) if (2) holds then a sequence is $g-$ statistically bounded if and only if it is f-statistically bounded, that is $BS_g = BS_f$.

Proof. Let the sequence (x_k) be $g-$ statistically bounded. Then there exists a real number $M > 0$ such that $\delta_g(\{k \in \mathbb{N} : |x_k| > M\}) = 0$. Taking $A = \{k \in \mathbb{N} : |x_k| > M\}$ the proof of (i) and (ii) follows from Theorem 2.2 (i) and (ii), respectively.

Note that the following example shows that the inclusion $BS_g \subseteq BS_f$ may be strict at least for some special modulus functions f and g. Indeed the sequence (x_k) defined by (4) is $f-$ statistically bounded but it is not $g-$ statistically bounded for the modulus functions $g(t) = \log(t+1)$ and $f(t) = t^{1/2}$, since $\delta_g(\{k \in \mathbb{N} : |x_k| > 3\}) = \frac{1}{2}$ and $\delta_f(\{k \in \mathbb{N} : |x_k| > 3\}) = 0$. Therefore $(x_k) \in BS_f - BS_g$ and so that the inclusion $BS_g \subset BS_f$ is strict.

Corollary 3.12. For any unbounded modulus f we have

(*i*) $BS_f \subseteq BS$;
(*ii*) if (3) holds $BS_f = BS$.

Proof. (*i*) follows from the fact "given a set $A \subseteq \mathbb{N}$, $\delta_f(A) = 0$ implies $\delta(A) = 0$ for any unbounded modulus f" and (*ii*) follows from Corollary 2.3.

Theorem 3.13. Let f and g be two unbounded modulus functions. If (1) holds then a $g-$ statistically convergent sequence is f-statistically bounded, that is $S_g \subseteq BS_f$ and the inclusion may be strict.

Proof. Suppose that the sequence (x_k) is $g-$ statistically convergent to l. Let $\varepsilon > 0$ be given and define $A(n) = \{k \leq n \ : \ |x_k - l| \geq \varepsilon\}$ and $B(n) = \{k \leq n \ : \ |x_k - l| \geq M\}$ for a number M $(> \varepsilon)$ large enough. Now since clearly $|A(n)| \geq |B(n)|$ for every $n \in \mathbb{N}$ we have that $\delta_g(A) \geq \delta_g(B)$ and so that $\delta_g(A) = 0$ implies $\delta_g(B) = 0$. If (1) holds then $\delta_g(B) = 0$ implies $\delta_f(B) = 0$ by Theorem 2.2 (i). This means that (x_k) is f-statistically bounded. This completes the proof.

From Theorem 3.13. and Corollary 3.12 we obtain the following result (see also Theorem 26 and Theorem 34 in [3]).

Corollary 3.14. An $f-$ statistically convergent sequence is statistically bounded, that is $S_f \subseteq BS$ for any unbounded modulus f.

Proof. If we take $g(t) = f(t)$ then the condition (1) will be provided directly for unbounded modulus functions f and g. From Theorem 3.13 we obtain $S_f \subseteq BS_f$ in this case. This and Corollary 3.12 (*i*) gives $S_f \subseteq BS$.

References

1. Aizpuru, A., Listán-García, M.C., Rambla-Barreno, F.: Density by moduli and statistical convergence. Questiones Math. **37**(4), 525–530 (2014)
2. Bhardwaj, V.K., Singh, N.: On some sequence spaces defined by a modulus. Indian J. Pure Appl. Math. **30**(8), 809–817 (1999)
3. Bhardwaj, V.K., Dhawan, S., Gupta, S.: Density by moduli and statistical boundedness. Abstr. Appl. Anal. **2016**, 6 (2016). http://dx.doi.org/10.1155/2016/2143018. Article ID 2143018
4. Connor, J.S.: The statistical and strong p-Cesàro convergence of sequences. Analysis **8**, 47–63 (1988)
5. Fast, H.: Sur la convergence statistique. Colloq. Math. **2**, 241–244 (1951)
6. Fridy, J.: On statistical convergence. Analysis **5**, 301–313 (1985)
7. Fridy, J.A., Orhan, C.: Statistical limit superior and limit inferior. Proc. Am. Math. Soc. **125**(12), 3625–3631 (1997)
8. Ghosh, D., Srivastava, P.D.: On some vector valued sequence spaces defined using a modulus function. Indian J. Pure Appl. Math. **30**(8), 819–826 (1999)
9. Kayan, E., Çolak, R., Altn, Y.: d- statistical convergence of order α and d- statistical boundedness of order α in metric spaces. U.P.B. Sci. Bull. Series A **80**(4), 229–238 (2018)

10. Maddox, I.J.: Sequence spaces defined by a modulus. Math. Proc. Camb. Philos. Soc. **100**, 161–166 (1986)
11. Maddox, I.J.: Inclusions between FK spaces and Kuttner's theorem. Math. Proc. Cambridge Philos. Soc. **101**(3), 523–527 (1987)
12. Nakano, H.: Concave modular. J. Math. Soc. Japan **5**, 29–49 (1953)
13. Ruckle, W.H.: FK spaces in which the sequence of coordinate vectors is bounded. Canad. J. Math. **25**, 973–978 (1973)
14. Salát, T.: On statistically convergent sequences of real numbers. Math. Slovaca **30**, 139–150 (1980)
15. Schoenberg, I.J.: The integrability of certain functions and related summability methods. Am. Math. Monthly **66**, 361–375 (1959)
16. Steinhaus, H.: Sur la convergence ordinaire et la convergence asymptotique. Colloq Math. **2**, 73–74 (1951)
17. Zygmund, A.: Trigonometric Series. Cambridge University Press, Cambridge (1979)

New Integral Inequalities for Product of Geometrically Convex Functions

Ahmet Ocak Akdemir[1(✉)] and Hemen Dutta[2]

[1] Faculty of Science and Letters, Department of Mathematics, Ağrı İbrahim Çeçen University, 04100 Ağrı, Turkey
ahmetakdemir@agri.edu.tr

[2] Department of Mathematics, Gauhati University, Guwahati, India
hemen_dutta08@rediffmail.com

Abstract. In the present paper, we have recalled some definitions and well known results for the field of inequality theory. Then, some new integral inequalities have been established for geometrically convex functions via some well-known inequalities such that Young and Hölder inequality. Our results involve new upper bounds for product of geometrically convex functions that have applications in different branches of mathematics and statistics including convex analysis.

Keywords: $GA-$convex function · $GG-$convex function · Integral inequalities · Young inequality · Hölder inequality

2010 Mathematics Subject Classication: Primary 26D10 · Secondary 26A51, 26D15

1 Introduction

The concept of convex function, which is widely used in inequality and is known with its applications in many fields of mathematics, has continued to be interesting even though its history dates back to ancient times. Differentiable or integrable convex functions have been the focus of many researchers and have been the subject of many articles in the field of inequality theory. Let's start with the definition of this aesthetics and useful functions.

Definition 1.1. *A function $f : I \subset \mathbb{R} \to \mathbb{R}$ is said to be convex on I if inequality*

$$f(tx + (1-t)y) \leq tf(x) + (1-t)f(y) \tag{1.1}$$

holds for all $x, y \in I$ and $t \in [0,1]$.

Some studies have been designed to form different types of convex function classes in order to carry the convex function classes to a new dimension. Many convex function types, especially the convex function definition, are closely related to the special means and contain the mean expressions in the definition.

H. Dutta et al. (Eds.): CMES 2019, AISC 1111, pp. 315–323, 2020.
https://doi.org/10.1007/978-3-030-39112-6_26

We will continue with some types of convex function types that come into prominence among convex function types and find application in statistics.

In [9], Niculescu mentioned definitions of geometrically convex functions as:

The arithmetic-geometric convex functions namely log −convex functions or $AG-$convexity are defined as $f : I \subseteq (0,\infty) \to (0,\infty)$ and holds the following inequality:

$$x, y \in I \text{ and } \lambda \in [0,1] \Longrightarrow f(\lambda x + (1-\lambda) y) \leq f(x)^{\lambda} f(y)^{1-\lambda}, \tag{1.2}$$

i.e., for which $\log f$ is convex.

The geometric-geometric convex functions namely multiplicatively convex functions or $GG-$convexity are defined as $f : I \subseteq (0,\infty) \to J \subseteq (0,\infty)$ and satisfies:

$$x, y \in I \text{ and } \lambda \in [0,1] \Longrightarrow f\left(x^{1-\lambda} y^{\lambda}\right) \leq f(x)^{1-\lambda} f(y)^{\lambda}. \tag{1.3}$$

The class of all geometric-arithmetic convex functions or namely $GA-$convexity are defined as $f : I \subseteq (0,\infty) \to (0,\infty)$ and holds:

$$x, y \in I \text{ and } \lambda \in [0,1] \Longrightarrow f\left(x^{1-\lambda} y^{\lambda}\right) \leq (1-\lambda) f(x) + \lambda f(y). \tag{1.4}$$

Besides, recall that the criterion of $GA-$convexity is $x^2 f'' + x f' \geq 0$ which implies all twice differentiable non-decreasing convex functions are also $GA-$convex.

Examples of functions that contain some important statistical definitions and closely related to many other mathematical concepts are presented below.

Remark 1.1. [4] Some examples of log-concave and concave functions:

(1) The normal probability density function $f(x) = \frac{1}{\sqrt{2\pi}} e^{-x^2/2}$ is log-concave on $(0,1)$, which is also concave on $(0,1)$ since $\log f(x) = -x^2/2 \ln \frac{1}{\sqrt{2\pi}}$ is a concave function.
(2) The probability density function of the beta distribution, for $0 \leq x \leq 1$, and shape parameters $\alpha_1, \alpha_2 > 0$, is a power function of the variable x and of its reflection $(1-x)$ like follows:

$$f(x; \alpha_1, \alpha_2) = \frac{1}{\beta(\alpha_1, \alpha_2)} x^{\alpha_1 - 1} (1-x)^{\alpha_2 - 1}$$

is concave and log-concave function for $\alpha_1 = \alpha_2 = 2$, where β is Euler-Beta function. Namely $f(x) = \frac{\Gamma(4)}{\Gamma(2)\Gamma(2)} x(1-x) = 6\left(x - x^2\right)$ is concave and log-concave function on $(0,1)$.

Hermite-Hadamard inequality, which suggests bounds for the mean value of a convex function, is given as follows.

Theorem 1.1. *(See [2, 3]) Let $f : I \subseteq \mathbb{R} \to \mathbb{R}$ be a convex function and $u, v \in I$ with $u < v$. The following double inequality:*

$$f\left(\frac{u+v}{2}\right) \leq \frac{1}{v-u} \int_u^v f(x)\, dx \leq \frac{f(u) + f(v)}{2} \tag{1.5}$$

is known in the literature as Hadamard's inequality (or Hermite-Hadamard inequality) for convex functions. If f is a positive concave function, then the inequality is reversed.

It is worth remembering some special means to be used in this study. The extended logarithmic mean L_p of $a, b > 0$ is given for $a = b$ by $L_p(a, a) = a$ and for $a \neq b$ by

$$L_p(a,b) = \begin{cases} \left[\frac{b^{p+1}-a^{p+1}}{(p+1)(b-a)}\right]^{\frac{1}{p}} & , p \neq -1, 0 \\ \frac{b-a}{\ln b - \ln a} & , \ p = -1 \\ \frac{1}{e}\left(\frac{b^b}{a^a}\right)^{\frac{1}{(b-a)}} & , \ p = 0 \end{cases}.$$

It is obivous that $L_{-1}(a,b)$ is called the logarithmic mean $L(a,b)$. Two important averages subject to arithmetic geometric inequality are defined as follows.

$$G(a,b) = \sqrt{ab},$$

$$A(a,b) = \frac{a+b}{2}$$

are the geometric mean and arithmetic mean, respectively.

For more detailed information, inequalities, generalizations, and interesting new results for special means and geometric convex function classes, please refer to the papers [1,5,6,8–17].

Many articles listed in the bibliography section have obtained results based on geometric convex function classes and special averages. Here we just want to remind you a few of these results. In [7] Yang et al. established the following results;

Theorem 1.2. *Let $f, g : I \to (0, \infty)$ be log-convex functions on I and $a, b \in I$ with $a < b$ and $\alpha, \beta > 0$ with $\alpha + \beta = 1$. Then the following inequality holds:*

$$\frac{1}{b-a}\int_a^b f(x)\,g(x)\,\mathrm{d}x \leq \alpha\left[L_{\frac{1}{\alpha}-1}(f(a), f(b))\right]^{\frac{1-\alpha}{\alpha}} L(f(a), f(b))$$
$$+\beta\left[L_{\frac{1}{\beta}-1}(g(a), g(b))\right]^{\frac{1-\beta}{\beta}} L(g(a), g(b)).$$

Theorem 1.3. *Let $f, g : I \to (0, \infty)$ be log-concave functions on I and $a, b \in I$ with $a < b$. Further, let $\alpha > 1$ with $\alpha + \beta = 1$ (or $\beta > 1$ with $\alpha + \beta = 1$). Then the following inequality holds:*

$$\frac{1}{b-a}\int_a^b f(x)\,g(x)\,\mathrm{d}x \geq \alpha\left[L_{\frac{1}{\alpha}-1}(f(a), f(b))\right]^{\frac{1-\alpha}{\alpha}} L(f(a), f(b))$$
$$+\beta\left[L_{\frac{1}{\beta}-1}(g(a), g(b))\right]^{\frac{1-\beta}{\beta}} L(g(a), g(b)).$$

The goal of this paper is to establish some new integral inequalities for log-convex and log-concave functions by using above-mentioned classical integral inequalities.

2 Main Results

Theorem 2.1. *Let $f, g : I \subseteq (0, \infty) \to (0, \infty)$ are $GA-$convex functions and f, g, fg are integrable functions on $[a, b]$ such that $a, b \in I$, $\frac{b}{a} \neq 1$. Then, we point out the following new result:*

$$\frac{1}{\ln b - \ln a} \int_a^b f(x) g\left(\frac{ab}{x}\right) dx \leq T_1 f(b)g(a) + T_2 f(a)g(b) + T_3 (f(a)g(a) + f(b)g(b)) \quad (2.1)$$

where

$$T_1 = \frac{b \ln^2 \frac{b}{a} - 2b \ln \frac{b}{a} + 2b - 2a}{\ln^3 \frac{b}{a}},$$
$$T_2 = \frac{a \ln^2 \frac{b}{a} - 2a \ln \frac{b}{a} + 2 \ln \frac{b}{a} - 2}{\ln^3 \frac{b}{a}},$$
$$T_3 = \frac{2b \ln^2 \frac{b}{a} - 3b \ln \frac{b}{a} + a \ln \frac{b}{a} + 2b - 2a}{\ln^3 \frac{b}{a}}.$$

Proof. By using the functions f and g and conditions of Theorem, we can write

$$\frac{1}{\ln b - \ln a} \int_a^b f(x) g\left(\frac{ab}{x}\right) dx = \int_0^1 b^t a^{1-t} f\left(b^t a^{1-t}\right) g\left(a^t b^{1-t}\right) dt.$$

Here we have used the change of the variables as $x = b^t a^{1-t}$ and $dx = b^t a^{1-t} \ln \frac{b}{a} dt$. By taking into account $GA-$convexity of f and g, we get

$$\begin{aligned} &\frac{1}{\ln b - \ln a} \int_a^b f(x) g\left(\frac{ab}{x}\right) dx \\ &\leq \int_0^1 b^t a^{1-t} [tf(b) + (1-t)f(a)] [tg(a) + (1-t)g(b)] dt \\ &= \int_0^1 b^t a^{1-t} \left[t^2 f(b)g(a) + t(1-t)(f(a)g(a) + f(b)g(b)) + (1-t)^2 f(a)g(b)\right] dt. \end{aligned}$$

By computing the above integrals, we obtain

$$\begin{aligned} &\frac{1}{\ln b - \ln a} \int_a^b f(x) g\left(\frac{ab}{x}\right) dx \\ &\leq \left[\frac{b \ln^2 \frac{b}{a} - 2b \ln \frac{b}{a} + 2b - 2a}{\ln^3 \frac{b}{a}}\right] f(b)g(a) \\ &+ \left[\frac{a \ln^2 \frac{b}{a} - 2a \ln \frac{b}{a} + 2 \ln \frac{b}{a} - 2}{\ln^3 \frac{b}{a}}\right] f(a)g(b) \\ &+ \left[\frac{2b \ln^2 \frac{b}{a} - 3b \ln \frac{b}{a} + a \ln \frac{b}{a} + 2b - 2a}{\ln^3 \frac{b}{a}}\right] (f(a)g(a) + f(b)g(b)). \end{aligned}$$

Thus, this complete the proof of the inequality (2.1). □

Similar to this result can be easily obtained for $GG-$convex functions.

Theorem 2.2. *Let $f, g : I \subseteq (0, \infty) \to (0, \infty)$ are $GG-$convex functions and f, g, fg are integrable functions on $[a, b]$ such that $a, b \in I$, $\frac{b}{a} \neq 1$. Then, we have:*

$$\frac{1}{\ln b - \ln a} \int_a^b f(x)\, g\left(\frac{ab}{x}\right) dx \leq L\left(af(a)g(b), bf(b)g(a)\right)$$

where $L(u, v)$ is the logarithmic mean of u and v.

Proof. Procedures similar to the proof of the previous theorem should be considered, but we only need to select the functions $GG-$convex. So, we can write

$$\begin{aligned} &\frac{1}{\ln b - \ln a} \int_a^b f(x)\, g\left(\frac{ab}{x}\right) dx \\ &= \int_0^1 b^t a^{1-t} f\left(b^t a^{1-t}\right) g\left(a^t b^{1-t}\right) dt \\ &\leq \int_0^1 \left[bf(b)g(a)\right]^t \left[af(a)g(b)\right]^{1-t} dt. \end{aligned}$$

By a simple calculation, we deduce

$$\frac{1}{\ln b - \ln a} \int_a^b f(x)\, g\left(\frac{ab}{x}\right) dx \leq \frac{bf(b)g(a) - af(a)g(b)}{\ln bf(b)g(a) - \ln af(a)g(b)}.$$

The proof is completed. □

Theorem 2.3. *Let $f : I \subseteq (0, \infty) \to (0, \infty)$ is $GG-$convex function and f is integrable function on $[a, b]$ such that $a, b \in I$, $\frac{b}{a} \neq 1$. Then, one has the following inequality:*

$$\frac{1}{\ln b - \ln a} \int_a^b f(x)\, f\left(\frac{ab}{x}\right) dx \leq G^2\left(f(a), f(b)\right) L(a, b)$$

where $L(u, v)$ is the logarithmic mean of u and v and $G(u, v)$ is the geometric mean of u and v.

Proof. By using the definition of $GG-$convexity, we have

$$\begin{aligned} &\frac{1}{\ln b - \ln a} \int_a^b f(x)\, f\left(\frac{ab}{x}\right) dx \\ &= \int_0^1 b^t a^{1-t} f\left(b^t a^{1-t}\right) f\left(a^t b^{1-t}\right) dt \\ &\leq f(a)f(b) \int_0^1 b^t a^{1-t} dt = f(a)f(b) \frac{b - a}{\ln b - \ln a}. \end{aligned}$$

Which completes the proof. □

Theorem 2.4. *Suppose that $f : I \subseteq (0,\infty) \to (0,\infty)$ be $GA-$convex function, $g : I \subseteq (0,\infty) \to (0,\infty)$ be $GG-$convex function and f, g, fg are integrable functions on $[a,b]$ such that $a, b \in I$, $\frac{b}{a} \neq 1$. Then, one has the following inequality:*

$$\frac{1}{\ln b - \ln a}\int_a^b f(x)\, g\left(\frac{ab}{x}\right) dx \leq \frac{L_p^p(f(a), f(b))}{p} + \frac{L\left((ag(b))^q, (bg(a))^q\right)}{q},$$

for $\frac{1}{p} + \frac{1}{q} = 1$ where $L(u,v)$ is the logarithmic mean of u and v and $G(u,v)$ is the geometric mean of u and v.

Proof. The classical Young's inequality can be represented by (see [2]):

$$\alpha\beta \leq \frac{\alpha^p}{p} + \frac{\beta^q}{q}, \text{ for } \frac{1}{p} + \frac{1}{q} = 1, \quad \alpha, \beta > 0.$$

Since f is $GA-$convex function and g is $GG-$convex function on I, we get

$$\begin{aligned} &\frac{1}{\ln b - \ln a}\int_a^b f(x)\, g\left(\frac{ab}{x}\right) dx \\ &= \int_0^1 b^t a^{1-t} f\left(b^t a^{1-t}\right) g\left(a^t b^{1-t}\right) dt \\ &\leq \int_0^1 b^t a^{1-t}\left[tf(b) + (1-t)f(a)\right]\left[g(a)\right]^t \left[g(b)\right]^{1-t} dt. \end{aligned}$$

By using the Young inequality, we can write

$$\begin{aligned} &\frac{1}{\ln b - \ln a}\int_a^b f(x)\, g\left(\frac{ab}{x}\right) dx \\ &\leq \int_0^1 b^t a^{1-t}\left[tf(b) + (1-t)f(a)\right]\left[g(a)\right]^t \left[g(b)\right]^{1-t} dt \\ &\leq \frac{\int_0^1 \left[tf(b) + (1-t)f(a)\right]^p dt}{p} + \frac{\int_0^1 b^{qt} a^{q(1-t)} \left[g(a)\right]^{qt} \left[g(b)\right]^{q(1-t)} dt}{q}. \end{aligned}$$

By computing the above integrals, we obtain

$$\begin{aligned} &\frac{1}{\ln b - \ln a}\int_a^b f(x)\, g\left(\frac{ab}{x}\right) dx \\ &\leq \left(\frac{[f(b)]^{p+1} - [f(a)]^{p+1}}{f(b) - f(a)}\right)\left(\frac{1}{p(p+1)}\right) + \frac{(bg(a))^q - (ag(b))^q}{q\left(\ln (bg(a))^q - \ln (ag(b))^q\right)}, \end{aligned}$$

which completes the proof. □

Theorem 2.5. *Suppose that $f, g : I \subseteq (0,\infty) \to (0,\infty)$ be two functions and $|f|, |g|, |fg|$ are integrable functions on $[a,b]$ such that $a, b \in I$, $a < b$. If $|f|$*

is GA−convex and $|g|$ is GG−convex function, then, one has the following inequality:

$$\frac{1}{\ln b-\ln a}\int_a^b |f(x)|\left|g\left(\frac{ab}{x}\right)\right| dx$$
$$\leq \left[\left(\frac{[|f(b)|]^{p+1}-[f(a)]^{p+1}}{|f(b)|-f(a)}\right)\left(\frac{1}{(p+1)}\right)\right]^{\frac{1}{p}}\left[\frac{(b|g(a)|)^q-(a|g(b)|)^q}{\ln(b|g(a)|)^q-\ln(a|g(b)|)^q}\right]^{\frac{1}{q}}.$$

for $\frac{1}{p}+\frac{1}{q}=1, p>1$.

Proof. By using the definition of $|f|$ and $|g|$ and changing of the variables, we have

$$\frac{1}{\ln b-\ln a}\int_a^b |f(x)|\left|g\left(\frac{ab}{x}\right)\right| dx$$
$$\leq \int_0^1 b^t a^{1-t}\left[t|f(b)|+(1-t)|f(a)|\right]\left[|g(a)|\right]^t\left[|g(b)|\right]^{1-t} dt.$$

By applying the well-known Hölder integral inequality, we get:

$$\frac{1}{\ln b-\ln a}\int_a^b |f(x)|\left|g\left(\frac{ab}{x}\right)\right| dx$$
$$\leq \left(\int_0^1 \left[t|f(b)|+(1-t)|f(a)|\right]^q dt\right)^{\frac{1}{p}}\left(\int_0^1 b^{qt}a^{q(1-t)}\left[|g(a)|\right]^{qt}\left[|g(b)|\right]^{q(1-t)} dt\right)^{\frac{1}{q}}.$$

By making use of necessary computations, we obtain;

$$\frac{1}{\ln b-\ln a}\int_a^b |f(x)|\left|g\left(\frac{ab}{x}\right)\right| dx$$
$$\leq \left[\left(\frac{[|f(b)|]^{p+1}-[f(a)]^{p+1}}{|f(b)|-f(a)}\right)\left(\frac{1}{(p+1)}\right)\right]^{\frac{1}{p}}\left[\frac{(b|g(a)|)^q-(a|g(b)|)^q}{\ln(b|g(a)|)^q-\ln(a|g(b)|)^q}\right]^{\frac{1}{q}}.$$

which completes the proof. □

Theorem 2.6. *Suppose that $f,g: I\subseteq(0,\infty)\to(0,\infty)$ be two functions and $|f|,|g|,|fg|$ are integrable functions on $[a,b]$ such that $a,b\in I$, $a<b$. If $|f|$ and $|g|$ are GG−convex function, then, one has the following inequality:*

$$\frac{1}{\ln b-\ln a}\int_a^b |f(x)|\left|g\left(\frac{ab}{x}\right)\right| dx$$
$$\leq \left[\frac{(|f(b)|)^p-(|f(a)|)^p}{\ln(|f(b)|)^p-\ln(|f(a)|)^p}\right]^{\frac{1}{p}}\left[\frac{(b|g(a)|)^q-(a|g(b)|)^q}{\ln(b|g(a)|)^q-\ln(a|g(b)|)^q}\right]^{\frac{1}{q}}$$
$$= L^{\frac{1}{p}}\left((|f(a)|)^p,(|f(b)|)^p\right)L^{\frac{1}{q}}\left((a|g(b)|)^q,(b|g(a)|)^q\right).$$

for $\frac{1}{p}+\frac{1}{q}=1, p>1$, where $L(u,v)$ is the logarithmic mean of u and v.

Proof. By using the definition of $|f|$ and $|g|$ and changing of the variables, we have

$$\frac{1}{\ln b-\ln a}\int_a^b |f(x)|\left|g\left(\frac{ab}{x}\right)\right|dx \leq \int_0^1 b^t a^{1-t}\left[|f(a)|\right]^t\left[|f(b)|\right]^{1-t}\left[|g(a)|\right]^t\left[|g(b)|\right]^{1-t}dt.$$

By applying the well-known Hölder integral inequality, we get:

$$\frac{1}{\ln b-\ln a}\int_a^b |f(x)|\left|g\left(\frac{ab}{x}\right)\right|dx \leq \left(\int_0^1 \left[|f(b)|\right]^{pt}\left[|f(a)|\right]^{p(1-t)}dt\right)^{\frac{1}{p}}\left(\int_0^1 b^{qt}a^{q(1-t)}\left[|g(a)|\right]^{qt}\left[|g(b)|\right]^{q(1-t)}dt\right)^{\frac{1}{q}}.$$

By making use of necessary computations, we obtain;

$$\frac{1}{\ln b-\ln a}\int_a^b |f(x)|\left|g\left(\frac{ab}{x}\right)\right|dx \leq \left[\frac{(|f(b)|)^p-(|f(a)|)^p}{\ln(|f(b)|)^p-\ln(|f(a)|)^p}\right]^{\frac{1}{p}}\left[\frac{(b\,|g(a)|)^q-(a\,|g(b)|)^q}{\ln(b\,|g(a)|)^q-\ln(a\,|g(b)|)^q}\right]^{\frac{1}{q}}.$$

which completes the proof. □

3 Conclusion

In this study, we have proved some new integral inequalities for product of $GA-$ and $GG-$convex functions by using some integration techniques and elemantery analysis. The results have been established via some classical inequalities such as Hölder integral inequality and Young inequality. We presented the results for special means of real numbers. The results can be extended to fractional calculus by using some new fractional Integral operators. Also, similar results can be found for more general convex function classes.

References

1. Dragomir, S.S., Pearce, C.E.M.: Selected Topics on Hermite-Hadamard Inequalities and Applications. RGMIA Monographs, Victoria University (2000)
2. Hardy, G.H., Littlewood, J.E., Pólya, G.: Inequalities, 2nd edn. Cambridge University Press, Cambridge (1988)
3. Mitrinović, D.S., Pečarić, J.E., Fink, A.M.: Classical and New Inequalties in Analysis. Kluwer Academic Publisher, Dordrecht (1993)
4. Niculescu, C.P.: The Hermite-Hadamard inequality for log-convex functions. Nonlinear Anal. **75**, 662–669 (2012)

5. Niculescu, C.P., Persson, L.E.: Convex Functions and Their Applications, A Contemporary Approach. In: CMS Books in Mathematics, vol. 23. Springer-Verlag, New York (2006)
6. Pearce, C.E.M., Pečarić, J., Šimic, V.: Stolarsky means and Hadamard's inequality. J. Math. Anal. Appl. **220**, 99–109 (1998)
7. Yang, G.-S., Tseng, K.-L., Wang, H.-T.: A note on integral inequalıtıes of Hadamard type for log-convex and log-concave functions. Taiwenese J. Math. **16**(2), 479–496 (2012)
8. Zhang, T.-Y., Ji, A.-P., Qi, F.: Sone inequalities of Hermite-Hadamard type for $GA-$convex functions with applications to means, Le Matematiche, vol. LXVIII, pp. 229–239 (2013). https://doi.org/10.4418/2013.68.1.17. Fasc. I
9. Niculescu, C.P.: Convexity according to the geometric mean. Math. Inequal. Appl. **3**(2), 155–167 (2000). https://doi.org/10.7153/mia-03-19
10. Niculescu, C.P.: Convexity according to means. Math. Inequal. Appl. **6**(4), 571–579 (2003). https://doi.org/10.7153/mia-06-53
11. Anderson, G.D., Vamanamurthy, M.K., Vuorinen, M.: Generalized convexity and inequalities. J. Math. Anal. Appl. **335**, 1294–1308 (2007)
12. İşcan, İ.: Some generalized Hermite-Hadamard type inequalities for quasi-geometrically convex functions. Am. J. Math. Anal. **1**(3), 48–52 (2013). https://doi.org/10.12691/ajma-1-3-5
13. İşcan, İ.: New general integral inequalities for some $GA-$convex and quasi-geometrically convex functions via fractional integrals, arXiv:1307.3265v1
14. İşcan, İ.: Hermite-Hadamard type inequalities gor $GA-s-$convex functions, arXiv:1306.1960v2
15. Zhang, X.-M., Chu, Y.-M., Zhang, X.-H.: The Hermite-Hadamard type inequality of $GA-$convex functions and its application. J. Inequal. Appl. **2010**, 11 (2010). https://doi.org/10.1155/2010/507560. Article ID 507560
16. Satnoianu, R.A.: Improved $GA-$convexity inequalities. J. Inequal. Pure Appl. Math. **3**(5), 1–6 (2002). Article 82
17. Latif, M.A.: New Hermite-Hadamard type integral inequalities for $GA-$convex functions with applications. Analysis **34**(4), 379–389 (2014). https://doi.org/10.1515/anly-2012-1235

Uniformly $(B, \lambda)-$Invariant Statistical Convergence

Rahmet Savaş(✉)

Department of Mathematics, İstanbul Medeniyet University, İstanbul, Turkey
rahmet.savas@medeniyet.edu.tr

Abstract. In the present chapter, we consider some properties of uniformly (B, λ)-invariant statistically convergent which is defined using the φ-function and invariant mean. Also we prove some inclusion theorems.

1 Introduction and Background

Let s denote the set of all real and complex sequences $x = (x_k)$. By l_∞ and c, we denote the Banach spaces of bounded and convergent sequences $x = (x_k)$ normed by $\|x\| = \sup_n |x_n|$, respectively.

If $x = (x_k)$ is a sequence and $B = (b_{nk})$ is an infinite matrix, then Bx is the sequence whose nth term is given by $B_n(x) = \sum_{k=0}^{\infty} b_{nk}x_k$. Thus we say that x is B-summable to L if $\lim_{n\to\infty} B_n(x) = L$. Let X and Y be two sequence spaces and $B = (b_{nk})$ an infinite matrix. If for each $x \in X$ the series $B_n(x) = \sum_{k=0}^{\infty} b_{nk}x_k$ converges for each n and the sequence $Bx = B_n(x) \in Y$ we say that B maps X into Y. By (X, Y) we denote the set of all matrices which maps X into Y, and in addition if the limit is preserved then we denote the class of such matrices by $(X, Y)_{reg}$.

Let $\lambda = (\lambda_i)$ be a non-decreasing sequence of positive numbers tending to ∞ such that

$$\lambda_{i+1} \leq \lambda_i + 1, \lambda_1 = 1.$$

The collection of such sequence λ will be denoted by Δ.

The generalized de la Valee-Poussin mean is defined by

$$T_i(x) = \frac{1}{\lambda_i} \sum_{k \in I_i} x_k$$

where $I_i = [i - \lambda_i + 1, i]$. A sequence $x = (x_n)$is said to be (V, λ)-summable to a number L, if $T_i(x) \to L$ as $i \to \infty$ (see [5]).

By a φ-function we understand a continuous non-decreasing function $\varphi(u)$ defined for $u \geq 0$ and such that $\varphi(0) = 0$, $\varphi(u) > 0$, for $u > 0$ and $\varphi(u) \to \infty$ as $u \to \infty$ (see [18]).

A φ-function φ is called non weaker than a φ-function ψ if there are constants $c, b, k, l > 0$ such that $c\psi(lu) \leq b\varphi(ku)$, (for all large u) and we write $\psi \prec \varphi$.

Let σ be a one-to-one mapping from the set of natural numbers into itself. A continuous linear functional ϕ on l_∞ is said to be an invariant mean or a σ-mean if and only if

H. Dutta et al. (Eds.): CMES 2019, AISC 1111, pp. 324–329, 2020.
https://doi.org/10.1007/978-3-030-39112-6_27

1. $\phi(x) \geq 0$ when the sequence $x = (x_n)$ has $x_n \geq 0$ for all n;
2. $\phi(e) = 1$ where $e = (1,1,1,...)$ and
3. $\phi(x_{\sigma(n)}) = \phi(x)$ for all $x \in l_\infty$.

For certain class of mapping σ every invariant mean φ extends the limit functional on space c, in the sense that $\varphi(x) = \lim x$ for all $x \in c$.

Consequently, $c \subset V_\sigma$ where V_σ is the subset of all bounded sequences whose σ-means are equal.

If $x = (x_n)$, set $Tx = (Tx_k) = (x_\sigma(n))$, it can be shown that (see, Schaefer [16])

$$V_\sigma = \left\{x \in l_\infty : \lim_m t_{m,n}(x) = L \text{ uniformly in } n, L = \sigma - \lim x\right\}$$

where

$$t_{m,n}(x) = \frac{x_n + x_{\sigma(n)} + ... + x_{\sigma^m(n)}}{m+1}, t_{-1,n}(x) = 0.$$

We say that a bounded sequence $x = (x_n)$ is σ-convergent if and only if $x \in V_\sigma$ such that $\sigma^m(n) \neq n$ for all $n \geq 0$, k ≥ 1.

Definition 1. Let $\Lambda = (\lambda_j)$ be same as above, φ be given φ-function and f be given modules function, respectively. Moreover, let $\mathbf{B} = (b_{nk})$ be the real matrix. Then we define,

$$V_\lambda^0((B,\varphi),f) = \left\{(x_k) : \lim_j \frac{1}{\lambda_j} \sum_{n \in I_j} f\left(\left|\sum_{k=1}^{\infty} b_{nk}\varphi(|x_{\sigma^m(k)}|)\right|\right) = 0, \text{ uniformly in m}\right\}.$$

If $x \in V_\lambda^0((B,\varphi),f)$, the sequence x is said to be $\lambda - strong(B,\varphi)$-invariant convergent to zero with respect to a modulus f.

2 Uniform $(B,\varphi)-$Invariant Statistically Convergent

The idea of convergence of a real sequence was extended to statistical convergence by Fast [1] (see also Schoenberg [17]). A sequence (x_k) of real numbers is said to be statistically convergent to L if for arbitrary $\varepsilon > 0$, the set $K(\varepsilon) = \{k \in \mathrm{N} : |x_k - l| \geq \varepsilon\}$ has natural density zero.

Over the years and under different names statistical convergence was discussed in the theory of fourier analysis, ergodic theory, number theory, measure theory, trigonometric series, turnpike theory and Banach spaces. Statistical convergence turned out to be one of the most active areas of research in summability theory after the work of Fridy [3] and Salat [15].

In another direction, Mursaleen [6] defined λ-statistical convergence as follows.

A sequence (x_k) of real numbers is said to be λ-statistically convergent to L (or S_λ-convergent to L) if for any $\varepsilon > 0$

$$\lim_{j\to\infty} \frac{1}{\lambda_j} \left|\{k \in I_j : |x_k - l| \geq \varepsilon\}\right| = 0$$

In [6] the relation between λ-statistical convergence and statistical convergence was established among other things.

Recently Savas [10] defined almost λ-statistical convergence by using the notion of (V,λ)-summability to generalize the concept of statistical convergence.

We can give the following definition.

Definition 2. A sequence $x=(x_k)$ is said to be uniformly σ-statistical convergent to the number L if for every $\varepsilon>0$,

$$\lim_n \frac{1}{n}\max_{m\geq 0}\left|\left\{k\leq n : \left|x_{\sigma^k(m)}-L\right|\geq\varepsilon\right\}\right|=0$$

In this case we write $S_\sigma-\lim x=L$ or $x_k\to L(S_\sigma)$ and we define

$$S_\sigma=\{x=(x_k): for \text{ some L}, \mathrm{S}_\sigma-\lim x=L\}.$$

Assume that B is a non-negative regular summability matrix. Then the sequence $x=(x_n)$ is called B-statistically convergent to L provided that, for every $\varepsilon>0$, (see [4])

$$\lim_j \sum_{n:|x_n-L|\geq\varepsilon} b_{jn}=0.$$

We denote this by $st_B-\lim_n x_n=L$.

Let $\mathbf{B}=(b_{nk})$ be real matrix and te sequence $x=(x_k)$, the φ-function $\varphi(u)$ and a positive number $\varepsilon>0$ be given. We write, for all m

$$K_\lambda^j\left((B,\sigma,\varphi),\varepsilon\right)=\left\{n\in I_j : \sum_{k=1}^{\infty} b_{nk}\varphi\left(\left|x_{\sigma^k(m)}\right|\right)\geq\varepsilon\right\}.$$

The sequence x is said to be uniformly (B,φ,λ)-invariant statistically convergent to a number zero if for every $\varepsilon>0$

$$\lim_j \frac{1}{\lambda_j}\max_{m\geq 0}\mu(K_\lambda^j((B,\sigma,\varphi),\varepsilon)=0,$$

where $\mu\left(K_\lambda^j\left((B,\sigma,\varphi),\varepsilon\right)\right)$ denotes the number of elements belonging to set $K_\lambda^j\left((B,\sigma,\varphi),\varepsilon\right)$. We denote by $S_\lambda^0(B,\sigma,\varphi)$, the set of sequences $x=(x_k)$ which are uniformly (B,σ,φ)-statistical convergent to zero.

If we take $B=I$ and $\varphi(x)=x$ respectively, then $S_\lambda^0(B,\sigma,\varphi)$ reduce to $S^0_{(\lambda,\varphi)}$-invariant.

$$S^0_{(\lambda,\sigma)}=\left\{x=(x_k):\lim_j \frac{1}{\lambda_j}\max_{m\geq 0}\left|\left\{k\in I_j : \left|x_{\sigma^k(m)}\right|\geq\varepsilon\right\}\right|=0\right\}.$$

We are now ready to state the following theorem.

Theorem 1. *If* $\Psi\prec\varphi$ *then* $\mathrm{S}_\lambda^0(B,\sigma,\Psi)\subset\mathrm{S}_\lambda^0(B,\sigma,\varphi)$.

Proof. By our assumptions we have $\psi\left(\left|x_{\sigma^k(m)}\right|\right) \leq b\varphi\left(c\left|x_{\sigma^k(m)}\right|\right)$ and we have for all m,

$$\sum_{k=1}^{\infty} b_{nk}\psi\left(\left|x_{\sigma^k(m)}\right|\right) \leq b\sum_{k=1}^{\infty} b_{nk}\varphi(c\left|x_{\sigma^k(m)}\right|) \leq K\sum_{k=1}^{\infty} b_{nk}\varphi\left(\left|x_{\sigma^k(m)}\right|\right)$$

for b,c >0, where the constant K is connected with properties of φ. Thus, the condition $\sum_{k=1}^{\infty} b_{nk}\Psi\left(\left|x_{\sigma^k(m)}\right|\right) \geq \varepsilon$ implies the condition $\sum_{k=1}^{\infty} b_{nk}\varphi\left(\left|x_{\sigma^k(m)}\right|\right) \geq \varepsilon$ and in consequence we get for all m,

$$\mu\left(K_{\lambda}^{j}(B,\sigma,\varphi),\varepsilon\right) \subset \mu\left(K_{\lambda}^{j}(B,\sigma,\psi),\varepsilon\right)$$

and

$$\lim_j \frac{1}{\lambda_j}\mu\left(\left(K_{\lambda}^{j}(B,\sigma,\varphi),\varepsilon\right)\right) \subset \lim_j \frac{1}{\lambda_j}\mu\left(\left(K_{\lambda}^{j}(B,\sigma,\Psi),\varepsilon\right)\right)$$

This completes the proof.

Theorem 2. *(a) If the matrix B, functions f and φ be given, then*

$$V_{\lambda}^{0}\left((B,\sigma,\varphi),f\right) \subset S_{\lambda}^{0}\left(B,\sigma,\varphi\right).$$

(b) If the φ-function $\varphi(u)$ and the matrix B are given, and if the modulus function f is bounded, then

$$S_{\lambda}^{0}\left(B,\sigma,\varphi\right) \subset V_{\lambda}^{0}\left((B,\sigma,\varphi),f\right).$$

Proof. (a) Let f be a modulus function and let ε be a positive numbers. We write the following inequalities for all m:

$$\begin{aligned}\frac{1}{\lambda_j}\sum_{n\in I_j} f\left(\left|\sum_{k=1}^{\infty} b_{nk}\varphi(\left|x_{\sigma^k(m)}-L\right|)\right|\right) &\geq \frac{1}{\lambda_j}\sum_{n\in I_j^1} f\left(\left|\sum_{k=1}^{\infty} b_{nk}(i)\varphi(\left|x_{\sigma^k(m)}-L\right|)\right|\right)\\ &\geq \frac{1}{\lambda_j} f(\varepsilon)\sum_{n\in I_j^1} 1\\ &\geq \frac{1}{\lambda_j} f(\varepsilon)\mu\left(K_{\lambda}^{j}\left(B,\sigma,\varphi\right),\varepsilon\right)\end{aligned}$$

where

$$I_j^1 = \left\{n\in I_j : \sum_{k=1}^{\infty} b_{nk}\varphi\left(\left|x_{\sigma^k(m)})\right|\right) \geq \varepsilon\right\}.$$

Finally, if $x\in V_{\lambda}^{0}\left((B,\sigma,\varphi),f\right)$ then $x\in S_{\lambda}^{0}\left(B,\sigma,\varphi\right)$.
(b) Let us suppose that $x\in S_{\lambda}^{0}\left(B,\sigma,\varphi\right)$. If the modulus function f is a bounded function, then there exists an integer M such that $f(x)<M$ for $x\geq 0$. Let us take, for all m

$$I_j^2 = \left\{n\in I_j : \sum_{k=1}^{\infty} b_{nk}(i)\varphi\left(\left|x_{\sigma^k}(m)\right|\right) < \varepsilon\right\}.$$

Thus we have,

$$\frac{1}{\lambda_j}\sum_{n\in I_j} f\left(\left|\sum_{k=1}^{\infty} b_{nk}\varphi(\left|x_{\sigma^k(m)}-L\right|)\right|\right)$$
$$\leq \frac{1}{\lambda_j}\sum_{n\in I_j^1} f\left(\left|\sum_{k=1}^{\infty} b_{nk}\varphi(\left|x_{\sigma^k(m)}-L\right|)\right|\right)$$
$$+\frac{1}{\lambda_j}\sum_{n\in I_j^2} f\left(\left|\sum_{k=1}^{\infty} b_{nk}\varphi(\left|x_{\sigma^k(m)}-L\right|)\right|\right)$$
$$\leq \frac{1}{\lambda_j} M \max_{m\geq 0} \mu\left(\mathrm{K}_{\lambda}^{j}\left((B,\varphi),\varepsilon\right)\right)+f(\varepsilon)$$

Taking the limit as $\varepsilon \to 0$, we obtain that $x \in V_{\lambda}^{0}(B,\sigma,\varphi,f)$.

Theorem 3. *If a sequence $x=(x_k)$ is $S(B,\sigma,\varphi)$-convergent to L and*

$$\lim inf_j\left(\frac{\lambda_j}{j}\right)>0$$

then it is $S_{\lambda}(B,\sigma,\varphi)$-convergent to L, where

$$S(\sigma,\varphi)=\left\{x=(x_k):\lim_j \frac{1}{j}\mu\left(K\left(B,\sigma,\varphi,\varepsilon\right)\right)=0\right\}.$$

Proof. For a given $\varepsilon>0$, we write, for all m

$$\left\{n\in I_j:\sum_{k=0}^{\infty} b_{nk}\varphi\left(\left|x_{\sigma^k(m)}-L\right|\right)\geq\varepsilon\right\}\subseteq\left\{n\leq j:\sum_{k=0}^{\infty} b_{nk}\varphi(x_{\sigma^k(m)}-L)\geq\varepsilon\right\}$$

Hence we have,

$$K_{\lambda}(B,\sigma,\varphi,\varepsilon)\subseteq K(B,\sigma,\varphi,\varepsilon).$$

Finally the proof follows from the following inequality:

$$\frac{1}{j}\max_{m\geq 0}\mu\left(K\left(B,\sigma,\varphi,\varepsilon\right)\right)\geq\frac{1}{j}\max_{m\geq 0}\mu\left(K_{\lambda}\left(B,\sigma,\varphi,\varepsilon\right)\right)=\frac{\lambda_j}{j}\frac{1}{\lambda_j}\max_{m\geq 0}\mu\left(K_{\lambda}\left(B,\sigma,\varphi,\varepsilon\right)\right)$$

This completes the proof.

3 Conclusion

In this work, we extended the previous concept of statistically convergence and gave some new definitions as uniformly σ-statistically convergence, β-statistically convergence and uniformly (B,φ,λ)–invariant statistically convergence where B is a real matrix and φ is a function. We also give some relationship between them. This subject also has some new open problems.

References

1. Fast, H.: Sur la convergence statistique. Colloq. Math. **2**, 241–244 (1951)
2. Freedman, A.R., Sember, J.J., Raphel, M.: Some Cesaro-type summability spaces. Proc. Lond. Math. Soc. **37**, 508–520 (1978)
3. Fridy, J.A.: On statistical convergence. Analysis **5**, 301–313 (1985)
4. Kolk, E.: Matrix summability of statistically convergent sequences. Analysis **18**, 77–83 (1993)
5. Malkowsky, E., Savaş, E.: Some λ-sequence spaces defined by a modulus. Arch. Math. **36**, 219–228 (2000)
6. Mursaleen, M.: λ-statistical convergence. Math. Slovaca **50**, 111–115 (2000)
7. Nuray, F., Savaş, E.: Invariant statistical convergence and A-invariant statistical convergence. Indian J. Pure Appl. Math. **25**, 267–274 (1994)
8. Salat, T.: On statistically convergent sequences of real numbers. Math. Slovaca **30**, 139–150 (1980)
9. Savas, E., Savaş, R.: Some λ-sequence spaces defined by Orlicz functions. Indian J. Pure. Appl. Math. **34**(12), 1673–1680 (2003)
10. Savas, E.: Strong almost convergence and almost λ-statistical convergence. Hokkaido Math. J. **24**(3), 531–536 (2000)
11. Savas, E.: Some sequence spaces and statistical convergence. Inter. J. Math. Math. Sci. **29**, 303–306 (2002)
12. Savas, E.: On some sequence spaces and A-statistical convergence. Inter. J. Math. Math. Sci. **29**, 303–306 (2002)
13. Savas, E.: On some sequence spaces and A-statistical convergence. In: 2nd Strathmore International Mathematics Conference, Nairobi, Kenya, 12–16 August 2013
14. Savaş, E.: On asymptotically λ-statistical equivalent sequences of fuzzy numbers. New Math. Nat. Comput. **3**, 301–306 (2007)
15. Savas, E., Nuray, F.: On σ-statistical convergence and lacunary σ-statistical convergence. Math. Slovaca **43**, 309–315 (1993)
16. Schaefer, P.: Infinite matrices and invariant means. Proc. Amer. Math. Soc. **36**, 104–110 (1972)
17. Schoenberg, I.J.: The integrability of certain functions and related summability methods. Amer. Math. Monthly **66**, 361–375 (1959)
18. Waszak, A.: On the strong convergence in sequence spaces. Fasciculi Math. **33**, 125–137 (2002)

Author Index

H. Dutta et al. (Eds.): CMES 2019, AISC 1111, pp. 331–332, 2020.
https://doi.org/10.1007/978-3-030-39112-6

The manufacturer's authorised representative in the EU is Springer Nature Customer Service Centre GmbH, Europaplatz 3, 69115 Heidelberg, Germany. If you have any concerns regarding our products, please contact ProductSafety@springernature.com

Printed and bound by CPI Group (UK) Ltd, Croydon, CR0 4YY
25/09/2026
02222891-0009